COAL MINE
STRUCTURES

COAL MINE STRUCTURES

Ian Farmer

LONDON   NEW YORK

CHAPMAN AND HALL

First published 1985 by
Chapman and Hall Ltd
11 New Fetter Lane, London EC4P 4EE
Published in the USA by
Chapman and Hall
733 Third Avenue, New York NY 10017
© 1985 I. W. Farmer

Softcover reprint of the hardcover 1st edition 1985

ISBN-13: 978-94-010-8643-1   e-ISBN-13: 978-94-009-4834-1
DOI: 10.1007/978-94-009-4834-1

British Library Cataloguing in Publication Data

Farmer, I.W.
   Coal mine structures.
   1. Coal mines and mining
   I. Title
   622'.334   TN802

Library of Congress Cataloging in Publication Data

Farmer, I. W. (Ian William)
   Coal mine structures.

   Bibliography: p.
   Includes index.
   1. Coal mines and mining.   2. Ground control
(Mining)   I. Title.
TN803.F37   1984   622'.334   84-7725

CONTENTS

PREFACE

Coal Mine Structures is based on a six-year study, carried out at the University of Newcastle upon Tyne between 1976 and 1982 and financed by the National Coal Board and the European Coal and Steel Community (Projects 7220 – AC/806 and 7220 – AC/814), into the behaviour of underground openings in British coal mines. The original work has been expanded to include other relevant British and international data. However, it remains, deliberately, a personal view based on a specific – albeit broad – research programme. It does not pretend to be a complete description of the behaviour of shafts, tunnels, rooms, pillars and longwall excavations. Nor does it set out to provide a manual for design. The specific aim is to show, often through consideration of quite detailed laboratory and field data, how the observed performance of some underground structures during mining, can be explained by the deformation characteristics of the rocks which surround them.

The work is based on observations by many engineers working for the National Coal Board's Mining Research and Development Establishment, and by research associates and postgraduate students at the University of Newcastle upon Tyne. I am grateful to them all, but I am particularly indebted to Dr A.H. Wilson and Mr M.J. Bell of the National Coal Board, to Dr P.F.R. Altounyan, Dr P. Garritty, Dr P. Holmes, and Dr P.D. Shelton formerly of the University of Newcastle upon Tyne, to Dr R.N. Gupta of C.M.R.S., Dhanbad, to Dr N.J. Kusznir of Keele University and to Dr A.M. Price of Warwick University.

Ian W. Farmer
Tucson, Arizona

1

BEHAVIOUR OF COAL MEASURES ROCKS

Design of engineering structures in sedimentary strata is complicated by the varying failure characteristics of individual lithological units. *Failure* is not, in an engineering sense, an easily definable state and can mean either reduction in strength or excessive deformation. For instance, stronger rocks, or rocks at low confining pressures will tend to fail through fracturing, with release of energy and dilation, accompanied by a reduction in load-bearing capacity from a peak to a residual strength. In coal mining the extreme manifestations of this type of failure will be associated with caving of strong sandstone beds overlying the coal seam during longwall mining, or possibly with collapse of isolated coal pillars in longwall, shortwall or room and pillar mining.

Weaker rocks or rocks at higher confining pressures will tend to deform homogeneously with minimum dilation and with continuing increase in load-bearing capacity. However, the deformation may be of such a large magnitude that the structure will cease to be usable and the overall effect will be equivalent to structural failure. In coal mining such extreme, near plastic, deformation will be associated particularly with weak or sheared fireclay seatearths which can cause large scale convergence in longwall face access roadways.

Failure of rock is not of course solely associated with deformation of the intact material. Sedimentary rocks contain bedding planes (often themselves containing interfacial and other shear zones), faults, joints and other discontinuities. These will affect the behaviour of most rocks, and particularly stronger rocks, in both the short term and the long term. Indeed the most common type of underground structural deformation results from loosening of blocks of rock bounded by discontinuities under the influence of gravity in a changing underground environment caused by excavation. It is often forgotten that the excavation of an underground structure, apart from changing the stress field and bringing gravitational forces into action, also changes the pattern of movement of air and water in the rock.

It is therefore important, before attempting to interpret case-history data in coal mines, to outline – albeit in a very simple way – some of the factors which control the deformation and failure of Coal Measures rocks.

1.1   Coal Measures rocks

The British Coal Measures comprise primarily mudstones, sandstones, seatearths and coals, arranged in a series of rhythmic or cyclic sequences. These sequences or cyclothems vary throughout the coalfield and through the succession. The central elements of the cycle are the coal seams and the underlying seatearths which can vary from soft organic fireclays to hard gannister. The coal seams are an average of 10 m apart and the sediments separating the seams usually contain varying proportions of clay minerals and fine-grained quartz, the former decreasing and the latter increasing through the upward succession. Thus, although sandstone layers are not uncommon above coal seams, the general succession comprises increasingly siliceous shales and mudstones above the seam, succeeded by siltstones, sandstones and grits beneath the seatearths of the overlying coal seam. Carbonates – principally ironstones – occur in the mudstones, and marine limestones occasionally occur as thin bands above the seam.

This type of stratigraphy, while specific to the British coalfield, is general to virtually all deposits of bituminous coal and anthracite. There are, however, occasional important variations in the thickness and succession of strata. For instance in the Indian coalfields the percentage of sandstone in the succession varies from 64 to 95 % (Saxena and Singh, 1982) with individual beds up to 50 m thick. In the Appalachian coalfield in Kentucky (Howell, Wright and Dearinger, 1976) thick beds of strong sandstone are a feature of the succession. Such sandstones also occur, more rarely, in the British Coal Measures.

The variable nature of the Coal Measures succession, and the existence of many indistinct lithological boundaries, means that it is difficult to relate accurately a lithological description to any engineering description, although Price (1966) has shown an inverse correlation between clay mineral content and strength. In an attempt to correlate lithology and simple test results, Davies (1977) collected together standard test data obtained by the Mining Research and Development Establishment of the National Coal Board on a suite of 927 Coal Measures rock samples from all parts of the British coalfield. These tests included compressive and tensile strength, scleroscope hardness number, durete,

Table 1.1 Compressive strengths of British Coal Measures rocks (MN m^{-2}) (after Davies, 1977)

Rock description	Mean	Variance	Standard deviation	Minimum value	Maximum value	Range	No. of sample
Sandstone	96.3	1582.0	39.8	27.1	303.0	235.9	123
Coarse sandstone	48.7	68.3	8.2	36.2	56.3	20.1	5
Medium sandstone	94.5	961.7	31.0	49.0	177.9	128.9	22
Fine sandstone	108.8	2130.0	46.2	34.9	221.2	186.3	47
Carbonaceous sandstone	73.6	918.6	30.3	19.1	146.6	127.5	17
Loose-grained sandstone	41.9	211.6	14.5	27.4	55.8	28.5	4
Silty sandstone	76.9	579.6	24.1	32.0	139.0	107.0	55
Siltstone sandstone	65.2	551.3	23.5	17.7	236.3	218.6	161
Sandy siltstone	67.8	462.4	21.5	35.3	121.3	86.0	42
Muddy siltstone	59.2	286.9	16.9	24.4	119.2	94.8	49
Sideritic siltstone	91.1	1458.1	38.2	42.1	115.7	73.6	14
Siltstone ironstone	68.2	280.5	16.7	47.6	93.0	45.4	5
Ironstone	114.1	1219.6	34.9	51.8	142.0	90.2	16
Mudstone	38.7	219.3	14.8	3.4	74.8	71.4	53
Silty mudstone	52.3	390.0	19.7	21.0	104.8	83.8	49
Shaly mudstone	31.1	304.6	17.4	7.6	75.1	67.5	13
Carbonaceous mudstone	43.4	690.1	26.3	19.9	149.7	129.8	33
Ferruginous mudstone	79.1	2563.1	50.6	40.7	152.7	112.0	4
Mudstone ironstone	52.8	85.1	9.2	42.8	63.8	21.0	5
Slickensided mudstone	25.2	10.9	3.3	19.7	28.6	8.9	6
Listric mudstone	28.1	73.1	8.5	15.8	41.4	25.6	9
Sandstone seatearth	63.5	155.6	12.5	48.8	89.9	41.1	10
Siltstone seatearth	48.7	230.5	15.2	26.7	81.3	54.6	41
Silty mudstone seatearth	39.1	62.7	7.9	25.3	58.9	33.6	25
Mudstone seatearth	30.5	109.6	10.5	4.7	56.5	51.8	47
Listric mudstone seatearth	26.6	23.6	4.9	21.8	33.1	11.3	4
Slickensided mudstone seatearth	19.3	178.7	13.4	5.5	37.8	32.3	5
Coal	30.4	154.3	12.4	11.9	55.0	43.1	53
Dull woody coal	48.3	58.5	7.6	41.0	60.6	19.6	5
Dull	26.0	165.1	12.8	15.8	43.8	28.0	4
Durain vitrain	29.7	51.8	7.2	20.0	37.9	17.9	6

specific energy index and abrasivity. In Table 1.1 compressive strength data are collected under the quoted lithological groups. Examination of the mean range and standard deviations of the data (calculated by Worsey, 1978) show that there is little consistency within or between lithological groups. For instance, the sandstones, with the exception of those classified as medium, all have members of their sample number with compressive strengths less than the mean strengths of the mudstones. There is, however, a general tendency for sandstones to be stronger than seatearths. The weakest materials are sheared, slickensided or listric mudstones and seatearths.

The most important conclusion is, however, that the data give a very unsatisfactory description of rock behaviour. What is really needed is less a quantitative description of strength, than a qualitative description of deformation characteristics under the types of stress which can exist in coal mining.

1.2   Deformation characteristics of Coal Measures rocks

The deformation characteristics of rocks in general, including some Coal Measures rocks, have been described in *Engineering Behaviour of Rocks* (Farmer, 1983). In order to define the deformation characteristics of a rock in a way which will be relevant to the design of mine structures, several features must be incorporated into any laboratory test programme:

(a) The testing machine must be capable of simulating the behaviour of an idealized rock mass or continuum surrounding the test specimen. Thus, if the specimen starts to fracture, forces which it can no longer transmit must rapidly be transferred (as they would be around an excavation) so that the residual strength of the rock can be examined. This requires a servo-controlled testing machine.
(b) The testing equipment must be capable of applying constraining forces similar to those which exist in coal mines. This requires a triaxial cell with a confining pressure range up to about $40\,\mathrm{MN\,m^{-2}}$, the horizontal pressure which would be expected at depths of about 1500 m in a geostatic stress field with equal horizontal and vertical components.

Fig. 1.1   Axial stress–axial strain and volumetric strain–axial strain curves for Coal Measures sandstone specimens tested in triaxial compression at confining pressures from 0 to $21\,\mathrm{MN\,m^{-2}}$. In the photographs of fractured 75 mm diameter specimens, test confining pressures were: bottom row, left to right 4, 7, $7\,\mathrm{MN\,m^{-2}}$; top row, left to right, 14, 14, $21\,\mathrm{MN\,m^{-2}}$ (from Farmer, 1983; Price, 1979).

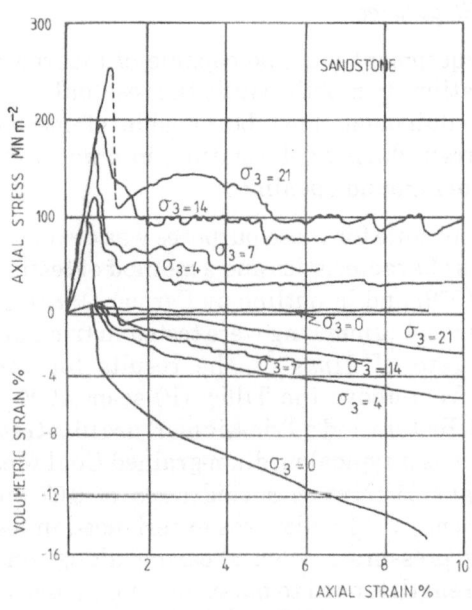

(c) The testing equipment must be capable of incorporating sufficiently large deformations to mobilize fully the residual strength of the rocks.

(d) The testing equipment must be capable of measuring dilation or volumetric strain during deformation, in order to allow estimates of closure in underground openings.

A triaxial cell suitable for these purposes was developed in the author's laboratory and its characteristics and a method of testing are described in detail by Price (1979) and in outline by Farmer (1983). Specimens of rock 75 mm in diameter by 150 mm long were tested in triaxial compression at a constant strain rate of $21\mu\varepsilon\,s^{-1}$. The results for three typical Coal Measures rocks from above the Tilley (P) seam at Woodhorn Colliery, Northumberland, Britain and a Triassic marl are illustrated in Figs 1.1–1.4.

The *sandstone* was a typical medium-grained Coal Measures sandstone with a unit weight $25\,kN\,m^{-3}$, a void ratio of 0.07 and a compressive strength of $97\,MN\,m^{-2}$. Difficulty was experienced in testing this rock at higher confining pressures, when shearing along single or conjugate planes at peak strength tended to burst the triaxial test membrane. At low confining pressures (up to $14\,MN\,m^{-2}$) failure was predominantly through vertical fractures (Fig. 1.1) parallel to the minor (confining) principal stress – typically Griffith-type tensile fracture.

At higher confining pressures shear planes were formed. The stress–strain characteristics (Fig. 1.1) showed a rapid fall from a peak to a residual stress level accompanied by rapid dilation at low confining pressures and representing quite marked brittle or strain-softening behaviour.

The *silty sandstone* was a medium-grained Coal Measures sandstone with silt inclusions, having a unit weight of $23.4\,kN\,m^{-3}$, a void ratio of 0.11 and a compressive strength of $61\,MN\,m^{-2}$. Its deformation behaviour was similar to that of sandstone, although shear and conjugate shear planes (Fig. 1.2) were much clearer at the higher confining pressures. The stress–strain characteristics (Fig. 1.2) showed a less rapid fall from peak to residual strength and the residual strength/peak strength ratio was higher. There was also reduced dilation at higher confining pressures.

All of these characteristics – albeit affected by the layered nature of the specimen – are more pronounced in the case of the *mudstone*, and there

Fig. 1.2 Axial stress–axial strain and volumetric strain–axial strain curves for Coal Measures silty sandstone specimens tested in triaxial compression at confining pressures from 0 to $42\,MN\,m^{-2}$. In the photographs of fractured 75 mm diameter specimens, test confining pressures were: bottom row, left to right 0, 4, 7, $14\,MN\,m^{-2}$; top row, left to right, 21, 29, 36, $42\,MN\,m^{-2}$ (from Farmer, 1983; Price, 1979).

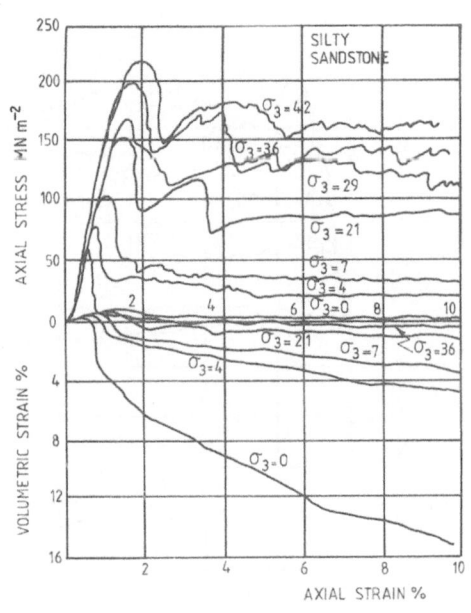

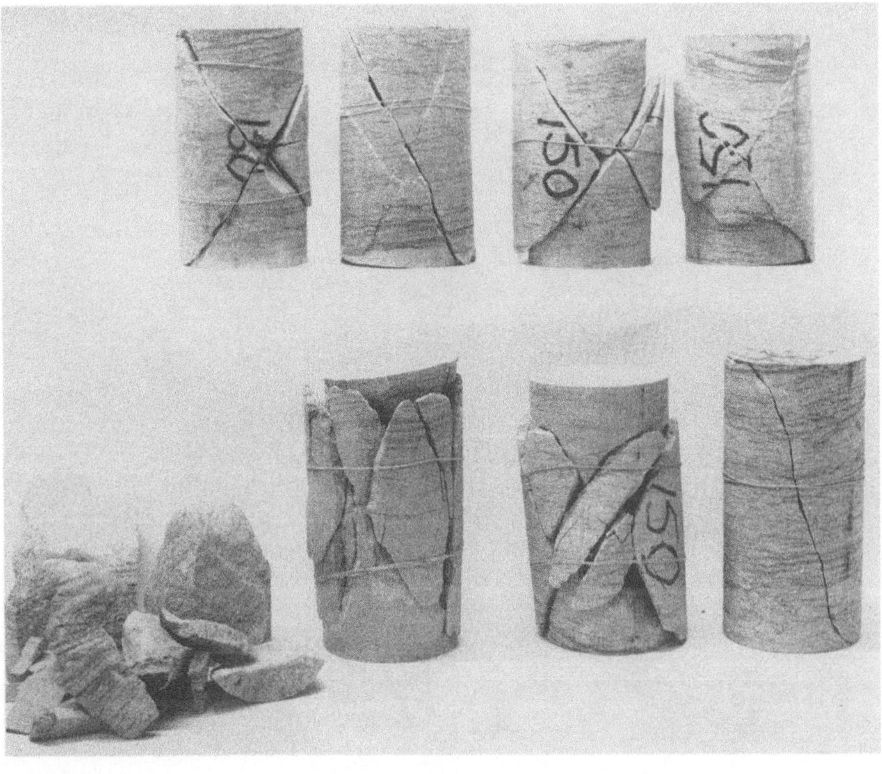

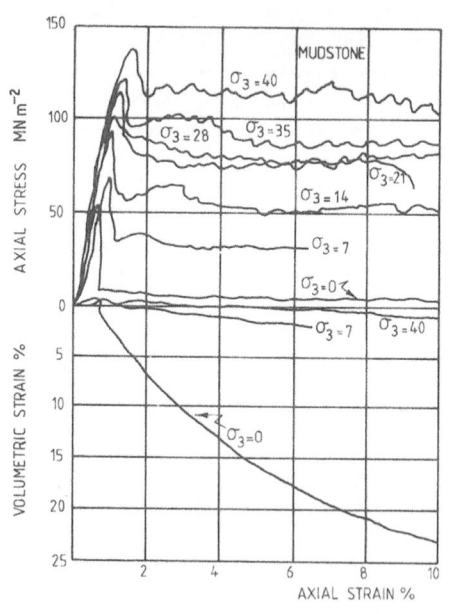

are indications of ductile deformation (Fig. 1.3). This was a fine-grained Coal Measures mudstone with occasional bands of silt and ironstone with a unit weight of $26.3\,\mathrm{kN\,m^{-3}}$, a void ratio of 0.02 and a compressive strength of $55\,\mathrm{MN\,m^{-2}}$.

The two common Coal Measures rocks which could not be obtained in a satisfactory form for testing were fireclay seatearths and bituminous coals – the former because they are invariably found in a sheared state, the latter because they contain cleat structures with narrow spacing. Both make it virtually impossible to prepare specimens of reasonable size for testing. In the case of the former, a *carnallite marl* – an anhydritic marl overlying the Permo–Triassic evaporites at Boulby Mine, North Yorkshire, Britain, was chosen as a substitute. This was mechanically similar to typical Coal Measures seatearths (which are considered further in Section 1.4) apart from the sheared structure, and had roughly the same liquid and plastic limits. It had a unit weight of $21.7\,\mathrm{kN\,m^{-3}}$, a void ratio of 0.05 and a compressive strength of $10\,\mathrm{MN\,m^{-2}}$. The deformation behaviour (Fig. 1.4) is characterized by a progressive change from overall strain-softening to strain-hardening behaviour with increasing confining pressure. There is also physical evidence – again hindered by the layered structure – of ductile behaviour.

As a substitute for bituminous coal, specimens of South Wales *anthracite* were chosen for testing. Although lacking the cleated structure of coal, it has the characteristic high peak and residual strengths (Fig. 1.5) exhibited by most bituminous coals in small samples (see Evans and Pomeroy, 1966). The fractures at higher confining pressures also exhibit typical single shear brittle fracture characteristics (Fig. 1.6). The brittle behaviour under typical engineering stresses has considerable importance in the design of coal pillars (see Chapter 2).

1.3   A general description of rock deformation

It is useful at this stage to comment generally on how the information in Figs 1.1–1.5 can be used to provide a simple description of the mechanics of rock deformation. For instance, confining pressure had a pronounced effect on the stress–strain characteristics of all the rocks tested. If the

Fig. 1.3 Axial stress–axial strain and volumetric strain–axial strain curves for Coal Measures mudstone specimens tested in triaxial compression at confining pressures from 0 to $40\,\mathrm{MN\,m^{-2}}$. In the photographs of fractured 75 mm diameter specimens, test confining pressures were: bottom row, left to right 7, 14, $21\,\mathrm{MN\,m^{-2}}$; top row, left to right, 28, 35, $40\,\mathrm{MN\,m^{-2}}$ (from Farmer, 1983; Price, 1979).

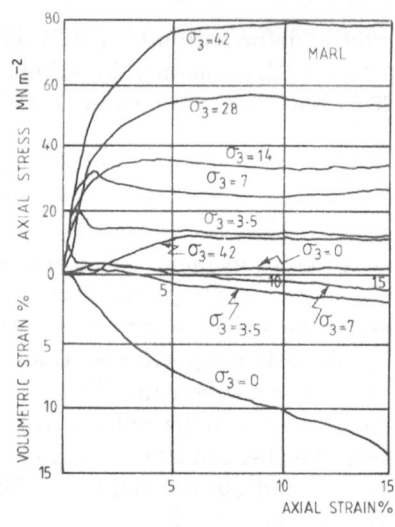

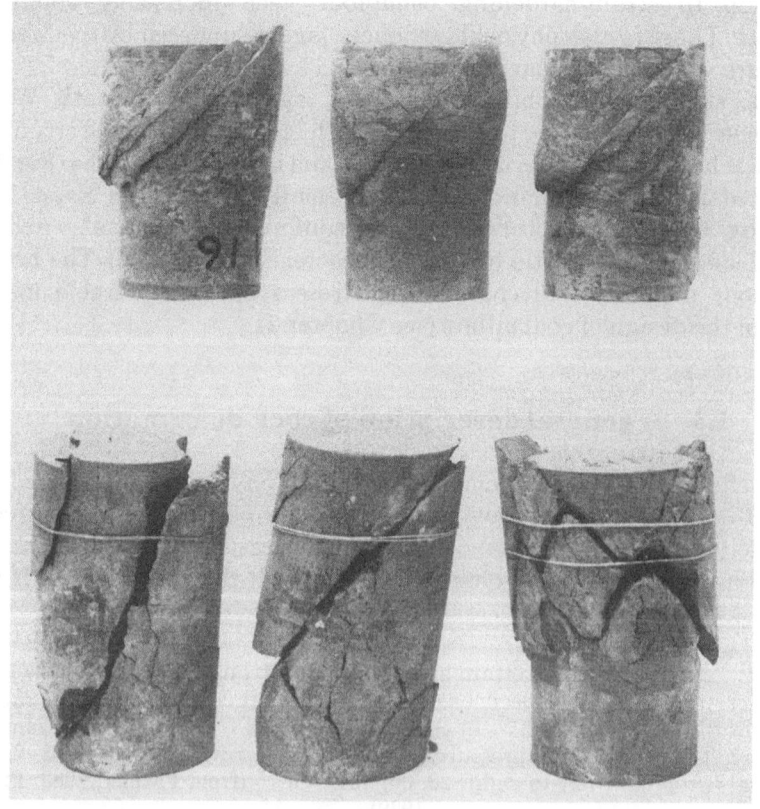

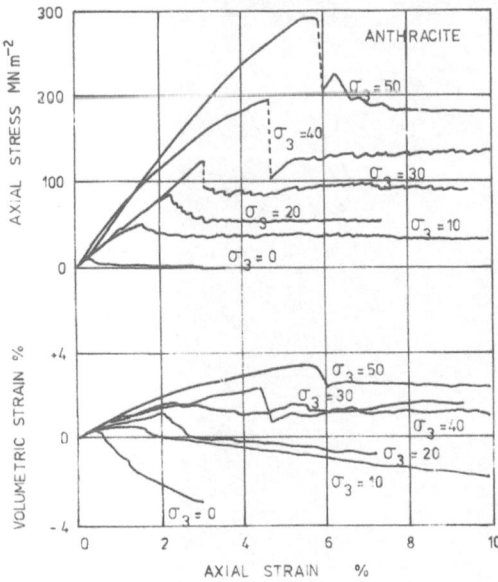

Fig. 1.5 Axial stress–axial strain and volumetric strain–axial strain curves for anthracite coal specimens tested at confining pressures from 0 to 50 MN m^{-2} (from Godden, 1982).

undulations of the stress–strain curves are attributed to uneven operation of a relief valve rather than stick–slip mechanisms, then it can be seen that increasing confinement increases both peak and residual strength. In the case of the marl this results in an eventual change from strain-softening to strain-hardening behaviour.

The effect of confining pressure can be illustrated most clearly by plotting the peak and residual strength envelopes for the sandstones, mudstone and marl in Figs 1.1–1.4 in terms of axial stress at fracture (σ_{1f}) and residual strength (σ_{1r}) against confining pressure (σ_3) in Fig. 1.7

This gives strength envelopes in the form:

$$\sigma'_{1f} = \sigma'_{cf} + k_p \sigma'_3 \tag{1.1}$$

$$\sigma'_{1f} = \sigma'_r + k_p \sigma'_3 \tag{1.2}$$

Fig. 1.4 Axial stress–axial strain and volumetric strain–axial strain curves for Triassic marl specimens tested in triaxial compression at confining pressures from 0 to 42 MN m^{-2}. In the photographs of fractured 75 mm diameter specimens, test confining pressures were: bottom row, left to right 0, 3.5, 7 MN m^{-2}; top row, left to right, 14, 29, 42 MN m^{-2} (Farmer, 1983; Price, 1979).

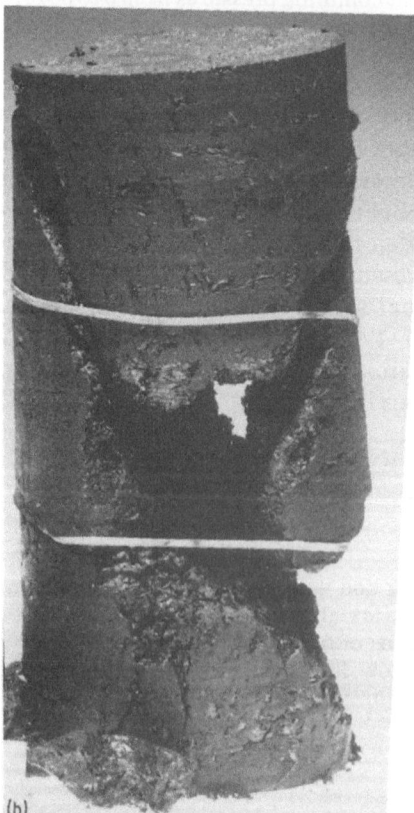

Fig. 1.6 Fractured specimens of anthracite coal tested at (a) 50, (b) 10, (c) 40 and (d) 50 MN m^{-2} (from Godden, 1982).

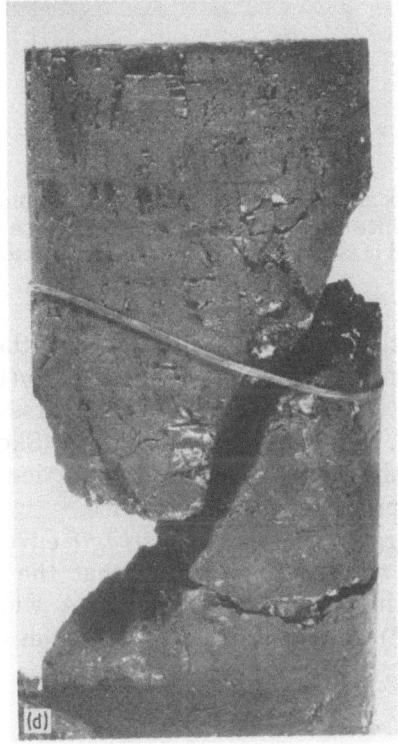

13

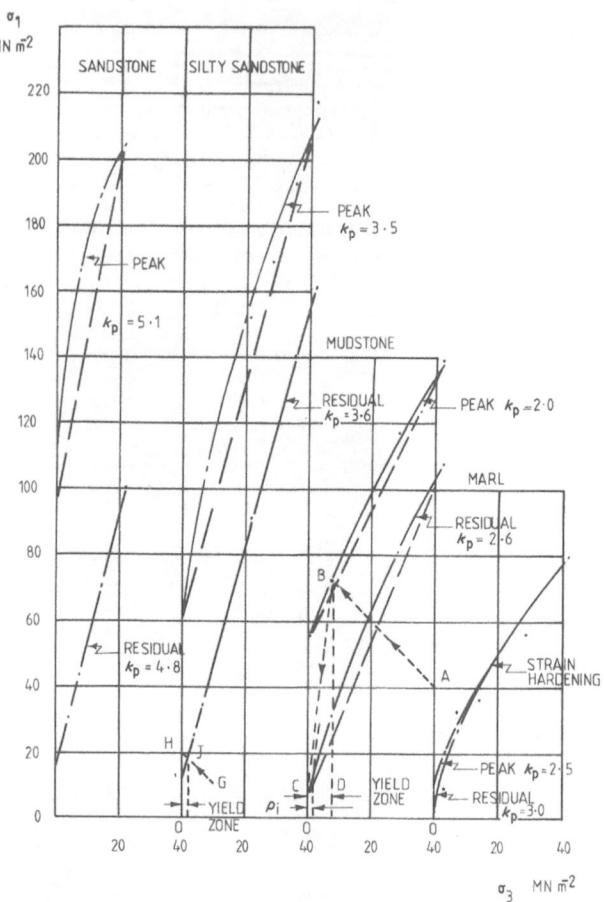

Fig. 1.7 Peak and residual strength envelopes plotted from the data in Figs 1.1 to 1.4. $k_p = (1 + \sin\phi)/(1 - \sin\phi)$ is the average slope of the envelope (see Equations (1.1), (1.2)) and is known as the triaxial stress factor or coefficient of passive earth pressure.

where σ_{cf} is the compressive strength, σ_r is the residual strength and k_p is the triaxial stress factor, equal to $(1 + \sin\phi)/(1 - \sin\phi)$.

The prime is used to denote effective stress, to conform with engineering practice, although it can be shown (Skempton, 1961) that in rocks which are stiff compared with other particulate materials, the effective stress concept has reduced importance.

It can be argued that the *peak strength* envelope represents the ultimate strength of the intact rock and that the *residual strength* envelope represents the frictional resistance. A wider selection of rocks (see Farmer, 1983) would provide evidence of an eventual coming together of the two curves as in the case of the marl, and this, together with the strength envelopes, can form a basis for a general description of rock deformation which can be particularly useful in engineering design in a succession of sedimentary rocks which behave in widely differing ways.

For instance, it has been shown by Brown and Hoek (1978) that at most depths below the immediate surface in sedimentary strata, the principal geostatic stresses are likely to be equal. Thus if a circular tunnel is driven in them and they are assumed to react in an elastic manner to short-term changes in compression the stress at the tunnel sidewall will change from:

$$\sigma_0 = \sigma_1 = \sigma_2 = \sigma_3 = \gamma z \qquad \text{before excavation}$$

where σ_0 is the stress at depth z and γ is the unit weight of the rock to:

$$\sigma_{\text{radial}} = 0, \sigma_{\text{tangential}} = 2\sigma_0 \qquad \text{before excavation.}$$

This is based on the thick cylinder solution found in any text on stress analysis (see Jaeger and Cook, 1979). The change in stress can be represented in Fig. 1.7 as an excavation stress path A–B, based on an assumption that at a depth of 1000 m, $\sigma_0 = 40 \, \text{MN m}^{-2}$ and that if a circular tunnel is constructed then at the surface of the tunnel, $\sigma_{\text{radial}} = \sigma_3 = 0$ and $\sigma_{\text{tangential}} = \sigma_1 = 2\sigma_0 = 80 \, \text{MN m}^{-2}$ which would be sufficient to fracture the mudstone and silty sandstone, but *not* the sandstone.

The effect of this fracturing will be considered later; but the mechanics of deformation can be considered immediately. When the stress path intersects the peak strength curve at point B the rock will fracture and a zone of fractured rock having zero confinement at the excavation boundary will be formed. In other words the excavation stress path will follow the line BC. Provided that some restraint is available at the tunnel surface, the residual strength mobilized at C will extend into the fractured rock along the residual strength envelope until it reaches a magnitude D, the confining stress required to suppress further fracture of the rock mass. This represents the boundary of the fracture zone, and if the radius of the tunnel is known the extent of the *fracture zone* can be evaluated very easily from the magnitude of D using the thick cylinder equation (for example, see Fig. 3.6). If the excavation is supported by supports capable of exerting an internal pressure p_i to the rock, then the extent of the fracture zone will be reduced. However, a support pressure equivalent to D would be required to prevent fracture completely and this is not feasible either economically or from the point of view of practical engineering. Supports in mines work essentially by providing the conditions under which the residual strength of the rock can be mobilized.

One further point which must be made in a general way, is that all simple elastic analyses such as the thick cylinder theory ignore body forces. In mining engineering these have supreme importance and support pressure due to rock loosening under gravity is more important than that attributed to rock fracture (see Ward, Coates and Tedd, 1976). Another point is the unfortunate use – deliberately perpetuated in Fig. 1.7 – of the term yield zone to describe a zone of fractured rock. The importance of yield can be more clearly illustrated by considering what would happen to the marl in Fig. 1.7 under the same stress path as the mudstone. In this case the stress path would intersect that part of the curve beyond the joining of the peak and residual envelopes. This represents the point (Fig. 1.4) at which deformation becomes stable – in other words a stress increment is always required to cause a deformation increment – and dilation is small. There are various names for this point or line in engineering, the most common being the stability line or *brittle–ductile transition*. It is sometimes assumed that constant volume plastic deformation occurs at the brittle–ductile transition, but this is not the case (see Price and Farmer, 1979, 1980, 1981). A useful test for this is to plot the data for the marl from Fig. 1.4 in terms of the ratio of deviatoric (q) and spherical (p) stress against the ratio of change in volumetric strain ($\delta \varepsilon_v$) and change in shear strain ($\delta \varepsilon_s$). If the deformation is plastic then the plastic potential function and the yield function defining the yield surface should coincide, giving (see Atkinson and Bransby, 1978) an associated flow rule of the form:

$$\frac{q}{p} = M - \frac{\delta \varepsilon_v}{\delta \varepsilon_s} \tag{1.3}$$

where M is the ordinate intercept.

In the case of the marl (Fig. 1.8) it can be seen that a multiplying factor of about 6 is required for the strain increment ratio, indicating a degree of non-plasticity over the test confining pressure range. This type of yield surface covering the range between the brittle–ductile transition and the critical state at which the material deforms plastically is sometimes called the Hvorslev surface (Schofield and Wroth, 1968). In mining engineering it is important in defining a dilatant deformation regime where resistance to deformation in weak rocks is greater than would be expected by the use of the term 'ductility' and which conditionally excludes the term 'creep' (see Farmer, 1983). The more important criterion is, however, the brittle ductile transition, defining as it does the point where the behaviour of the rock changes from strain-softening to strain-hardening.

It is worthwhile attempting to explain the basic mechanics of these two types of deformation, particularly because they highlight the similarities between the behaviour of rock and other engineering materials. In the case of strain-hardening behaviour, strain will tend towards homogeneity throughout the confined specimen, since those elements of the rock which

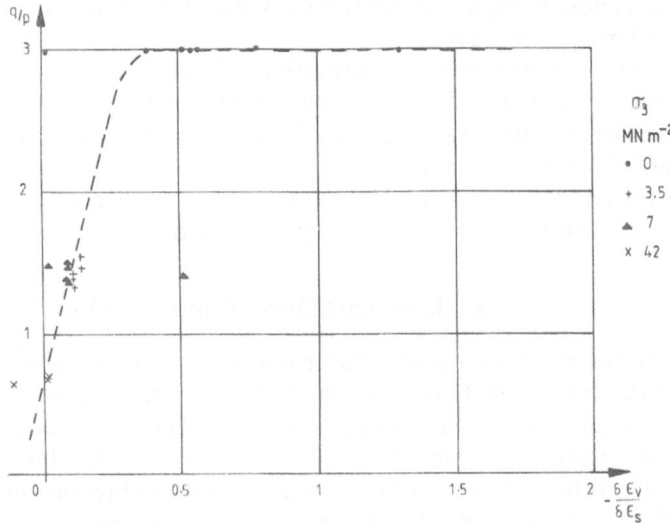

Fig. 1.8 A Hvorslev surface for marl computed from the data in Fig. 1.4. Note $q = \sigma_1 - \sigma_3$, $p = 1/3\,(\sigma_1 + 2\sigma_3)$ so that $q/p = 3$ corresponds to the uniaxial state where $\sigma_3 = 0$ (from Farmer, 1983; Price and Farmer, 1981).

have been strained most will be stronger than those which have been strained less. Then at failure, defined as the point of maximum deviatoric stress, or at the end of the test, a uniformly deformed specimen would be expected. In the case of the marl specimens tested at higher confining pressures (Fig. 1.4) there is a combination of homogeneous (bulging) and shear deformation although the stress–strain characteristics exhibit a change to strain-hardening behaviour.

In the case of strain-softening behaviour, the specimens become weaker with increased strain. Thus strain will tend to be inhomogeneous and further strain will be concentrated in the weaker elements of the rock, which have already been subjected to the most strain. Thus following peak stress, thin zones of concentrated strain – or shear planes – will be expected to develop. This is illustrated most clearly by the conjugate shear planes in the silty sandstone of Fig. 1.2.

The importance of the data from Figs 1.1–1.5 can be summarized briefly from the general point of view of coal mining design as follows:

(a) The peak strength of all the rocks tested except marl is sufficiently high to resist fracture at most coal mining depths whatever the stress redistribution.

(b) In dry rocks, ductile behaviour only occurs at relatively high confining pressures in weak rocks such as seatearths. These are the only rocks which approach the critical state and deform on a large scale at coal mining pressures.

(c) Even in these rocks deformation of a strain-softening type will probably occur under conditions of low confinement – as found in an excavation sidewall. The resultant residual strength can then be mobilized to assist support.

(d) Rocks have only been considered in a dry state. When wet, seatearths may be expected to deform in a plastic manner.

1.4 Deformation of seatearths

In Coal Measures strata, the materials most likely to deform excessively are seatearths – whether sheared or intact – and the key to predicting strata deformation often lies in predicting the behaviour of the seatearths. Unfortunately, the structure of seatearths is such that laboratory specimens are difficult to obtain in a suitable state for testing, and the results would probably be of doubtful value in any case. This is partly due to the structure but mainly due to the very wide variability in seatearth composition and in the reaction of seatearths to the presence of water.

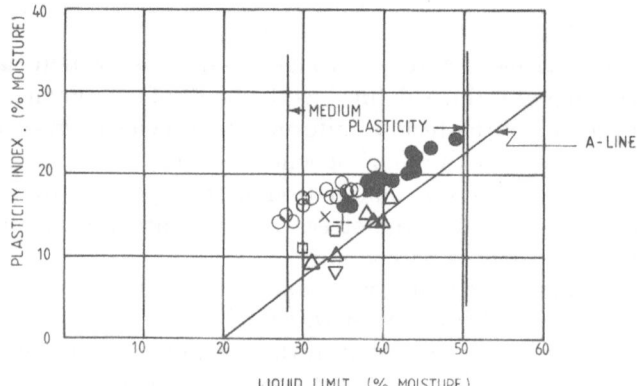

Fig. 1.9 Plasticity chart of plasticity index against liquid limit (both having units of percentage moisture content) including data from Coal Measures seatearths:

○ Clays from unsheared horizons in Northumberland (Salahey, Money and Dearman, 1977);

● Clays from sheared horizons, Northumberland (Salahey *et al.*, 1977);

+ Sheared and unsheared clays, average of 13 samples from South Wales (Stimpson and Walton, 1970);

☐ Unsheared clays, Barnsley seam seatearth;

△ Unsheared clay, Midlands, (Lain, 1974);

▽ Silty seatearth, Durham (Lain, 1974);

× Anhydritic marl (see Fig. 1.4).

Mineralogical analysis of British seatearths shows the presence of chlorites, montmorillonites, illites, kaolinites and quartz in varying quantities, and a similarly variable composition is found in most of the coalfields of the world. Depending on the mineralogical composition, seatearths are usually characterized as swelling materials, but this is not, strictly speaking, correct. Like most rocks with a high clay mineral content they have a capacity to adsorb water and this causes some swelling, but it is likely to be limited in the case of most seatearths where calcium cations tend to be dominant and quartz content is high. This can be illustrated by several examples from various sources.

In Fig. 1.9 data obtained from various reconstituted British seatearths on liquid limit (the maximum percentage moisture content at which a clay can remain plastic) and plastic limit (the minimum percentage moisture content at which clay can remain plastic) are compared through a plot of plastic index (the difference between liquid and plastic limits) against liquid limit. Compared with Casagrande's A-line classification this indicates inorganic clays of low to medium plasticity, low activity and low organic content and containing some silt or sand size particles.

Various relations have been proposed between plasticity index (PI) and the properties of clays. For instance according to Casagrande's classification (Wagner, 1957) inorganic clays of medium plasticity would be expected to have a dry unconfined compressive strength of the order of $2\text{--}10\,MN\,m^{-2}$ with relatively low dilatancy and swelling. Average strength when saturated of such clays will be:

	ϕ'	$C'(kN\,m^{-2})$
Peak	16°	1000–1500
Residual	11°	–
Cu	0°	200–300

At a slightly more speculative level, Kenny (1967) quotes an approximate range of ϕ' values for normally consolidated clays from 35° for a PI of 10, 30° for a PI of 25, 25° for a PI of 50 and 20° for a PI of 100. This confirms Skempton's (1964) observation that residual ϕ' decreases and plasticity increases with increasing clay mineral content in all types of clay. This is an important observation to make in the case of seatearths. Confirmation can be found in the data of Smart, Rowlands and Isaac (1982) reproduced in Fig. 1.10, which relates strength, cohesion and particularly swelling (in terms of elongation of core samples), to quartz content for British Coal Measures seatearths. It can be seen here that swelling is relatively low. Surprisingly Smart *et al.* (1982) observe that ϕ' values do not appear to be affected by quartz content.

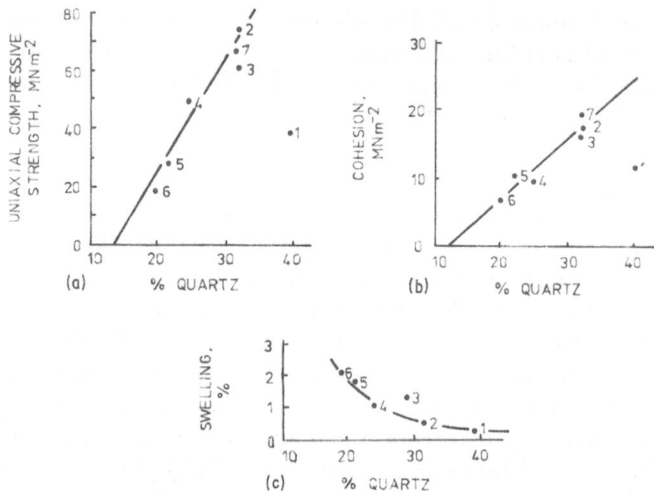

Fig. 1.10 Effect of quartz content on (a) uniaxial compressive strength, (b) cohesion and (c) swelling of oven-dried specimens of Coal Measures seatearths from South Wales collieries. Markham M21 (1 floor, 2 roof) Taff Merthyr B3 (3 floor) Merthyr Vale B20 (4 roof) Marine BL11 (5 roof, 6 floor) Cwm 713 (7 roof) (after Smart *et al.*, 1982).

It would be possible, although difficult, to relate rock strength to moisture content. Van Eekhout and Peng (1975) note that plasticity increases rapidly at relative humidities in the mine atmosphere greater than 50% and a reduction in compressive strength by up to an order of magnitude in saturated rocks. In mines in the United States subjected to seasonal changes in humidity it can be shown that directly related increases in roadway convergence (Chugh and Missavage, 1981) and increases in falls of roof (Stateham and Radcliffe, 1976) occur with increases in relative humidity.

Moisture absorption is, however, a surface phenomenon and the rate of absorption is much greater along bedding planes than across them. It follows therefore that sheared seatearths (see Section 1.5) absorb water much more rapidly than intact seatearths. The nature of absorption is illustrated by Holmes (1983) in scanning electron micrographs in Fig. 1.11. These are samples from the Barnsley seam seatearth at Markham Main Colliery, Yorkshire. Figures 1.11(a), (c) are of dry seatearth and 1.11(b), (d) of samples immersed in water for 2 hours. At low magnification (Figs 1.11(a), (b)) the effects of immersion can be clearly seen, water penetrating the surfaces and causing partial disintegration. At high magnification (Figs 1.11(c), (d)) it can be seen that the matrix away from the surface has not been affected.

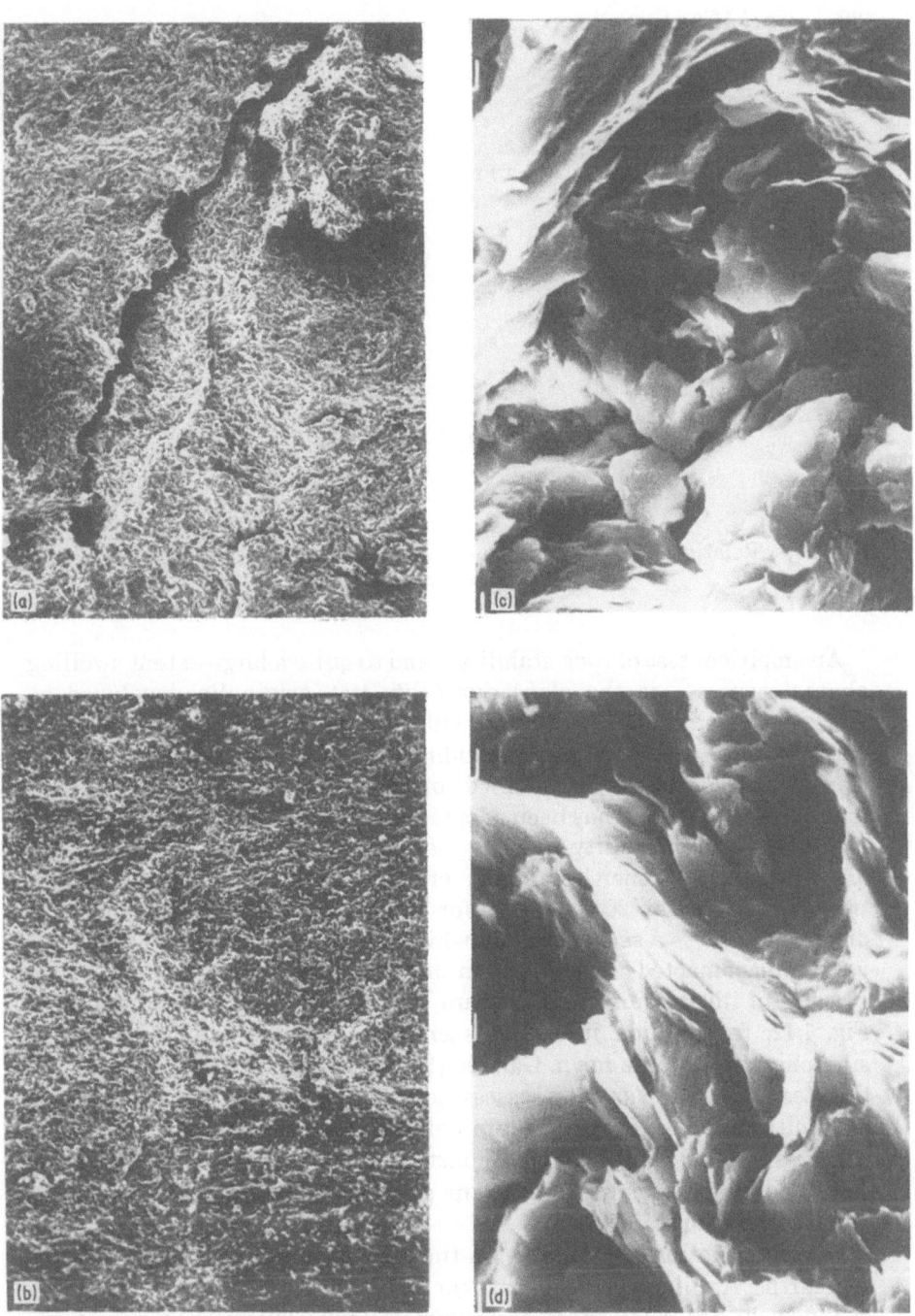

Fig. 1.11 Scanning electron micrographs of Barnsley seam seatearths from Markham Main Colliery, Yorkshire: (a) × 75 dry; (b) × 75 immersed; (c) × 5000 dry; (d) × 5000 immersed (photograph P. Holmes).

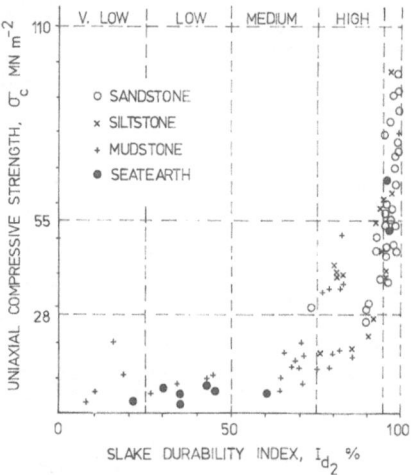

Fig. 1.12 Relation between slake durability index I_{d2} and rock strength (after Hassani and Scoble, 1981).

An empirical test of rock stability – and to quite a large extent swelling characteristics – is the *slake durability test*, originally developed by Badger, Cummings and Whitmore (1956) to assess the likely disintegration of Coal Measures shales, including seatearths, during coal preparation. It was subsequently modified for core logging by Franklin, Broch and Walton (1971) and has been adopted (Brown, 1981) as a standard ISRM test. The slake durability index I_{d2} is a measure of the proportion of a 0.5 kg broken specimen remaining after rotation in a sieve through a trough of water at 20 rev min^{-1} for 20 minutes with an intermediate drying cycle. It is a searching test which leaves only non-clay mineral or very well compacted clay mineral in the sieve at the end of the test.

A useful illustration from Hassani and Scoble (1981) is reproduced in Fig. 1.12, from which it can be seen that only sandstones and some siltstones resist breakdown. Gamble (1971) has suggested a way in which this test might be used to assess seatearths, by plotting I_{d2} against plasticity index. In Fig. 1.13, data collected by Lain (1974) and Holmes (1983) are included on Gamble's chart. These are primarily British seatearths which once again confirm the general trend of medium plasticity and low slake durability.

Less information is available on the mechanics of deformation of wet seatearths. An illustration from shear box tests performed by Lain (1974) is shown in Fig. 1.14. These are 60 mm × 60 mm × 20 mm specimens from sample C in Fig. 1.13 with a plasticity index of 15 and negligible slake durability. They were tested in a shear box at a strain rate of 2 % per minute

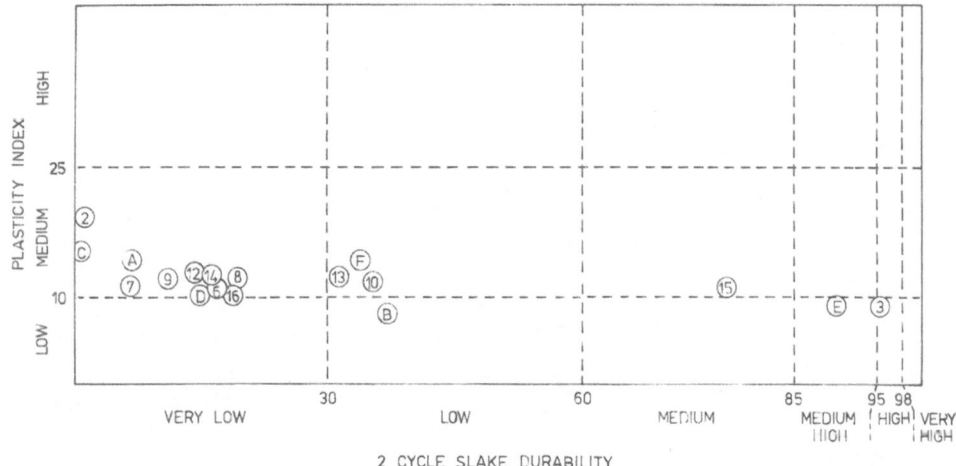

Fig. 1.13 A durability–plasticity classification suggested by Gamble (1971) with data from various sources collected by Lain (1974), A–F and by Holmes (1983), 1–16.

in a *saturated undrained* state – possibly a slightly higher rate than would be expected in coal mine deformation conditions.

The results which are typical of slightly over-consolidated clays, are reproduced in Fig. 1.15(a) as peak and residual strength envelopes and may be taken to represent the worst seatearth conditions likely to be encountered. They lead to certain interesting observations. It can be seen that even in a saturated state, seatearths exhibit peak and residual

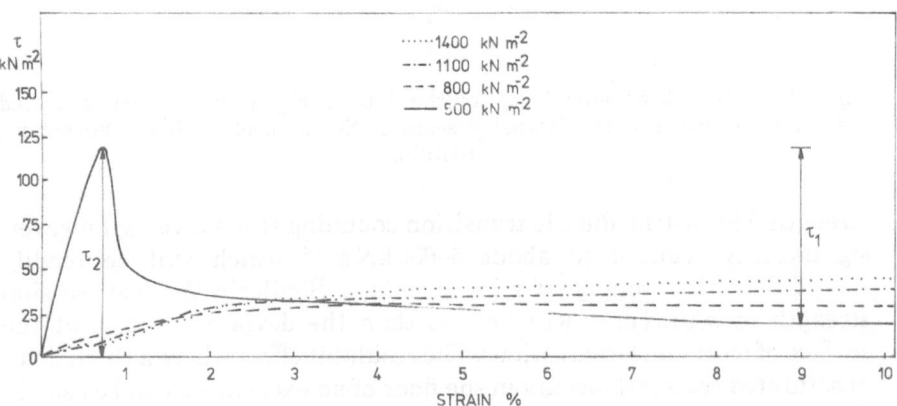

Fig. 1.14 Shear box tests on specimens of saturated seatearths from Birch Coppice Colliery, Staffordshire, Britain (after Lain, 1974).

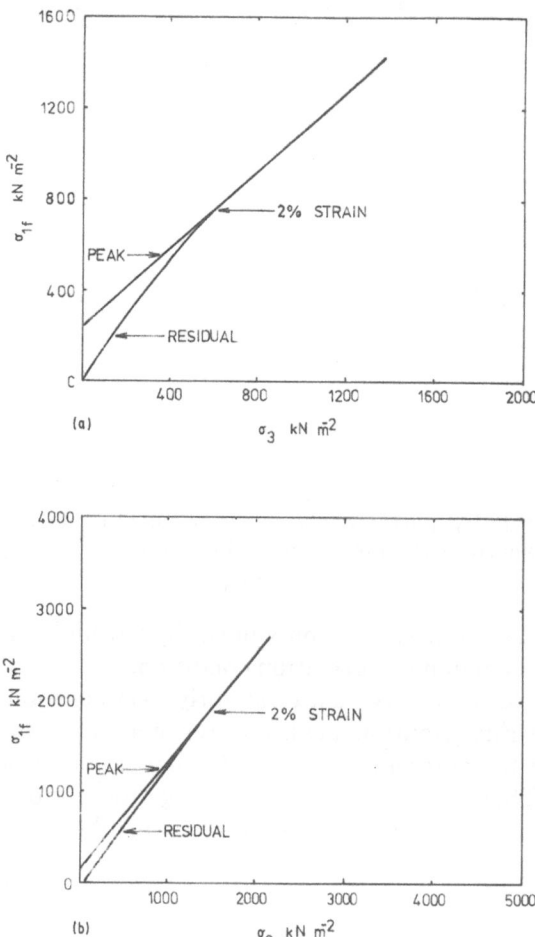

Fig. 1.15 Strength envelopes: (a) for the data in Fig. 1.14; (b) from saturated seatearths from below the Barnsley seam at Stillingfleet Colliery, Yorkshire, Britain.

strength. The brittle–ductile transition confining stress level is, however, significantly reduced to about 5–600 kN m^{-2} which will be readily exceeded in the vicinity of mine openings. Similarly the compression strength of 200 kN m^{-2} will be less than the deviatoric stress at the surface of most mine excavations. The resultant effect where a thick layer of saturated seatearth occurs in the floor of an excavation can be seen in Fig. 1.16.

In Fig. 1.15(b) results from triaxial tests on a similar saturated seatearth are reproduced, showing again the potential weakness of the material.

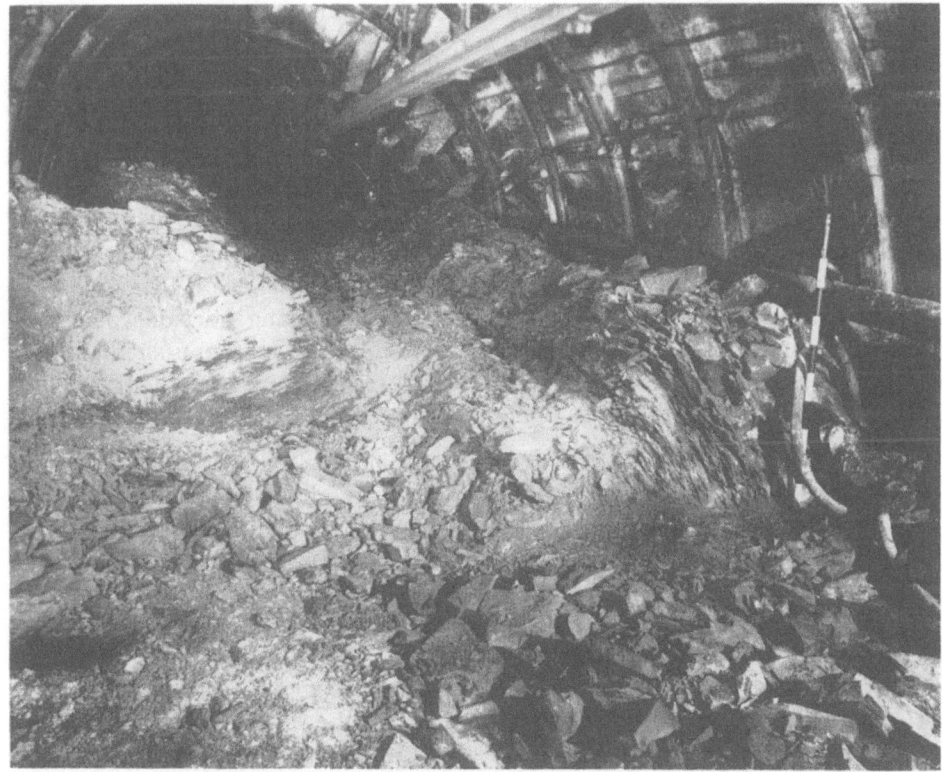

Fig. 1.16 Typical floor heave in a mine roadway resulting from seatearth deformation at Cotgrave Colliery, Nottinghamshire, Britain, Deep Hard seam (photograph by L.J. Thomas).

1.5 Shear zones

In a seminal paper to the 4th Congress of the International Society of Rock Mechanics at Montreux in which he reflected on two decades of involvement in rock mechanics, Deere (1979) recanted many of the rather esoteric approaches to rock mass classification (see for instance Deere, Hendron, Patton and Cording, 1966; Deere and Miller, 1966) he had previously encouraged. Although he re-emphasized the importance of verification of rock mass design parameters through observation of structural behaviour, he emphasized that the majority of engineering problems encountered in rock could be related to two adverse geological features:

(a) the thin but continuous zones of weakness found in sedimentary rocks along certain bedding surfaces – the so-called *bedding plane shear zone* and foliation shear zone; and

(b) the transition zone in profiles of weathered rock of all types.

Bedding plane shear zones are of course a particular feature of Coal Measures rocks. They are found characteristically in beds with a high clay mineral content such as seatearths or shales where these occur in sequences of sandstone or limestone. They constitute zones which are often difficult to detect, which are weak in shear and tension, and which may have continuity over a considerable distance. They occur mainly at the contact of seatearths and shales with overlying horizons of coal, sandstones, ironstones and mudstones. Not all Coal Measures shear zones are bedding plane shears and four types of genesis have been identified by various workers (see particularly Salahey *et al.*, 1977 and Stimpson and Walton, 1970); these are:

(a) *Intraformational shear zones* which are closely associated with faulting both in occurrence and orientation. The shear planes are formed by non-homogeneous straining (similar to the strain-softening mechanism described in Section 1.3) and can have large areal extent and thickness in excess of 1 m. Intraformational shears are typically highly polished and striated.

(b) *Bedding plane shear zones of tectonic origin*, usually developed on the upper contact of a rock with high clay mineral content, and a more competent overlying rock. These are formed by folding and are concentrated on the limbs of folds. The thickness is usually less than 100 mm. There is often some association between intraformational and bedding plane shears of tectonic origin.

(c) *Bedding plane shears of sedimentary origin* which are similar to those of tectonic origin, occurring typically beneath the contact of coarse sandstones overlying rocks with high clay mineral content. They result from prelithification differential compaction of the coarse sands (low compressibility) and clays (high compressibility). This results in penetration of the clays by the sands during subsequent deposition, causing interfacial shear displacement with limited areal extent and thickness. These shears are commonly encountered at the base of scour channel deposits above coal seams.

(d) *Lystric surfaces* which occur as a zone of randomly orientated shear planes usually in clay rich seatearths but also in carbonaceous mudstones. They comprise individually slickensided and polished curvilinear surfaces with no preferred orientation and very limited areal extent and thickness which recur frequently in the horizon. They result from localized volume changes during diagenesis, caused by consolidation, syneresis and organic decomposition. They are particu-

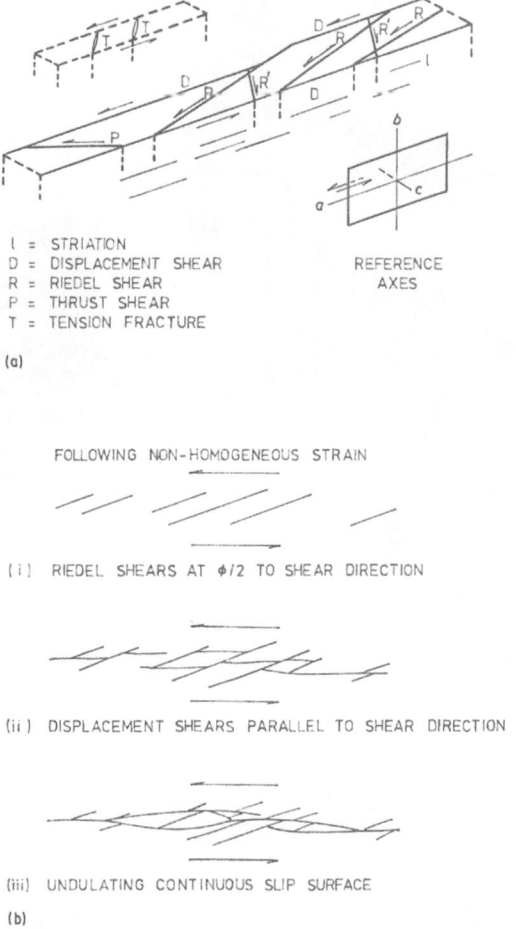

l = STRIATION
D = DISPLACEMENT SHEAR
R = RIEDEL SHEAR
P = THRUST SHEAR
T = TENSION FRACTURE

REFERENCE
AXES

(a)

FOLLOWING NON-HOMOGENEOUS STRAIN

(i)   RIEDEL SHEARS AT $\phi/2$ TO SHEAR DIRECTION

(ii) DISPLACEMENT SHEARS PARALLEL TO SHEAR DIRECTION

(iii) UNDULATING CONTINUOUS SLIP SURFACE

(b)

Fig. 1.17 Schematic representation of (a) the characteristics of a shear zone formed by tectonic movement; (b) the successive stages of development of slip surfaces in clay subjected to simple shear (after Skempton, 1966).

larly marked in seatearths containing significant proportions of montmorillonites.

There are therefore two main mechanisms of shear zone deformation – one involving considerable tectonic movement either during faulting or folding, the other primarily from localized compaction and consolidation. The former are the more interesting and Skempton (1966) describes their particular characteristics in detail, by reference to a system of rectangular axes, a, b, c (Fig. 1.17) in which a is the direction of movement,

b lies in the plane of shear and *c* is at right angles to this plane. Then three types of slip surface can be identified:

1. *Displacement shears* which lie parallel or sub-parallel to the *a*, *b* plane and are similar to the slip surface formed during a laboratory shear box test.
2. *Riedel shears* which lie *en echelon* inclined at 10°–30° (equivalent to $\phi/2$) to the *a*, *b* plane with the acute angle pointing against the direction of relative movement.
3. *Thrust shears* which are effectively a mirror image of the Riedel shears with the acute angle with the *a* axis pointing in the direction of movement. They have the same attitude as fracture cleavage.

The effect, illustrated in Fig. 1.17, of first displacement, then Riedel and thrust shears – the latter having a tendency to rotate during continued shearing – is to divide the shear zone into numerous lenses of a rhombic or occasionally trapezoidal shape, with slickensided, polished and curved surfaces. Typical tectonic shear material is illustrated in Fig. 1.18. This contrasts the appearance of typical tectonic shear zone material and typical compaction/consolidation material.

Shear zones can vary in thickness from 100 mm or less to several metres, depending partly on the mode of genesis. A typical example from Dawdon Colliery, Durham, Britain is illustrated in Fig. 1.19. In Fig. 1.20 scanning electron micrographs contrast sheared and fractured surfaces of seat-earths from Birch Coppice Colliery, Staffordshire, Britain.

That typical shiny slickensided appearance of shear zones is reflected in British coal mining terminology by the use of various local names ranging from 'backslips', 'bind', 'clod' and 'clunch' to 'duns', 'spavin' and 'rammel'. There is little to be gained through detailed examination of the shear resistance of shear zones. Tensile strength and cohesion are negligible and the friction angle (9°–19° depending on clay mineral content) is close to residual. Because of this the small displacements which occur during stress redistribution associated with excavation should be sufficient to mobilize the full shear resistance of sheared rocks. This can lead to several predictable phenomena. The most obvious is the problem of slope stability in open pit mines when planes including sheared strata daylight

---

Fig. 1.18 Shear zone structures collected by Holmes (1979): top left, intraformational shear in mudstone below Busty seam at Dawdon Colliery, Durham, Britain; top right, bedding fault (inverted) mudstone overlain by ironstone band above Harvey seam at Woodhorn Colliery, Northumberland, Britain; bottom left, seatearth with lystric surfaces in addition to intraformational structure in Busty seam seatearth at Dawdon Colliery; bottom right, seatearth with lystric surfaces, Tilley seam seatearth at Acklington opencast site, Northumberland, Britain.

Fig. 1.19 A zone of sheared seatearth about 1.5 m thick in the vicinity of the Seaham fault at Dawdon Colliery, Durham, Britain (from Farmer and Glossop, 1983).

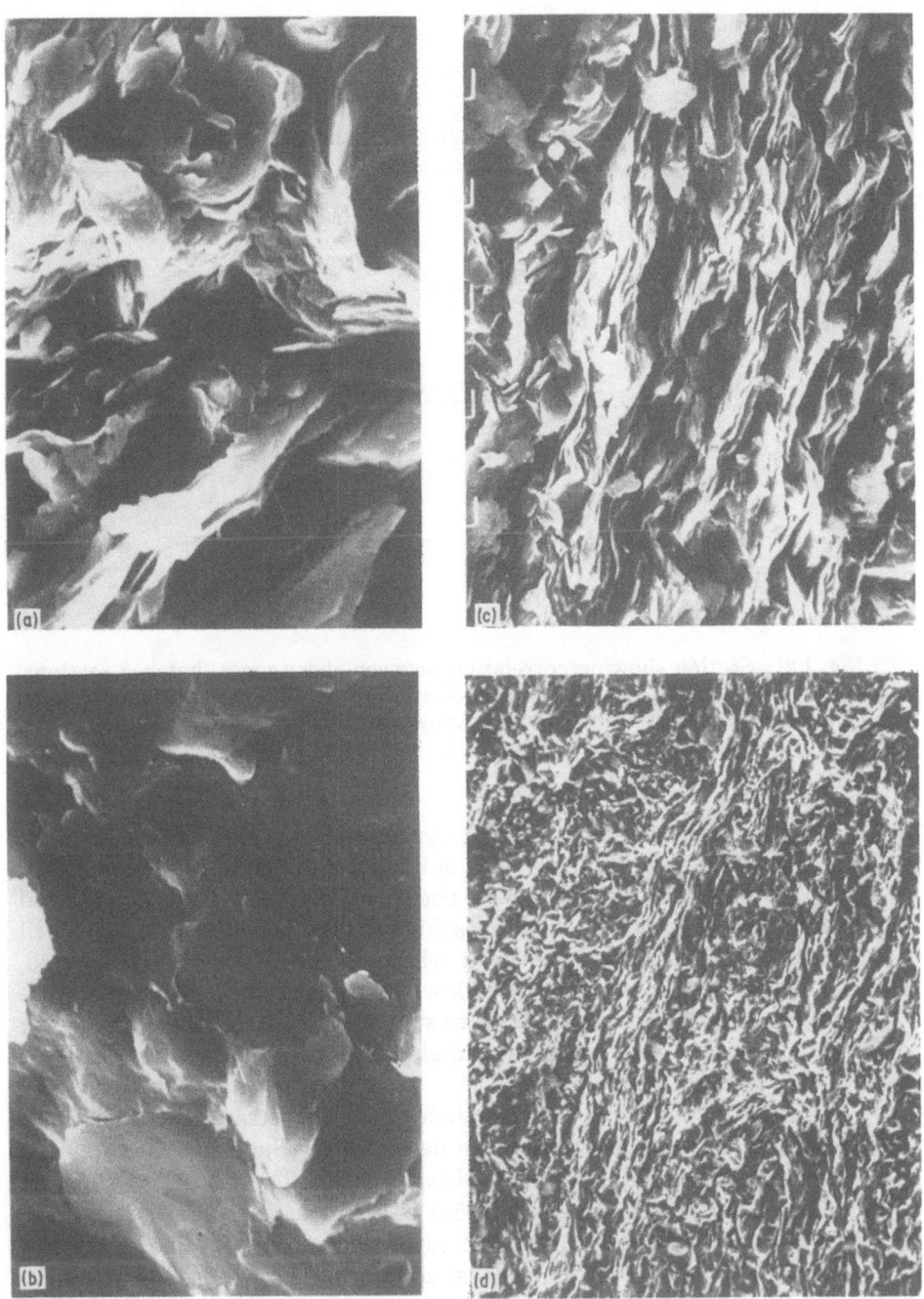

Fig. 1.20 Scanning electron micrographs of slickensided or sheared seatearths (a, b) and fractured seatearths (c, d) from beneath the Birch Coppice seam, Birch Coppice Colliery, Staffordshire, Britain (from Lain, 1974).

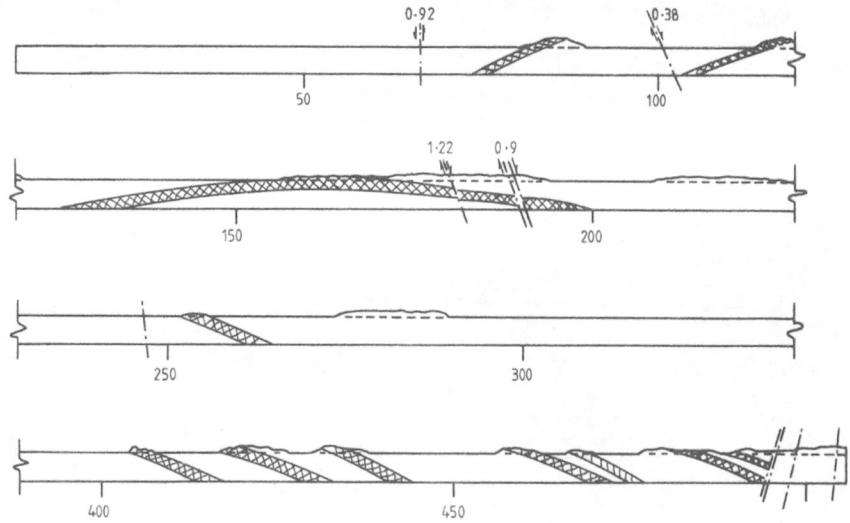

Fig. 1.21 Section showing correlation between shear zones (hatched sections) and overbreak above Sea Drift tunnel at Dawdon Colliery, Durham, Britain. Chainages and fault throws, where known, are in metres and horizontal and vertical scales are equal (after Farmer and Glossop, 1983).

in bench or sidewall faces. In underground mining, it must be assumed that at all bedding planes with a substrate rock containing some clay mineral, the substrate rock will be sheared. This means that virtually all bedding planes in coal mines separating rocks of different lithologies will form a potential parting and that where the rock is strongly sheared, rocks forming a roof structure below the parting and a floor structure above the parting will be difficult to support or control. The engineering problems will be overbreak in tunnels and on longwall faces and floor heave in floor rocks.

An example of a cross-measure drift in sheared strata which might be considered a classic case has been described by Snowdon, Glossop and Farmer (1983) and Farmer and Glossop (1983). The tunnel at Dawdon Colliery, Durham, Britain, was 3.65 m in diameter, circular and machine driven, and was located in a faulted area containing numerous intra-formational and bedding plane shear zones, similar to those illustrated in Fig. 1.19, at a depth of 500 m. After advancing 179 m in 8 months the machine was withdrawn. One of the major factors affecting machine progress was the relative weakness of the shear zones. This allowed penetration of the machine into the floor and penetration of thrust pads into the sidewalls, and made roof support difficult. A tunnel section in Fig.

1.21 shows the close correlation between overbreak in the tunnel roof and the presence of shear zones.

On the coal face there are two basic problems: the release of blocks bounded by compaction shears, and the overbreak of layers up to 2 m thick of tectonically sheared rock beneath a strong parting. The former is less of a problem with powered supports. The latter is a major problem with powered supports. Collapse of such rocks, usually beneath sandstone layers, can lead to loss of support efficiency and extensive roof cavity formation. Methods of control will be discussed in Chapter 7.

In access roadways or tunnels, sheared floor rocks can lead to excessive floor heave. The mechanics of this are discussed further in Chapters 3 and 6, but it is evident that redistributed stresses around an excavation will tend to induce peak shear stress in the peripheral rocks. The sheared floor rocks with lower shear resistance and little or no radial support will tend to deform more than the surrounding rocks. A typical end result has been illustrated in Fig. 1.16.

The presence of shear zones in fireclay seatearths also has an important secondary effect, since most of these contain plastic clays. Surface water can gain ready access through shear planes to the seatearths, enhancing the tendency to plasticity (see Fig. 1.11). In addition, since movement can be more easily accommodated along existing planes, deformation and extrusion from roadway peripheries will be exacerbated.

1.6 Faults, joints and cleat

Although shear zones are the most important geological structures in Coal Measures rocks, the rocks also contain other structures which can interact in a highly significant way with coal mine structures. It is difficult to discuss these in a general way, but useful to consider some implications.

In relatively undisturbed Coal Measures strata the majority of faults tend to be *normal faults* hading at 60–70° to the vertical. These result partly from the *up-warping* and *down-warping* associated with the depositional cycles of the Coal Measures formations and partly from similar post-Coal Measures depositional cycles. Where erosion has taken place, thrust faulting at 20°–30° to the vertical can also be found. In *mountain building* areas, the main structures tend to be folds and *thrust faults*. The two types of movement are termed by structural geologists, respectively *epeirogenic* and *orogenic*.

In the British Coal Measures, the eastern coalfields tend to be undisturbed, and the western coalfields, particularly South Wales, tend to fall in the mountain building phase of the Hercynian orogeny. In both

cases the faults and all other structures tend to follow the pattern of
major movement. Thus, in the major eastern coalfields of Britain the
structures are determined by later Carboniferous/Permian movements
which formed broad shallow basins separated by narrow faulted anti-
clines (Kent, 1966) which strike in a north-east and north-west direction.

The engineering significance of faults varies. In the case of normal
faults, there is the likelihood of passage of water along the fault plane.
There may, depending on the degree of movement along the fault, be
sheared ground ranging from a simple parting to a major fault shear zone
and associated intraformational shears. Because the major principal
stress direction inducing shear is vertical there may also be high vertical
stresses in the vicinity of some normal faults and low horizontal stresses
which can lead to roadway damage. This is exacerbated in the case of
longwall workings since these are invariably planned, for maximum
economy, to run parallel to fault directions and access roadways will
invariably abut fault zones. The major importance of normal faults,

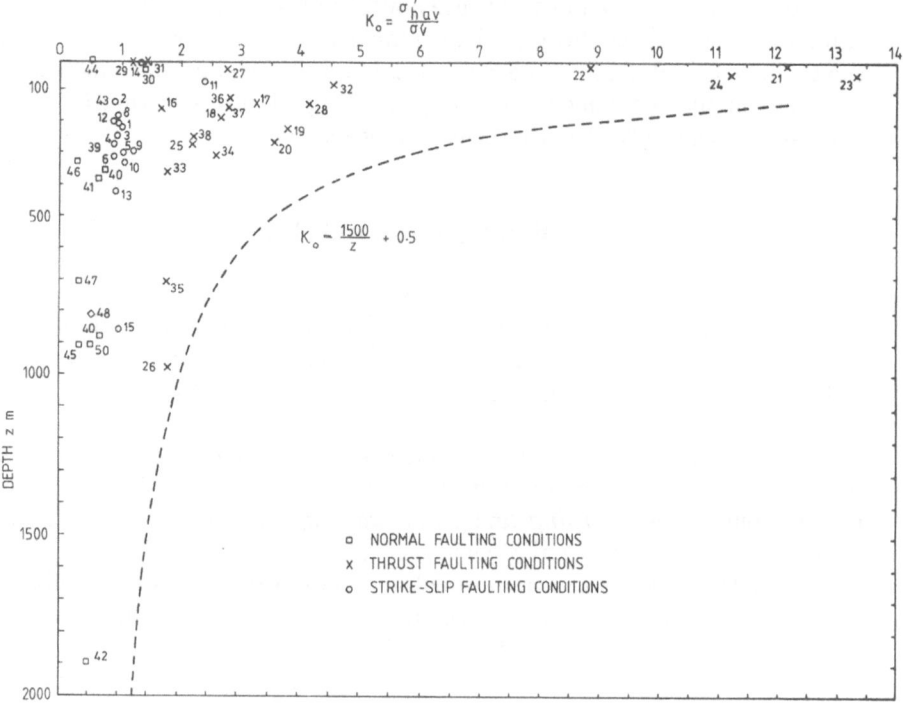

Fig. 1.22 Effect of geological structures on the horizontal/vertical stress ratio
(after Jamison and Cook, 1979). The numbers refer to case histories identified in
the reference.

Fig. 1.23 An example of distortion encountered in a roadway in the shaft pillar at Abernant Colliery, South Wales, in the vicinity of a thrust fault at a depth of 730 m (Wilson, 1960) (photograph by permission of the N.C.B.).

however, tends to be their presence and the effect they have on the mine system. Their occasional association with bumps and bursts is discussed in Chapter 8 – as is the more important association of thrust faults with these phenomena.

Thrust faults tend to have a more important effect in mining engineering – principally because they are formed by a major principal stress in a horizontal direction. By its nature and the necessity to overcome a high passive vertical stress, such a stress must be large. The ratio can be illustrated by considering failure in the Rankine passive condition when $\sigma_h/\sigma_v = k_p = 3$ where $\phi = 30°$ and the rock has minimum strength (see Equation (1.2)). It is interesting to note in Fig. 1.22 that many of the high horizontal–vertical stress ratios identified in practice are associated with thrust faulting. If such high ratios were to exist at depth, then they could constitute a considerable engineering problem. It is probable that horizontal stresses of considerable magnitude did exist in the vicinity of a thrust fault in the shaft pillar at Abernant Colliery in South Wales (Fig. 1.23), where difficulty was encountered (Wilson, 1960) in constructing shaft bottom roadways.

A joint pattern exists in most sedimentary rocks roughly parallel to the direction of faulting. This is the case in the British Coal Measures. Joints are fractures or cracks which differ from faults in that there has been no discernible movement along them. Their frequency is several orders of magnitude greater than that of faults and Price (1966) shows that this frequency is related to rock lithology, bed thickness and the degree of tectonic deformation.

Price justifies the former by suggesting that the number of joints which develop in a rock is related to the strain energy which the rock is capable of storing. This will be related to the strength/modulus ratio or to the inverse of the commonly quoted *modulus/strength ratio* (see Farmer, 1983). The latter is exceptionally low in the case of coal (< 50), between 50 and 150 for shales and mudstones and above 200 for sandstones. Since beds of sandstones are generally thicker than shales or coals, and Price (1966) proposes a near linear relation between bed thickness and joint spacing, this goes some way to explaining the much closer joint spacing in coal than sandstones.

Since the Coal Measures cyclothem comprises cycles of beds of widely varying thickness and lithology it can be expected that the joint frequency will vary widely. This can be illustrated through a log (Fig. 1.24) from an anchor extensometer borehole at Lynemouth Colliery, Northumberland, Britain, in which one of the measures of joint frequency, rock quality designation (RQD), is related to lithological descriptions and strength. This constitutes the minimum data required to assess the possible deformation behaviour of strata, and since RQD is

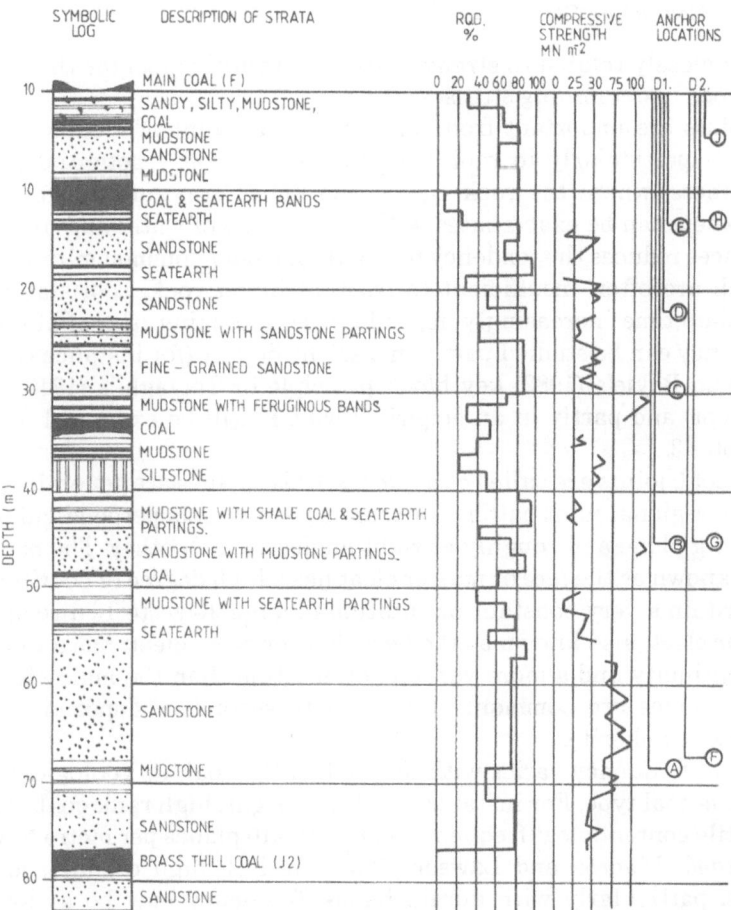

Fig. 1.24 A log of borehole core obtained from an instrumentation borehole between the Main and Brass Thill seams at Lynemouth Colliery, Northumberland, Britain. The information on RQD and strength using a point load tester is the minimum geotechnical data necessary for strata assessment. RQD (Deere, 1968; Deere *et al.*, 1966) is the percentage of core recovered in lengths greater than 100 mm during drilling, expressed as a percentage of the total core. It can be related roughly to frequency:

RQD (%)	Quality classification	Fracture frequency (m^{-1})
90–100	Excellent	1
75–90	Good	1–5
50–75	Fair	5–10
25–50	Poor	10–20
0–25	Very poor	20

often closely related to strength gives an intuitive feel for the quality of the rock under mining stresses.

Joints are important from an engineering point of view in that they form – particularly in roof structures in tension – potential planes of weakness, where the rock can loosen under gravitational forces. This weakness can be exacerbated with time as flow of water and oxidation of surfaces reduces the tendency to stability. Under common mining stresses which are often insufficient to fracture intact rock, roof stability and stand-up time increasingly depend on the loosening of roof blocks. This tendency can be studied partly in a scientific way (for instance Goodman, Shi and Boyle's (1982) key block principle or Terzaghi's (1946) arching concept) and partly in an empirical way. It will be examined further in Chapter 3.

In coal, joints are called *cleat planes*. Cleat planes are generally parallel to the regional joint pattern – occurring in two sets perpendicular to the bedding plane and roughly at right angles to each other. The *main* cleat also known as the *face* or master cleat has a high degree of continuity and in Britain a very constant orientation in a north-westerly direction. The *minor* cleat, also known as the *bord*, butt or cross cleat, has a low degree of continuity and a more variable orientation than the main cleat. Both cleat planes are commonly infilled with secondary deposits – usually calcite or ankerite.

Cleat frequency varies with the rank of the coal, its geological history and the coal type. For instance, in clean, bright, high rank coals with low volatile content cleat frequencies of up to 210 planes per metre have been recorded (Macrea and Lawson, 1954). In dull, high organic, low rank coals, particularly with durain bands, frequency may be as low as 15 planes per metre.

There are various theories as to the genesis of cleat ranging from diagenetic processes of compaction and coalification to tectonic forces. The strong correlation between cleat direction and other structures suggests the latter. It is this strong correlation which makes cleat so important in mining operations – and particularly modern mining operations. Longwall faces, for maximum economy, tend to be set out in directions parallel to major or minor faulting. This means that the face will tend to be parallel to the cleat direction. The abutment pressures which exist ahead of longwall faces are generally of sufficient magnitude to induce shear and tensile fracturing of the roof, seam and floor. Termed by Faulkner and Phillips (1935) *induced cleavage*, these fractures tend to complement cleat when this is at a low angle to the face line. The resultant spalling of face and roof, which will be discussed at greater length in Chapter 7, is a major factor in determining face stability.

DESIGN OF ROOMS AND PILLARS

The simplest coal mine structures and the ones which most closely relate to conventional laboratory testing are *pillars*. In coal mining, pillars can be part of a room and pillar mine layout, or can be designed as a support element in a different type of mine layout. In Europe, these pillars tend to be used between longwall faces to protect access roads, while in the United States the multiple entry system utilizes two or more support pillars.

In room and pillar mining where ideally the mine layout should comprise square or rectangular pillars (see for instance Figs 2.2, 2.3) supporting the roof spans of *rooms* between pillars and across junctions, there is another simple engineering design problem. This concerns the stability of the roof, and the maximum span which can safely be maintained over the working life of the panel. This is a different type of problem from pillar design. In the former the relation between compression stresses in the pillar and compressive strength of the pillar rock – usually coal – is paramount. In the latter the relation between tensile and shear stresses in the roof and the discontinuous nature of the roof rock is the most important factor. Both, however, introduce fundamental problems of mine design, and are an ideal introduction to engineering design concepts in coal mining.

2.1 Stresses acting on pillars

Despite the simplicity of the structure, and the detailed knowledge of rock behaviour obtained over the past few years, pillar design – where it is practised – has barely changed during the present century. It is based on the assumption that the stress in a pillar is evenly distributed and equal to the original vertical geostatic stress divided by the pillar area/original area ratio; and that pillar failure occurs when this stress exceeds the compressive strength of the pillar rock. It would be a naive assumption for any engineering structure in any material. It is particularly so in the case of pillars with high width–height ratios in rock.

The major work on stresses acting on pillars was carried out by Coates (1970). He starts with the simplest statement of average pillar stress, known as the 'tributary area theory'. This assumes that each of the pillars left during excavation, supports all the overlying strata which are 'tributary' to their location. Then the average pillar stress for square pillars with rooms of consistent width is:

$$\sigma_p = \sigma_z \left(\frac{B + B_o/2}{B^2} \right)^2 = \frac{\sigma_z}{1 - R} \tag{2.1}$$

where B is the pillar width, B_o is the opening or room width, R is the extraction ratio equal to $4(B + B_o/2)^2/B_o$ and σ_z is the vertical stress.

For rectangular pillars this becomes:

$$\sigma_p = \frac{\sigma_z B_o^2}{B_o^2 - (B + B_o)(l + B_o)} \tag{2.2}$$

where l is the pillar length.

This approach assumes essentially that the mined area is extensive and shallow, that the mined rock is horizontally stratified and that the pillars are equidimensional. It specifically ignores the relative extent and depth of the mined area, the stress component parallel to the seam, the relative deformation properties of pillar, roof and floor rocks, and the positions of the pillars in the mining zone. Taking some of these into account, Coates obtained a solution principally for deep, long mine pillars, but applicable generally by solving the statically indeterminate net deflection of the roof and floor rocks resulting from mining. Then the general solution for average pillar stress becomes:

$$\sigma_p = \sigma_z \left\{ \frac{\left[2R - K_0 \dfrac{H}{L} \dfrac{(1 - 2v_w)}{(1 - v_w)} - \dfrac{v_p}{(1 - v_p)} K_0 \dfrac{H}{L} \dfrac{E_w}{E_p} \right]}{\left[\dfrac{H}{L} \dfrac{E_w}{E_p} + 2(1 - R)\left(1 + \dfrac{1}{N} \right) + \dfrac{2RB}{L} \dfrac{(1 - 2v_w)}{(1 - v_w)} \right]} \right\} \tag{2.3}$$

where H is the seam height, L is the extent of the mined area, K_0 is the ratio between σ_h and σ_z or the coefficient of geostatic stress, E_w, E_p and v_w, v_p are the elastic constants of the wall (roof and floor) and pillar materials.

This is a two-dimensional solution in plane strain and requires, strictly speaking, a length/width ratio of about 3 or more to be applicable. A simple three-dimensional approach is not feasible, although numerical

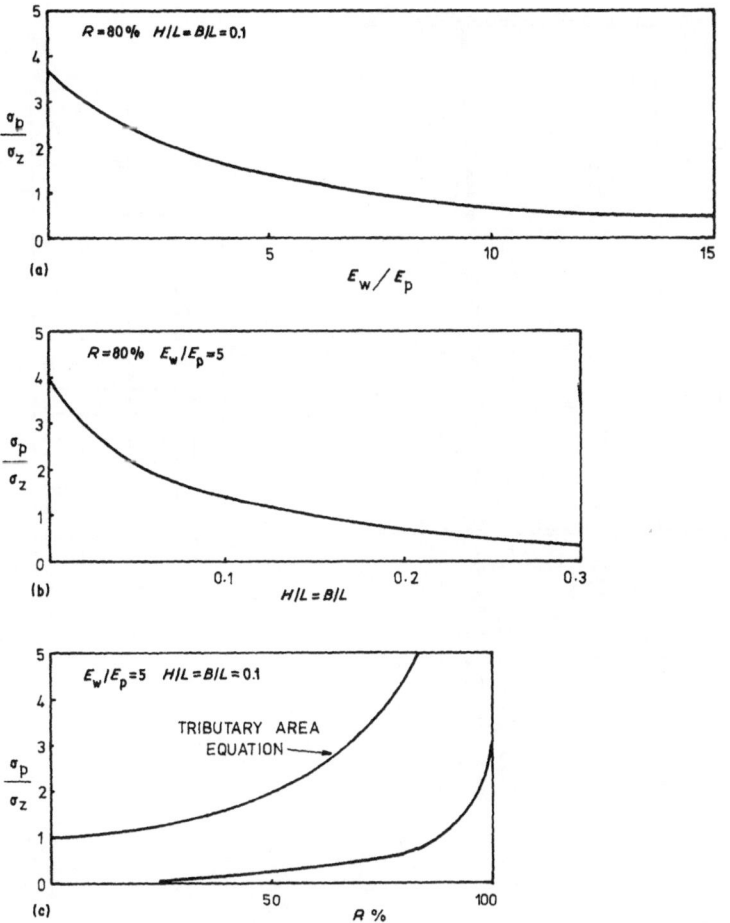

Fig. 2.1   Estimates of pillar stress σ_p as a proportion of vertical stress σ_z based on the variables in Equation (2.3), putting $K_0 = 1$, $\nu_p = \nu_w = 0.33$ and N large, so that

$$\frac{\sigma_p}{\sigma_z} = \frac{(2R - 0.5H/L) - (0.5H/L\ E_w/E_p)}{H/L\ E_w/E_p + 2(1 - R) + RB/\overline{L}}$$

and analogue solutions have been proposed by Orawecz (1977) amongst others. Coates' approach is useful in that it can be used simply to illustrate several of the fundamental characteristics of strata and geometry which affect pillar stresses. Some of these are illustrated in Fig. 2.1. For instance as the E_w/E_p ratio rises (Fig. 2.1(a)), so the pillar stress is reduced from a magnitude close to $4\sigma_z$ (the extraction ratio has been chosen as 80 %) to a level of $0.5\sigma_z$ for $H/L = B/L = 0.1$. This illustrates the bridging effect of the stiffer roof and floor layers and the tendency to

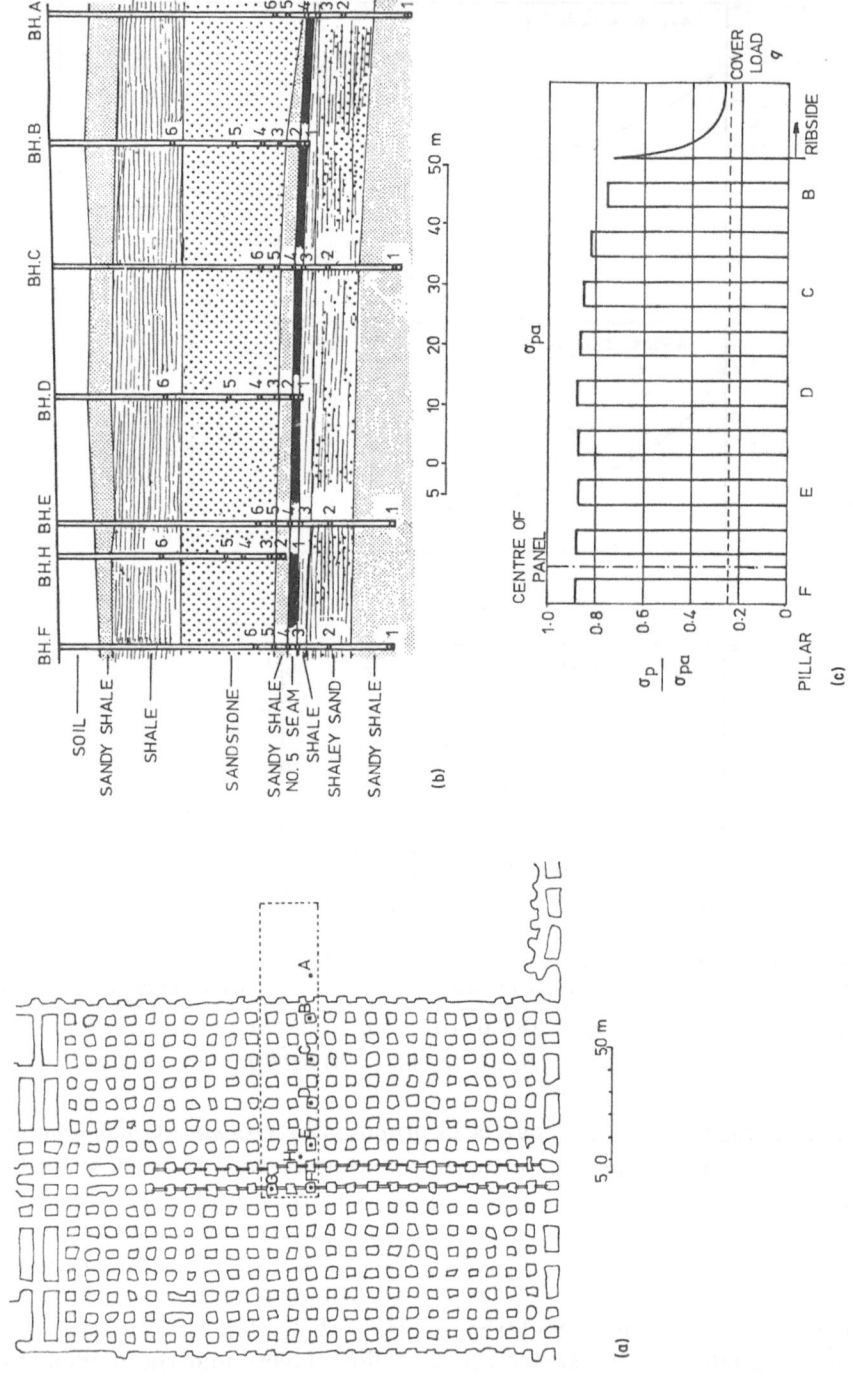

(b)

SOIL
SANDY SHALE
SHALE
SANDSTONE
SANDY SHALE
NO. 5 SEAM
SHALE
SHALEY SAND
SANDY SHALE

(c)

(a)

transfer stress to the side abutment. Similarly as L is decreased (Fig. 2.1(b)) the pillar stress is reduced from a maximum magnitude of $4\sigma_z$ to zero at $H/L = 0.4$ for a E_w/E_p ratio of 5. Again this can be attributed to bridging at low spans. As a further illustration (Fig. 2.1(c)), using fixed values for E_w/E_p, H/L, B/L, there is considerable variation between the tributary area calculation and Coates' equation for stress at increasing extraction ratios.

It should be stressed that this is used as an illustration, and that measurements of *average* pillar stresses are very infrequent. In fact a review of the literature shows virtually no reliable measurements of average stress, principally because such measurements are virtually impossible to obtain – this is discussed in the following sections. One of the more interesting sets of data is by Orawecz (1977) from work in South Africa. He describes two case histories in which surface settlements and underground displacements were measured using levelling and borehole anchors drilled from the surface to the seam level and below. The seams were at average depths of 40 m and 68 m. The purpose of the measurements was to test an analogue model, and satisfactory simulation allowed computation of pillar stresses from observed seam deformations.

The pillar geometries and data on the mining and instrumentation layouts are illustrated in Figs 2.2 and 2.3 together with the average pillar stresses σ_p, computed from seam deformations in Figs 2.2(c) and 2.3(c). These are quite close to the pillar stresses σ_{pa} computed from the 'tributary area' equation. In these cases the E_w/E_p and H/L ratios were respectively 3 and 0.01 and 2 and 0.05, and it can be seen from Fig. 2.1 that such a result would be expected. It is interesting to note the reduced pressure on the pillars adjacent to the ribside and also the relatively low level of the abutment stress. The former would be expected; the latter is rather surprising.

2.2 Stress distribution in pillars

The concept of average pillar stress is not a good one, since pillar stresses are not evenly distributed. This can be illustrated from case history data. Wagner (1974) describes some particularly useful tests in South Africa in

Fig. 2.2 Estimation of average pillar stress σ_p as a proportion of pillar stress computed from tributary area theory σ_{pa}, from experiments by Oravecz (1977) in No. 5 seam at Colliery A, South Africa. Data: average depth to mid-seam 40.3 m; seam height 1.5 m; pillar width 5.2 m; room width 5.5 m; percentage extraction 76.4 %; panel width 176.2 m (est.); deformation modulus, seam (est.) 1.54 GN m^{-2}; deformation modulus strata (est.) 4.43 GN m^{-2}; Poisson's ratio (est.) 0.15.

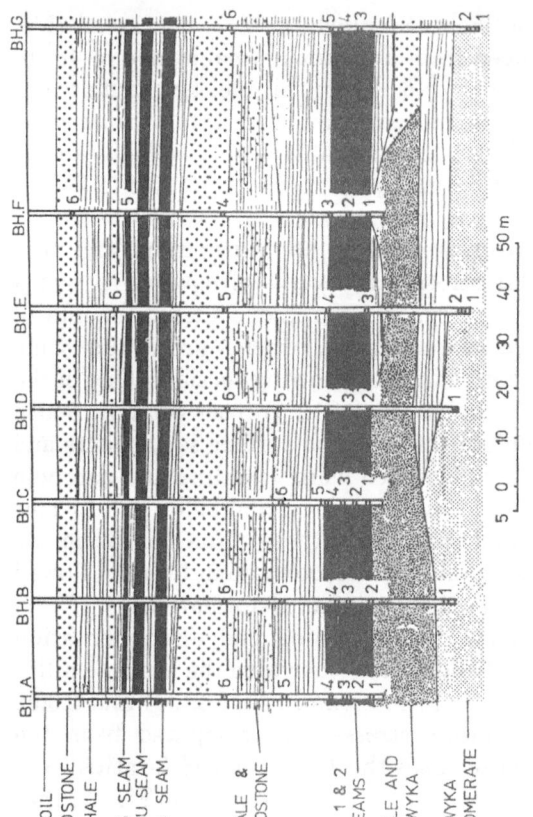

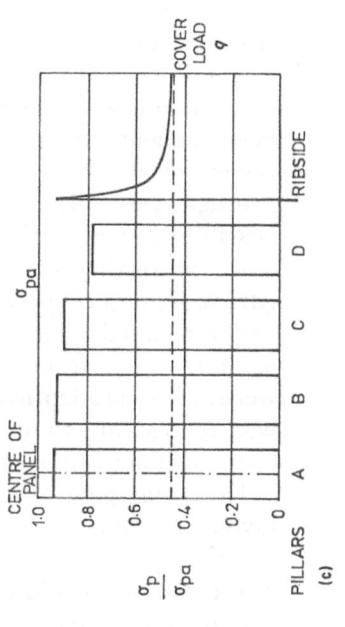

SOIL
SANDSTONE
SHALE

NO. 5 SEAM
NO. 4U SEAM
NO. 4 SEAM

SHALE &
S.ANDSTONE

NO. 1 & 2
SEAMS

SHALE AND
DWYKA

DWYKA
CONGLOMERATE

5   0   5   10   20   30   40   50 m

BH.A   BH.B   BH.C   BH.D   BH.E   BH.F   BH.G

(b)

$\frac{\sigma_p}{\sigma_{pa}}$

CENTRE OF
PANEL

σ_{pa}

COVER
LOAD
q

1.0

0.8

0.6

0.4

0.2

0

PILLARS   A   B   C   D   RIBSIDE

(c)

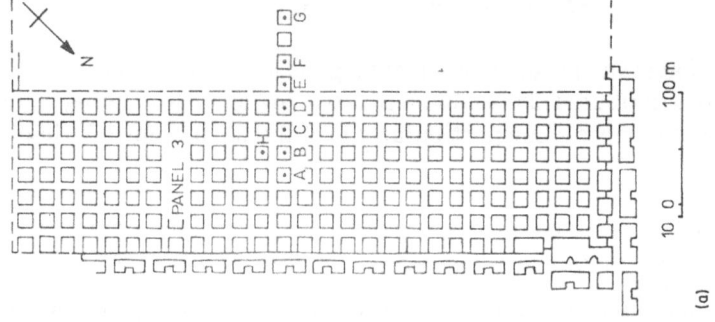

N

PANEL 3

A B C D E F G
H

10 0   100 m

(a)

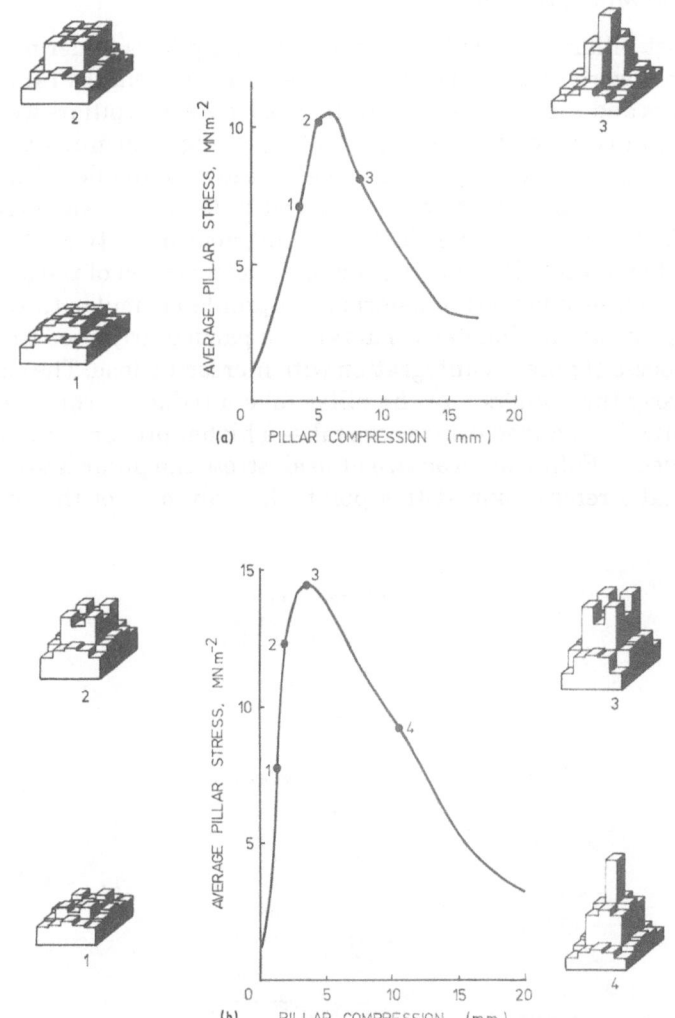

Fig. 2.4 Typical stress distributions at various stages of the deformation process of 1 m square coal pillars with (a) width–height ratio of 1; (b) width–height ratio of 2 (after Wagner, 1974).

Fig. 2.3 Estimation of average pillar stress σ_p as a proportion of pillar stress σ_{pa} computed from tributary area theory, from experiments by Oravecz (1977) in No. 2 seam at Colliery B, South Africa. Data: average depth to mid-seam 66.7 m; seam height 5.5 m; pillar width 13.7 m; room width 6.1 m; percentage extraction 52.1 %; panel width 144.8 m; deformation modulus, seam (est.) 3.92 GN m^{-2}; deformation modulus, strata (est.) 6.27 GN m^{-2}; Poisson's ratio (est.) 0.15.

which test pillars were cut from existing coal pillars underground. The test pillars were rectangular and square with side lengths ranging from 0.6 m to 2 m and width–height ratios from 0.6 to 2.2. The pillars were loaded to failure in compression by a hydraulic jacking system installed along the central plane of each pillar parallel to the roof and floor. In this way the modulus mismatch between wall and pillar rocks was maintained. Figure 2.4 illustrates the vertical stress distribution in two pillars – also measured hydraulically – during complete deformation of the pillars. The average stress–compression deformation profile is similar to the strain-softening profiles in Chapter 1. This gives a particularly good illustration of the process of pillar disintegration with increasing load. The sides – and particularly the corners – of the pillar take a reduced proportion of the load, while the confined centre core takes a higher proportion as deformation proceeds. Following fracture at peak stress the pillar is still capable of residual strength, but at this point, the only part of the pillar with

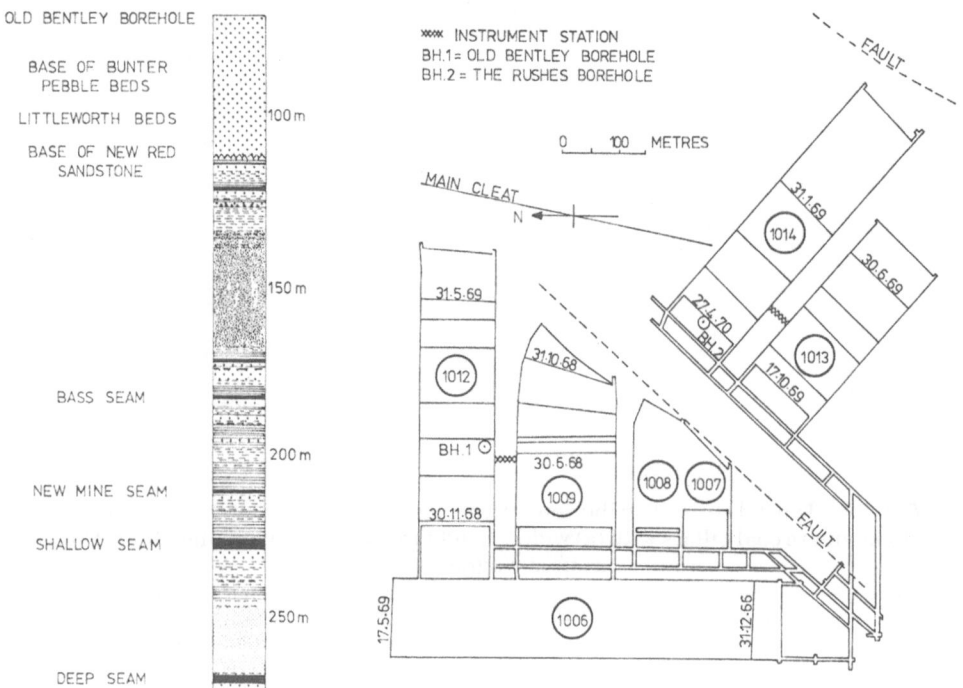

Fig. 2.5 Plan of workings in the Shallow seam at Lea Hall Colliery, Staffordshire, Britain, showing the position of instrumentation between 1009 and 1012 advancing longwall faces (230 m depth) and between 1013 and 1014 retreating longwall faces (280 m depth) (after Wilson and Ashwin, 1972).

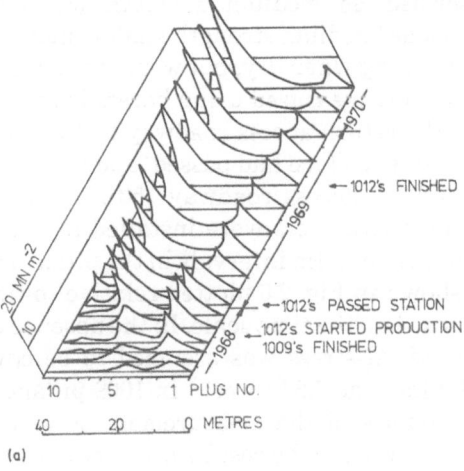

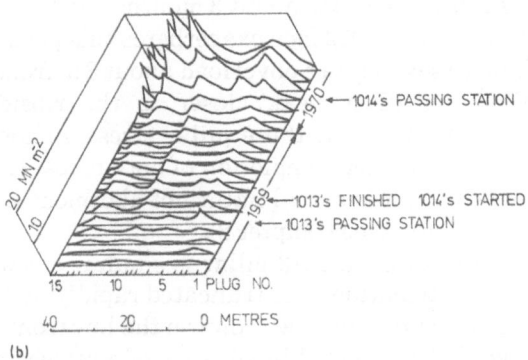

Fig. 2.6 Stress change with time across (a) 1009 ribside and 1009–1012 pillar and (b) 1013 ribside and 1013–1014 pillar at instrumentation stations in Fig. 2.5 (after Wilson and Ashwin, 1972).

sufficient integrity or confinement to accept significant load is the centre core. It is a particularly good illustration of the similarity between pillar deformation and rock specimen deformation, and of the importance of the confined centre of the pillar.

The stress distribution in pillars between two advancing and two retreating longwall faces has been measured by Wilson and Ashwin (1972) at Lea Hall Colliery, Britain. The position of the pillars is illustrated in Fig. 2.5. The seam was 3 m thick with an extraction height of 2.2 to 2.3 m at a depth of between 230 and 280 m. A borehole section is included in Fig. 2.5. Both advancing and retreating faces were separated by a pillar 41 m wide.

The coal is described as medium strength ($\sigma_{cf} = 26\,\mathrm{MN\,m^{-2}}$, $E = 4.8\,\mathrm{GN\,m^{-2}}$), the roof as 'medium strength shaley mudstone' and the floor as seatearth, 'fairly strong when dry, but weak when wet'.

Changes in stress were measured using borehole stressmeters installed in the ribside from the return access roadway to No. 1009 advancing face as soon as possible after the face had passed, and in the solid coal forming the ribside of the return access roadway 150 m in advance of No. 1013 retreating face. The stressmeters were installed in individual boreholes about 1 m apart and at about 3 m intervals in the pillar. The stress changes in the pillars are shown in Fig. 2.6 plotted relative to time; with the face positions superimposed on the time scale. In the absence of absolute stress measurements the initial stress was assumed to be cover stress (about $7\,\mathrm{MN\,m^{-2}}$ in 1013 pillar and $5.8\,\mathrm{MN\,m^{-2}}$ in 1009 pillar). Bearing in mind some of the inadequacies of the measurement system discussed in the paper and at the presentation, the results are very convincing. In the case of 1009 pillar, a ribside abutment reaches a peak about 4 m from the ribside of two to three times coverload about 3 months and 200 m after passage of the face. The working of 1012 face exacerbates this peak and also induces another peak up to seven times coverload about 8 m from 1012 ribside and a later smaller subsidiary peak closer to the ribside. A feature of particular importance is that these pillar stresses showed little sign of abating at least a year after completion of both faces. This is particularly valuable evidence of the residual nature of abutment and pillar stresses which is discussed later (see Chapter 8).

The pattern of stresses in 1013 pillar is similar, but with lower peaks, and a more even distribution. It is truncated rapidly by the loss of access, after passage of 1014 face, but does confirm the low front corner abutment stress which would be expected in advance of a retreating longwall face, but which is not always achieved. Certainly bearing in mind the greater depth of 1013 pillar, it would be feasible to consider a much shorter pillar or even elimination of the pillar in the retreating face compared with the advancing face.

It is not the place here to comment on the relative merits of advancing and retreating longwall faces, but the case history illustrates the much lower stress regime on working access roadways in the latter case. Particularly though, it illustrates the existence of residual pillar stresses, which are a lasting interaction problem in multi-seam longwall layouts.

2.3   Strength of pillars

Most of the work on pillars has been carried out on pillar strength. It is mainly empirical and of dubious quality. The most complete work is by

Salamon and Monro (1967); the best summaries are by Bieniawski (1981) and Tsur-Lavie and Denekamp (1982).

The strength of pillars, determined by experiment, depends on three things, the *size* of the pillar, the *shape* of the pillar material and the unconfined compressive *strength* of the pillar material. The most common way in which these can be represented takes the form:

$$\sigma_{pf} = \sigma_{cf}\left(a + b\frac{B}{H}\right) \tag{2.4}$$

$$\sigma_{pf} = K\frac{B^\alpha}{H^\beta} \tag{2.5}$$

where σ_{pf} is the pillar strength, σ_{cf} is the uniaxial compressive strength of a cube of specified dimension, B, H are pillar width and height, a, b are dimensionless constants usually chosen so that $a + b = 1$, and $\sigma_{pf} = \sigma_{cf}$ when $B/H = 1$, α, β are dimensionless constants and $K = f(\sigma_{cf})$ is a constant such that $K = \sigma_{cf}$ when $\alpha = \beta$ and $B = H$.

Variations on these equations have been suggested, but they have been found effective for reasonably equidimensional pillars, in most coalfields in the world. There is a surprising measure of agreement about the

Table 2.1   Constants in Equations (2.4), (2.5)

Source	a	b	α	β	Comments
Bunting (1911)	0.7	0.3	–	–	Laboratory data
Obert, Windes and Duvall (1946)	0.78	0.22	–	–	Laboratory data
Bieniawski (1968)	0.64	0.36	–	–	*In situ* – S. Africa
Van Heerden (1974)	0.70	0.30	–	–	*In situ* – S. Africa
Wang, Skelley and Wolgamott (1977)	0.78	0.22	–	–	W. Virginia mines, United States
Sorensen and Pariseau (1978)	0.69	0.31	–	–	Statistical – United States
Greenwald, Howarth and Hartman (1939)	–	–	0.5	0.83	*In situ* – Pittsburgh mines, United States
Streat (1954)	–	–	0.5	1	Statistical – S. Africa
Holland (1964)	–	–	0.5	1	Statistical – United States
Salamon and Monro (1967)	–	–	0.46	0.66	Statistical – S. Africa
Bieniawski (1968)	–	–	0.16	0.55	Statistical – S. Africa
Hazen and Artler (1972)	–	–	0.5	0.5	Statistical – United States
Zern (1926)	–	–	0.5	0.5	Empirical – United States
Morrison, Corlett and Rice (1975)	–	–	0.5	0.5	*In situ* – Canada

constants a, b, α, β in Equations (2.4) and (2.5). Some representative values from early times to recent research are quoted in Table 2.1. Essentially a and b are shape characteristics and α, β are size characteristics – the relative importance of each will be discussed in the next section. The greatest difficulty in estimating strengths is in estimating values of K. In Equation (2.4), σ_{cf} is usually quoted, in the case of coal, for a 25 mm cube, and accurate data are available. In Table 2.2, compressive strengths of British coals show that both low (anthracite) and high volatile content coals tend to be strong and intermediate rank coals tend to be weaker. There is no reason why these data should not be equally valid for other countries. The very empirical nature of Equation (2.5) means that the values of K quoted are less reliable. They are usually based on a larger cube size than the compressive strengths in Table 2.2 and are usually statistically derived for a particular district and system of units. For this reason there is a reluctance to quote values for K in the literature, and where they are quoted (see Table 2.3) they are usually significantly lower than the small cube strengths. A general consensus quoted by Wilson (1980) is that the large scale strength of coal is about one-fifth of the values in Table 2.2.

Some attempts (see for instance Babcock, Morgan and Haramy, 1981) have been made to rationalize Equations (2.4) and (2.5). Tsur-Lavie and Denekamp suggest a simple modification of Equation (2.5) in the form:

$$F_{\mathrm{p}} = K\left(\frac{B}{H}\right)^{\beta} A^{\frac{1}{2}(2+\alpha-\beta)} \tag{2.6}$$

where F_{p} is the force acting on the pillar and where the exponent $(2+\alpha-\beta)/2$ describes the size effect and applies to the cross-section A, and

Table 2.2   Compressive strength of British coals (after Evans and Pomeroy, 1966)

Coal rank number	Volatile content (%)	Average compressive strength of twenty 25 mm cubes (MN m^{-2})	
		Perpendicular to beds	Parallel to beds
100	5	38.2	36.4
201	12	18.2	16.1
301	22	7.9	6.1
3/501	30	18.1	12.0
4/502	39	32.8	31.5
702	37	36.5	26.2
7/802	37	57.3	39.2
801	36	55.7	34.1
901	39	54.8	28.6

Table 2.3   Quoted values of K in Equation (2.5)

Source	K (MNm^{-2})	Comments
Greenwald *et al.* (1939)	19.3	Originally in psi for B, H values in inches. Pittsburgh, United States coals
Salamon and Monro (1967)	9.1	Originally in psi for B, H values in feet. S. Africa
Bieniawski (1968)	6.9	Originally in psi for B, H values in feet. S. Africa
Jenkins and Szeki (1964)	12.4	Originally in psi for B, H values in feet. Britain
Wagner (1974)	11.0	B, H values in metres. Based on in-situ tests. S. Africa

the exponent β describes the shape effect or end constraint. It is a useful starting point to a discussion – which is fundamental to pillar design – of the mechanics of pillar breakdown.

2.4   Mechanics of pillar breakdown

Figure 2.7 illustrates the breakdown of two pillars in a potash mine at a depth of 1100 m. Both have deformed and, to a certain extent, failed. The mechanism in each case is, however, quite different. The pillar with the smaller width–height ratio has fractured with the type of strain-softening deformation illustrated in Figs 1.1–1.3. The pillar with the greater width–height ratio, although it has fractured at the edges, has probably followed the strain-hardening deformation path of Fig. 1.4. The reason for the change in deformation characteristics in this case is that a unique combination of stress and constraint has combined to effect a degree of stability in the centre of the pillar. Tsur-Lavie and Denekamp (1982) have discussed the effect of this change on overall pillar stability.

They consider the strain energy, U, stored in a pillar:

$$U = \sigma \varepsilon V = S \frac{\sigma^2}{E} l^3 \tag{2.7}$$

where ε, E, V are strain and elastic modulus and pillar volume and S, l are a shape factor and linear dimension.

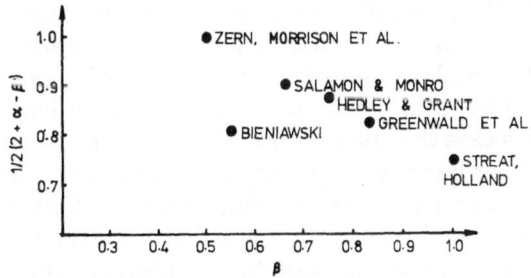

Fig. 2.8 Size effect $(2+\alpha-\beta)/2$ plotted against shape effect β for the data in Table 2.1 (after Tsur-Lavie and Denekamp, 1982).

Then if failure occurs through shear and formation of a failure plane, this energy will be dissipated over an area proportional to l^2. But if failure occurs through ductile or plastic deformation, it will be evenly distributed throughout the volume, proportional to l^3. In the former case the energy at failure per unit volume of rock will be proportional to $S\sigma_{cf}^2 l/E$ and in the latter case to σ_{cf}^2/E. This means that if the energy at failure per unit volume of rock is assumed constant, in the former case:

$$\sigma_{cf} \propto l^{-0.5} \tag{2.8}$$

and in the latter case σ_{cf} will not be affected by dimension.

This explains the effect of size on pillar strength in the empirical equations – it is a feature of brittle rather than ductile behaviour. Tsur-Lavie and Denekamp plot the size effect exponent $(2+\alpha-\beta)/2$ from Equation (2.6) against the shape effect exponent β for the data in Table 2.1 and other data from coal pillars and pillars from other rocks. This is reproduced in Fig. 2.8 and it leads to an interesting conclusion. $(2+\alpha-\beta)/2$ has a magnitude close to unity where the exponent has been derived from a study of pillar failures in mines. In other words there is a limited size effect. It is only in laboratory experiments that the size effect is significant. In addition where the size effect is less important as $(2+\alpha-\beta)/2 \to 1$, so the shape or end constraint effect becomes more important, β reducing to a magnitude of 0.5.

The end constraint on the pillar will depend on the relative strengths and stiffness of the pillar, roof and floor rocks. If the roof and/or floor

Fig. 2.7 Types of deformation observed in potash pillars at a depth of about 1100 m. Note that the pillar with the large width–height ratio (a) while showing signs of fracture at the sides also exhibits signs of ductile behaviour. The pillar with a low width–height ratio (b) deforms in a typically brittle manner as in Figs 1.1–1.3 (photograph C. Wiggett).

rocks are stiffer than the pillar, then lateral restraint during load transfer will inhibit expansion of the pillar and impose constraint. It is possible to visualize a pillar with a high roof rock/pillar rock modulus ratio and a high width–height ratio being so constrained as to be virtually incompressible. In this case the apparent strength of the pillar will be extremely high. On the other hand if the pillar rocks are stiffer than the roof/floor rocks, lateral expansion of the roof/floor rocks will, in an extreme case, induce lateral tension in the pillar and reduce its strength.

An example of this, although not due to direct loading, can be found in shallow undersea workings at Lynemouth Colliery, Northumberland, Britain (Hodkin, 1978; Tubby and Farmer, 1981). The investigation was also designed to measure strata deformation (see Chapter 5 and Fig. 5.5) above a retreating longwall face (K4 face) 160 m below the sea bed and pillar workings in an overlying seam 75 m above the face. An interseam strata section is given in Fig. 1.25. The face was 180 m long with an extraction height of 1.6 m, and part of the investigation involved examining deformation of the pillars during undermining.

The average depth of the pillar workings was 85 m below the sea bed and

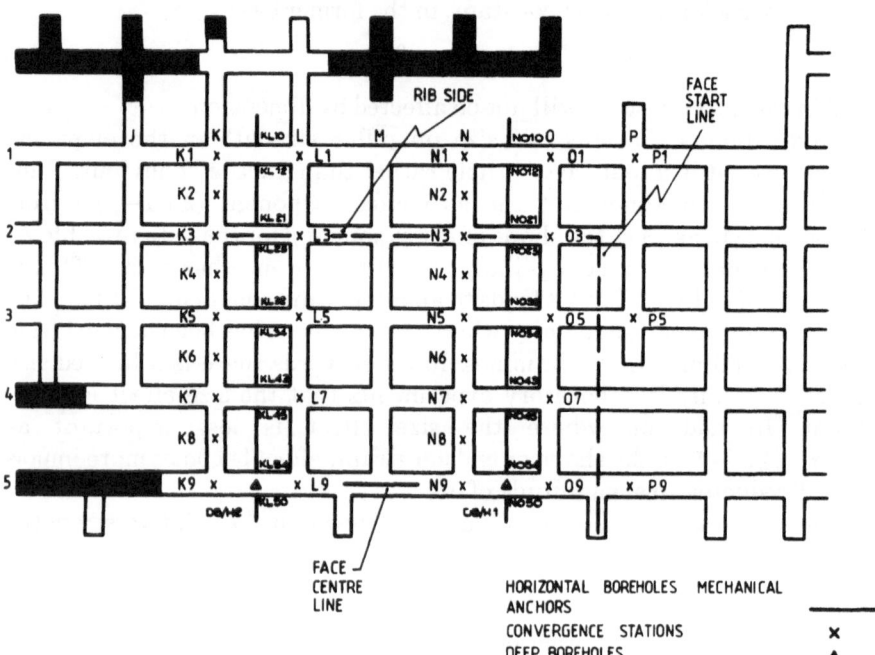

Fig. 2.9 Position of pillar instrumentation boreholes in Main seam pillars above K4 face in the Brass Thill seam at Lynemouth Colliery, Northumberland, Britain. A geological section is included in Fig. 5.3.

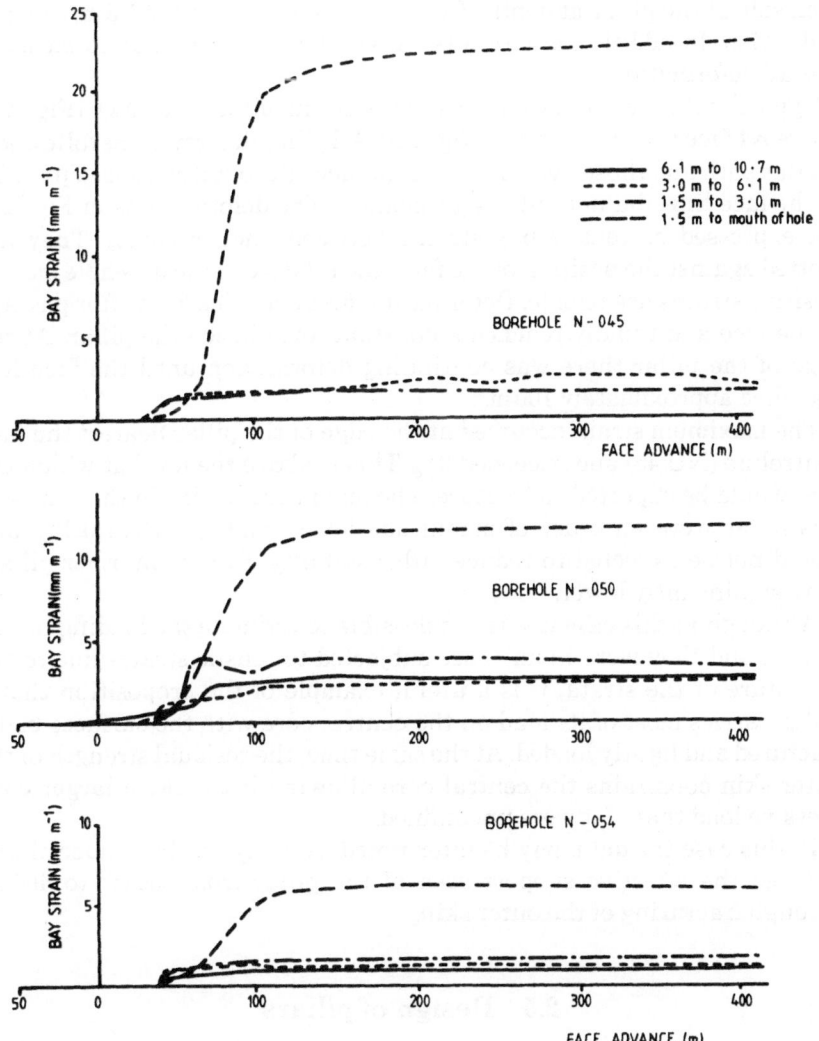

Fig. 2.10 Typical horizontal tensile bay strains related to face position and computed from relative deformations between anchors in Main seam pillars. The locations refer to Fig. 2.9. Note the much larger strain at the edges of the pillar.

the extraction ratio of 35 % from 5.5 m rooms and 23 m pillars to a height of 2.5 m was relatively conservative. Boreholes (Fig. 2.9) were drilled horizontally into the pillars normal to the direction of advance over half the face width. On an adjacent face, boreholes were drilled parallel to the face direction. Extensometer wire anchors were installed in boreholes from each side of the pillar at depths from the borehole mouth of 1.5 m, 3 m, 6 m and 10.7 m. In addition a wire was passed through the pillar to measure overall deformation.

Typical pillar deformation observations in pillar line NO (Fig. 2.9) above K4 face are illustrated in Fig. 2.10. All pillar deformations followed a similar path and those over the adjacent face, although measured parallel to the face line, did not differ significantly. The deformations in Fig. 2.10 are expressed as tensile bay strains between anchor points. They are plotted against the position of the face line relative to the borehole mouth. Positive strains are tensile. Deformation first occurred 35 m after passage of the face and rapidly reached a constant level inside the pillar. At the edge of the pillar there was continuing deformation until the face had travelled approximately 100 m.

The maximum strain occurred at the edge of the pillar nearest the face centreline (NO 45) and exceeded 2%. This is above the level at which the coal would be expected to fracture. The maximum strains in the centre of this pillar were an order of magnitude lower, and less than 0.2%, and would not be expected to reduce pillar stability. Strains in other pillars were significantly lower.

Although in this case it was not possible to estimate the loading on the pillars, and they were in any case subjected to tensile strains due to the curvature of the strata, it is a useful example of the proposition that a pillar carries most of its load on the central core with the surfaces being fractured and lightly loaded. At the same time, the residual strength of the outer skin constrains the central core allowing it to take a larger compressive load than if it were unconfined.

In this case the data may be interpreted as implying that undermining reduced the effective support area of the pillar from $530 \, \text{m}^2$ to $460 \, \text{m}^2$ through fracturing of the outer skin.

2.5   Design of pillars

The stress/strain distribution in a pillar can be used as a basis for pillar design which is superior to the empirical strength equations of the previous sections.

Pillars are simple structures, and with information on the relative stiffness of the pillar roof and floor rocks, it is relatively easy to estimate the stress distribution in the pillar and adjacent rocks. Wilson (1980) has demonstrated methods of estimating pillar stress distributions based on observation and hypothesis, but the best approaches are through boundary or finite element analysis.

A simple two-dimensional boundary element programme, developed by

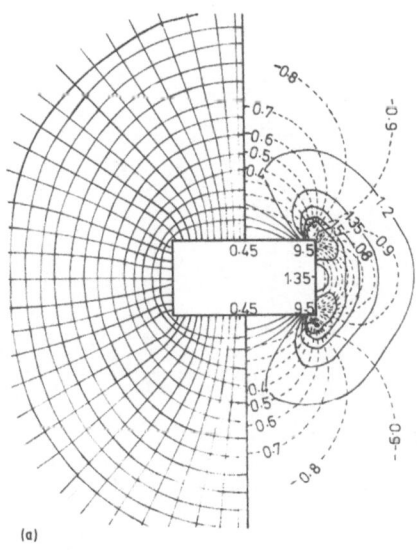

(a)

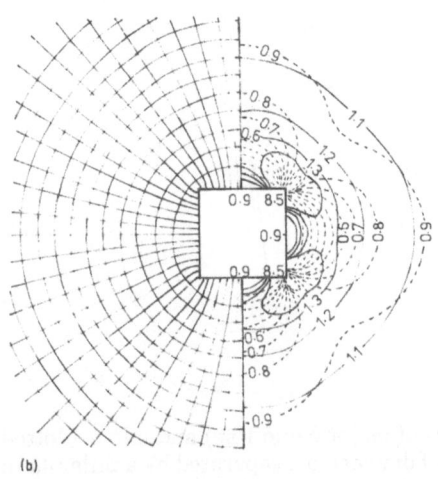

(b)

Fig. 2.11 Principal stress trajectories (LHS) and contours (RHS) of the ratio of major principal stress to applied stress (solid line) and minor principal stress to applied stress (dotted line) for (a) a rectangular and (b) a square opening in an infinite medium subject to a uniform stress field (using Bray and Hocking's two-dimensional boundary element analysis, in Hoek and Brown, 1981).

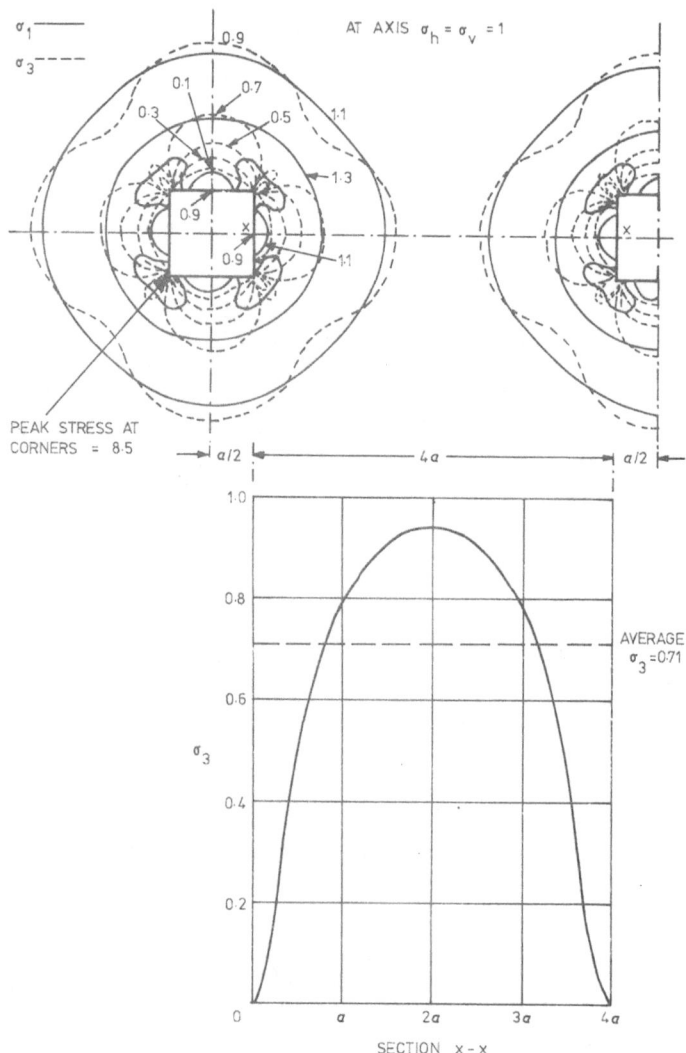

Fig. 2.12 Contours of major (solid line) and minor (dotted line) principal stress around two rooms of dimension *a* separated by a pillar 4*a* in width, and plotted to give minor principal stress (expressed as a proportion of applied stress) distribution in the pillar.

Bray and Hocking amongst others is included in Hoek and Brown (1981). This can be used after modification to calculate stresses around an opening or openings in a homogeneous, isotropic, linearly elastic material under conditions of plane strain in an infinite medium subjected

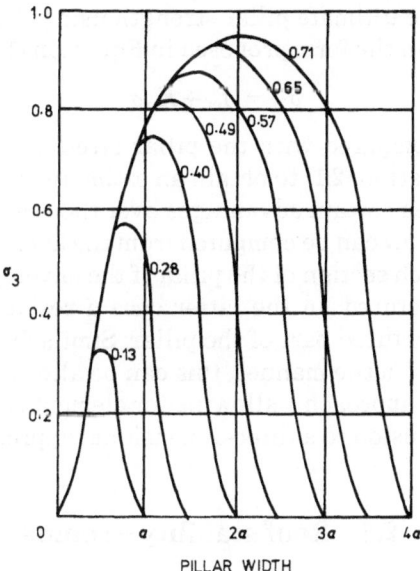

Fig. 2.13 Relation between minor principal stress expressed as a proportion of uniform applied stress and pillar width – for pillars of varying width. Values for average σ_3 are given for each curve.

to various combinations of uniform field stresses or external loadings. Typical solutions are given in Hoek and Brown, and the solutions for square and rectangular openings in a uniform stress field are reproduced in Fig. 2.11.

It can be argued that the boundary conditions may be a little extreme for coal mining. Nevertheless, the range of moduli for coals (see Evans and Pomeroy, 1966) of 2–$4\,\text{GN}\,\text{m}^{-2}$ is of the same order as for Coal Measures shales and seatearths and not less than half the range of most sandstones and mudstones. The effect of assumptions of isotropy and of two dimensions is also unlikely to affect the analysis as much as the assumptions made in some empirical approaches.

A simple example of how these computed stress distributions can be used in pillar design is given in Fig. 2.12. This takes the stress distribution in Fig. 2.11(b) and assumes initially two square rooms of dimension a at a distance $4a$ apart. Then the minor principal stress or confining stress in the pillar between the two can be projected on to a graph of minor principal stress against pillar width, to give the minor principal stress distribution and the average minor principal stress. This can be computed for pillars of any width (see Fig. 2.13) and the resultant distribution can be used to compute the ultimate pillar strength using the strength envelope of the rock or coal in the form proposed in Equation (1.1):

$$\sigma_{1f} = \sigma_{cf} + K_p \sigma_3$$

Then σ_{1f} can be compared with the pillar stress σ_p computed from the tributary area Equation (2.1) to obtain an estimate of safety factor.

This approach has certain advantages over those examined in previous sections. Although σ_{1f} can be computed from the average of σ_3, it can also be computed for each section of the pillar if the envelope is curved or part of the pillar is fractured. In the latter case a residual strength can be allocated to the fractured part of the pillar. Similarly if part or all of the pillar behaves in a ductile manner, this can be allowed for in design. It is therefore a flexible approach – allowing for elements of limit state design, and capable of extension to a three-dimensional approach.

2.6   Roof stability – rooms

In a perfect world, it would be possible to extend the pillar analysis to examine the stability of the roof and to a lesser extent the floor of the rooms. This is a more important problem than the stability of pillars. There have been disasters (Bryan, Bryan and Fouche, 1964) and near disasters (Mottahed and Szeki, 1982) due to multiple pillar failures, usually when a combination of deteriorating pillars and a weakness of the overall roof or floor strata – which prevents bridging to ribside abutments – leads to overall collapse. Nevertheless, the majority of accidents in room and pillar workings occur through falls of roof – and roof design is considerably more difficult than pillar design.

This can be illustrated most simply in Fig. 2.14, where it can be seen that in the centre of both the sidewall and roof of a square room in a uniform stress field, quite high lateral compression stresses will be sufficient to retain in compression any discontinuities in the roof. At the same time at moderate depths, deviatoric stresses will be insufficiently high to induce fracture. If this does occur, it is likely to be associated with peak stresses at the corners of the excavation. This itself may lead to reduction of lateral compression.

The real weakness of the approach, however, is that it ignores *body forces* – as do most methods of stress analysis based on continuum mechanics. This can be illustrated very simply in Fig. 2.15 by plotting a normalized stress at the centre of the roof of an excavation against the excavation width–height ratio – ostensibly its *span*. It is an exponential type of relation, σ_1 asymptoting to infinity as the span reduces to zero and vice versa. It is evident that that will not happen and that as the span

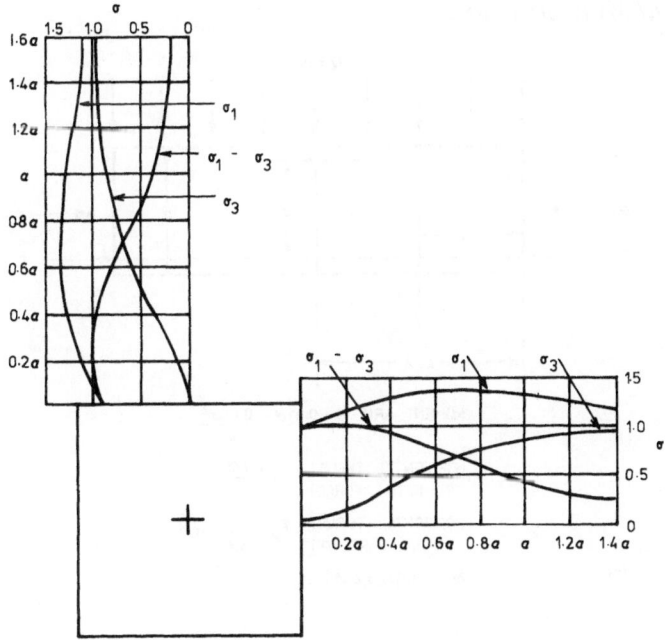

Fig. 2.14 Major, minor and deviatoric stresses at roof centre line and side wall axis level of a square room in a uniform stress field expressed as a proportion of applied stress.

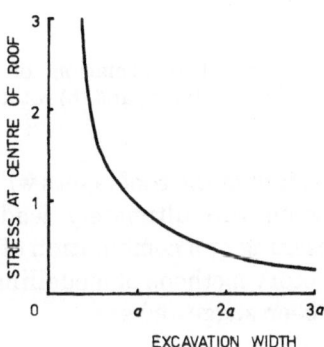

Fig. 2.15 Relation between the horizontal (originally major) principal stress at the roof centre line expressed as a proportion of a uniform applied stress field and as a ratio of room height and room width.

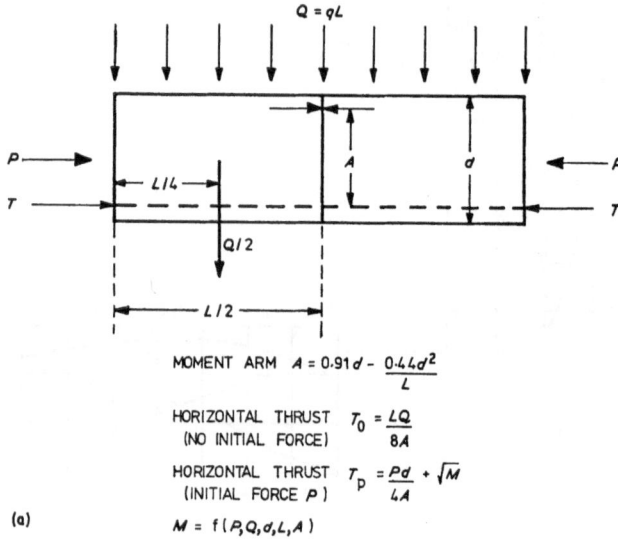

$$\text{MOMENT ARM} \quad A = 0.91d - \frac{0.44d^2}{L}$$

$$\text{HORIZONTAL THRUST} \quad T_0 = \frac{LQ}{8A}$$
$$\text{(NO INITIAL FORCE)}$$

$$\text{HORIZONTAL THRUST} \quad T_p = \frac{Pd}{4A} + \sqrt{M}$$
$$\text{(INITIAL FORCE } P \text{)}$$

(a)

$$M = f(P, Q, d, L, A)$$

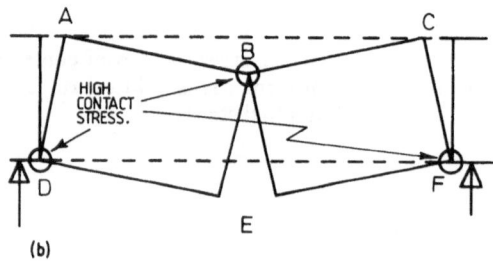

(b)

Fig. 2.16 Possible mechanisms of deformation of a cracked roof section involving (a) sliding and (b) rotation.

increases, so the body weight of the roof layers will induce tensile stresses at the roof centre, which will ultimately lead to breakdown, either through fracture or loosening or a combination of both.

There are few satisfactory methods of modelling this type of structural unit. Some which have been suggested are:

(a) To categorize the roof as a *beam* or *plate* subjected to assumed constraints and load distributions. This is a simplistic approach and relies on very difficult estimates of roof strata thickness, continuity and body and hydrostatic load distribution. The basics are covered by Obert and Duvall (1967) and Wright (1973).

(b) To categorize the roof as a *cracked flat arch*, loaded transversely and axially, which may fail through sliding along or rotation of blocks bounded by cracks which exist normal to the beam axis. Two versions of this can be summarized in Fig. 2.16. Wright (1973) suggesting the flat arch analogy proposes that sliding will occur when the downward thrust along the crack ($Q/2$ in Fig. 2.16(a)) exceeds the shearing resistance $T \tan \phi$ along the crack. Calculation of T involves a series of complex assumptions about the moment arm distance A, the thickness d and horizontal thrust P which make the analysis extremely complex. There are additional assumptions about load distribution. It is also doubtful whether the beam actually fails in this way. A more likely mechanism suggested by Ward (1978) is that hinges will tend to be accentuated at vertical joints reducing the contact area at the joints and thus increasing the contact pressure and the tendency to collapse. This mechanism is illustrated in Fig. 2.16(b) which shows that if there are joints AB, BE, CF, sagging will reduce contact areas at D, B and F. The resultant crushing under high contact stresses will eventually lead to failure. The analysis of this mechanism would be extremely complex. In roofs where jointing does not follow a horizontal–vertical pattern typical of sedimentary rocks, wedge analysis of the type demonstrated by Hoek and Brown (1981) or by Goodman *et al.* (1982) can be used to isolate areas of potential instability.

(c) To use *empirical* methods to compare the performance of different types of roof and different roof support systems. The simplest of the empirical methods is Terzaghi's (1946) classification system. This summarizes the main rock types which might be found in a tunnel sidewall, wall or roof into 10 categories. Experiments in tunnels varying from Alpine to urban were then carried out by Terzaghi to determine initial and final loads on square sets and arch supported tunnels. Terzaghi's load cells were calibrated wooden crush pads under the uprights. The estimated loads were then related to the classification categories to give a simple method of estimating rock loads in terms of metres of rock acting on the tunnel roof. The classifications and rock loads are reproduced in Table 2.4. Terzaghi's method is not as empirical as it seems. It is based firmly on his arching theory (Terzaghi, 1943). This in turn is based on the assumption that a frictional material above a tunnel roof (represented by a trap door beneath a mass of material) when allowed to move vertically downwards mobilizes sufficient shear resistance along vertical planes at the roof boundary, to form ultimately a stable arch.

In Table 2.4 an initial rock load and final rock load are given. For classes 1–5 the initial load is zero or thereabouts, changing with time in the case of classes 2–5 to a final rock load of increasing magnitude.

Table 2.4 Description of rock conditions (after Terzaghi, 1946)

1	Hard and intact rock	Contains neither joints nor hair cracks. If fractured, it breaks across intact rock. After blasting, spalls may drop off the roof for several hours or days. At high stresses, spontaneous and violent spalling of rock slabs from sides or roof may occur.
2	Hard stratified or schistose rock	Consists of individual strata with little or no resistance against separation along the boundaries between strata. The strata may or may not be weakened by transverse joints. In such rock, spalling is quite common.
3	Massive moderately jointed rock	Contains joints and hair cracks, but the blocks between joints are intimately interlocked so that vertical walls do not require lateral support. Spalling may occur.
4 and 5	Moderately and very blocky and seamy rock	Consists of chemically intact or almost intact rock fragments which are entirely separated from each other and imperfectly interlocked. In such rock, vertical walls may require support.
6 and 7	Crushed rock and sand	Comprises chemically intact rock having the character of a crusher run and capable of exerting considerable side pressure on tunnel supports. If most or all of the fragments are as small as fine sand grains and no recementation has taken place, crushed rock below the water table will exhibit the properties of a water bearing sand.
8 and 9	Squeezing rock at moderate and great depth	Slowly advances into the tunnel without perceptible volume increase. A prerequisite for squeeze is a high percentage of microscopic and submicroscopic particles of micaceous minerals or of clay minerals with a low swelling capacity.
10	Swelling rock	Advances into the tunnel chiefly on account of expansion. The capacity to swell seems to be limited to those rocks which contain clay minerals such as montmorillonite, with a high swelling capacity.

Classification number	Initial rock load (m rock)	Final rock load (m rock)
1	0	0
2	0	$0.25S$
3	0	$0.5S$
4	0	$0.25S–0.35(S+H)$
5	$0–0.6\ (S+H)$*	$(0.25–1.1)(S+H)$
6	$(0.5–1.2)(S+H)$	$1.1\ (S+H)$

Table 2.4 (continued)

Classification number	Initial rock load (m rock)	Final rock load (m rock)
7	$(1.0–1.2)(S+H)$	$(1.1–1.4)(S+H)$
8	–	$(1.1–2.1)(S+H)$
9	–	$(2.1–4.5)(S+H)$
10	–	up to 80

* S is room span, H is room height.

Table 2.5 Relation between stand-up time and rock classification systems

Terzaghi classification	Rock behaviour and possible causes of instability	Approximate stand-up time	Deere classification
1 Hard and intact	Stable excavation unless induced stress greater than rock strength	Many years	Excellent: RQD 90–100
2 Hard stratified and schistose	Bed separation with time; surface spalling	1 year	
3 Massive, moderately jointed	Immediately stable. Detachment of blocks under gravity with time	1 month	Good: RQD 75–90
4 Moderately blocky and seamy	Immediately stable. Detachment of blocks, progressively releasing further blocks	1 week	Fair: RQD 50–75
5 Very blocky and seamy and shattered	Immediately fairly stable. Surface dilation of rock due to rapid block detachment	1 day	Poor: RQD 25–50
6 Completely crushed	Local roof falls during excavation. Rapid peripheral dilation	1 hour	Very poor: RQD 0–25
7 Sand and gravel	Immediate collapse	0	
8 Squeezing: moderate depth	Rapid yielding and deformation		Squeezing and swelling ground
9 Squeezing: great depth			
10 Swelling			

Based on Terzaghi (1946) and Deere *et al.* (1970).

Table 2.6 Summary of parameter descriptions and ratings for computation of rock mass quality (after Barton *et al.*, 1974)

1	Rock quality designation	(RQD)	
A	Very poor	10–25	
B	Poor	25–50	
C	Fair	50–75	
D	Good	75–90	
E	Excellent	90–100	

2	Joint set number	(J_n)	
A	Massive, no or few joints	0.5–1.0	
B	One joint set	2	
C	One joint set plus random	3	
D	Two joint sets	4	
E	Two joint sets plus random	6	
F	Three joint sets	9	
G	Three joint sets plus random	12	
H	Four or more joint sets, random, heavily jointed, sugar-cube, etc.	15	
J	Crushed rock, earth-like	20	

3	Joint roughness number	(J_r)	
A	Discontinuous joints	4	
B	Rough or irregular, undulating	3	
C	Smooth, undulating	2	
D	Slickensided, undulating	1.5	
E	Rough or irregular, planar	1.5	
F	Smooth, planar	1.0	
G	Slickensided, planar	0.5	
H	Zone containing clay minerals	1.0	
J	Sandy, gravelly or crushed zone	1.0	

4	Joint alteration number	(J_a)	(σ_r) approx.
A	Tightly healed, hard	0.75	(–)
B	Unaltered joint walls	1.0	(25–35°)
C	Slightly altered joint walls	2.0	(25–30°)
D	Silty-, or sandy-clay coatings	3.0	(20–25°)
E	Softening or low friction clay mineral coatings	4.0	(8–16°)
F	Sandy particles, clay-free disintegrated rock, etc.	4.0	(25–30°)
G	Strongly over-consolidated non-softening clay mineral fillings	6.0	(16–24°)
H	Medium or low over-consolidated, softening, clay mineral fillings	8.0	(12–16°)
J	Swelling-clay fillings	8–12	(6–12°)

Table 2.6 (continued)

5	*Joint water reduction factor*	(J_w)	Approximate water pressure (kg cm^{-2})
A	Dry excavations or minor inflow	1.0	< 1
B	Medium inflow or pressure, occasional outwash of joint fillings	0.66	1–2.5
C	Large inflow or high pressure in competent rock with unfilled joints	0.5	2.5–10
D	Large inflow or high pressure, considerable outwash of joint fillings	0.33	2.5–10
E	Exceptionally high inflow or water pressure	0.2–0.1	> 10

6	*Stress reduction factor*	SRF
A	Multiple occurrences of weakness zones containing clay	10.0
B	Single weakness zones containing clay – shallow rock	5.0
C	Single weakness zones containing clay – deep rock	2.5
D	Multiple shear zones in competent rock (clay free)	7.5
E	Single shear zones in shallow competent rock (clay free)	5.0
F	Single shear zones in deep competent rock (clay free)	2.5
G	Loose open joints, heavily jointed	5.0
H	Competent rock – low stress, near surface	2.5
J	Medium stress	1.0
K	High stress, very tight structure	0.5–2.0
L	Mild rock burst (massive rock)	5–10
M	Heavy rock burst (massive rock)	10–20
N	Mild squeezing rock pressure	5–10
O	Heavy squeezing rock pressure	10–20
P	Mild swelling rock pressure	5–10
R	Heavy swelling rock pressure	10–15

Classes 1–5 will cover most Coal Measures roof rocks. The time required for the intact rock to collapse – whether through the mechanism of Fig. 2.16 or some other mechanism – to the final state is sometimes referred to as *stand-up time*.

In Table 2.5 the stand-up time for a 5 m wide span of roof is related to Terzaghi's and to Deere's RQD classification, together with comments on the rock behaviour. The concept of stand-up time is useful in coal mining – both in assessing room stability and possibly caveability. A more sophisticated empirical method of categorizing rock in terms of stand-up time has been suggested by Barton, Lien and Lunde (1974). Known as the

Norwegian Geotechnical Institute (NGI) Classification it is based on Deere's system but with added sophistication. Six parameters in three groups are used to define rock mass quality, Q:

$$Q = \frac{\text{RQD}}{J_n} \frac{J_r}{J_a} \frac{J_w}{\text{SRF}} \tag{2.9}$$

The parameter description and rating ranges are described in abbreviated form in Table 2.6. The factor Q is obtained by substituting the equivalent ratings from Table 2.6 in Equation (2.9) and then classifying the rock in groups. The groups chosen are:

Very good rock	> 100
Good rock	10–100
Fair rock	1–10
Poor rock	0.1–1
Very poor rock	< 0.1

It is of course similar to Deere's classification, but the method of multiplication gives a wider spread and the effects of discontinuities and of the stress conditions acting on the structure are amplified.

The nominal properties which the three groups in the equation represent can be summarized as:

1. Block size (RQD/J_n)
2. Interblock frictional or shear resistance (J_r/J_a)
3. Active stress (J_w/SRF)

It is an interesting observation that the same approximate parameters can be shown (see Farmer and Attewell, 1973) to control the compression of rock fill or broken rock and might be applied to post-caving or pack consolidation. The application in mining roof support is illustrated in Fig. 2.17 which is based on case history experience in rather stronger rocks than the Coal Measures by Bieniawski (1976), although data obtained from caving longwall excavations fit the curve reasonably well.

For instance a strong, massive sandstone roof such as occurs in the roof of the Brass Thill seam at Lynemouth Colliery in Fig. 1.25 might have an RQD of 80, J_n 4, J_r 4, J_a 0.75, J_w 1 and SRF 1 giving a Q factor of about 100. This indicates a maximum unsupported span of about 20–50 m with a stand-up time of 1–4 weeks, which is confirmed by the first cave monitored during observations (see Fig. 5.2) at Lynemouth Colliery which occurred after a face advance of 24 m in a period of 16 days.

The roof of the main seam at Lynemouth Colliery comprises a layered mudstone having an RQD of 70, J_n 4, J_r 3, J_a 2, J_w 5 and SRF 2.5 giving a Q

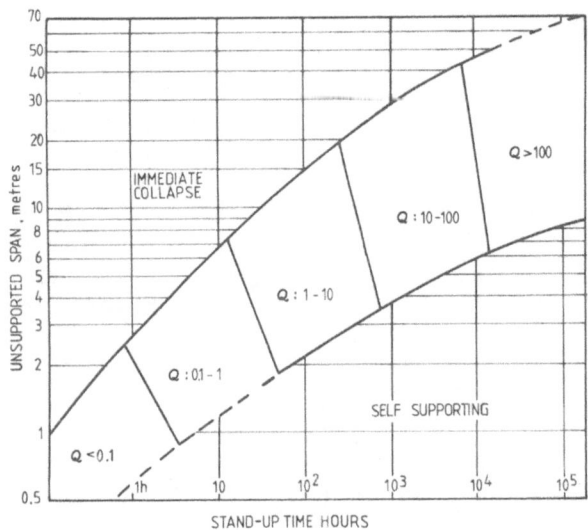

Fig. 2.17 Relation between unsupported span, stand-up time and rock mass quality (after Bieniawski, 1976).

value of 4.7. Unsupported spans of 5.5 m (support had been withdrawn) were still standing after 15 years (5×10^3 days) indicating, if anything, a degree of conservatism in Fig. 2.17. However, after very little undermining disturbances (see Tubby and Farmer, 1981) there was widespread collapse of rooms.

2.7 Support of mine rooms

Traditionally in Britain, where mine roofs tend to be weak, rooms are supported by RSJs, with a section of about 150 mm by 125 mm, and wooden legs. RSJs across junctions are usually supported by double girders on chocks. It is a safe arrangement – apart from the occasions when junction supports are dislodged by shuttle cars – but an expensive and occasionally inconvenient one. It may also be unnecessary.

An example of deformation during conventional junction excavation and support is given by Hodkin (1982). This was in a 2.5 m seam at a depth below sea bed at Ellington Colliery, Northumberland of 120 m. Extraction was 35 % and workings comprised 3.5 m rooms and 23 m pillars. Anchors were installed at three junctions up to 3.5 m into the roof at the centre of the junction. Results of relative movement between the roof surface and

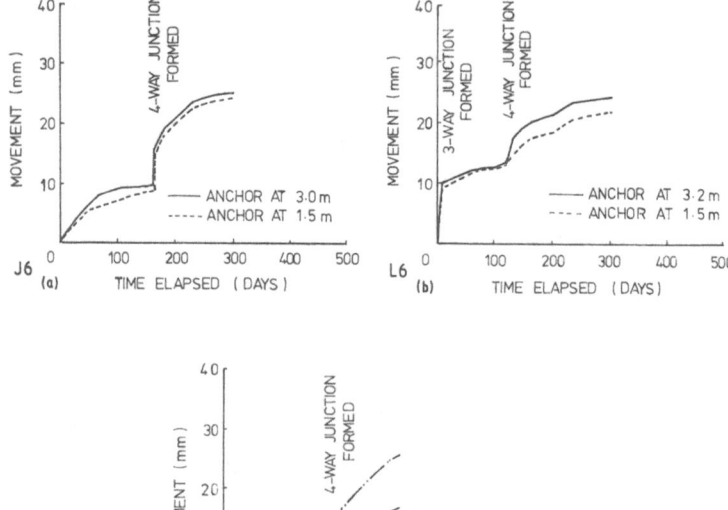

Fig. 2.18 Plots of relative vertical downward movement between the mouth of the hole (at the roof surface) and anchors at specified distances into the roof at the centre of three junctions at Ellington Colliery (after Hodkin, 1982); junction G/6 also includes roof to floor convergence. Junctions are 5.5 m by 5.5 m; pillars 23 m × 23 m. The strata comprise 0.3 m coal in the roof beneath 1.5 m of mudstone grading into 2 m of seatearth beneath a further coal seam.

the anchor position (Fig. 2.18) indicate that all the deformation occurred in the first metre or so. There was virtually no deep-seated roof deformation and little indication of bed separation. It is interesting to note the accelerated deformation as the fourth leg of the junction was formed in each case. It is also interesting to note the continuing deformation with time of the junction – a feature of arch or girder support.

The limited amount of deformation in this case, and the occurrence of this deformation close to the surface, suggests that a more suitable form of support would have been *rockbolts*. These are the more conventional form of support for rooms – particularly in stronger rock, and may usefully be considered here.

Rockbolts are discussed at greater length by various authors (see for instance Hobst and Zajic (1983)). Design often entails a complexity which is not necessary in such a simple support method. For a preliminary assessment of whether a rock is suitable for rockbolting, the only information required (Farmer and Shelton, 1980) is whether rockbolting is necessary or feasible. Rocks where reinforcement is unnecessary are those with few or tightly closed discontinuities normally described as hard, intact rocks (see Table 2.4). They are not common in Coal Measures strata. Rocks where bolting is not feasible are:

(a) Those which are insufficiently strong to provide an anchorage location – these are surprisingly few.
(b) Those with a tendency to ductile deformation at low stresses or to swelling or slaking in the presence of water.
(c) Those with high water flows or high pore or fissure water pressures.

Between the two limits there is a wide range of strata – including most Coal Measures strata – where reinforcement has varying degrees of attractiveness. It is most attractive in relatively strong rock with widely spaced bedding and jointing and low peripheral stresses. It is least attractive in weaker rocks with closely spaced bedding and discontinuities and peripheral stresses close to the strength of the rock. In such cases, deformation with time may occur partly through loosening and partly through fracturing and yielding of intact rock. An example in Fig. 2.19 shows a typical example of this phenomenon.

In rocks where stable anchorages can be obtained, and where there are a reasonable number of clean discontinuities (say Terzaghi classes 2–5 in Table 2.4) then justification of bolting is quite simple. Consider, for instance, the case of homogeneously fissured rock having a residual shear strength along discontinuities given by:

$$\tau = \sigma_n \tan \phi$$

In the case of randomly orientated discontinuities it can be shown that the preferred direction for sliding will be when the angle α between the discontinuity and the major principal stress is equal to:

$$\frac{\pi}{4} + \frac{\phi}{2}$$

Then if $\phi = 30°$, the Coulomb–Navier failure criterion gives conditions for sliding when $\sigma_1 > 3\sigma_3$.

At the surface of an underground excavation the major and minor principal stresses (usually expressed as radial and tangential stresses relative to the sidewall), which may be compressive or tensile, depending on the depth, in-situ stress distribution and excavation geometry, invariably approach or exceed these conditions, except in an intact rock. Depending again on the geometry and in-situ stress distribution, there will be quite rapid stress convergence away from the surface so that $\sigma_1 <$

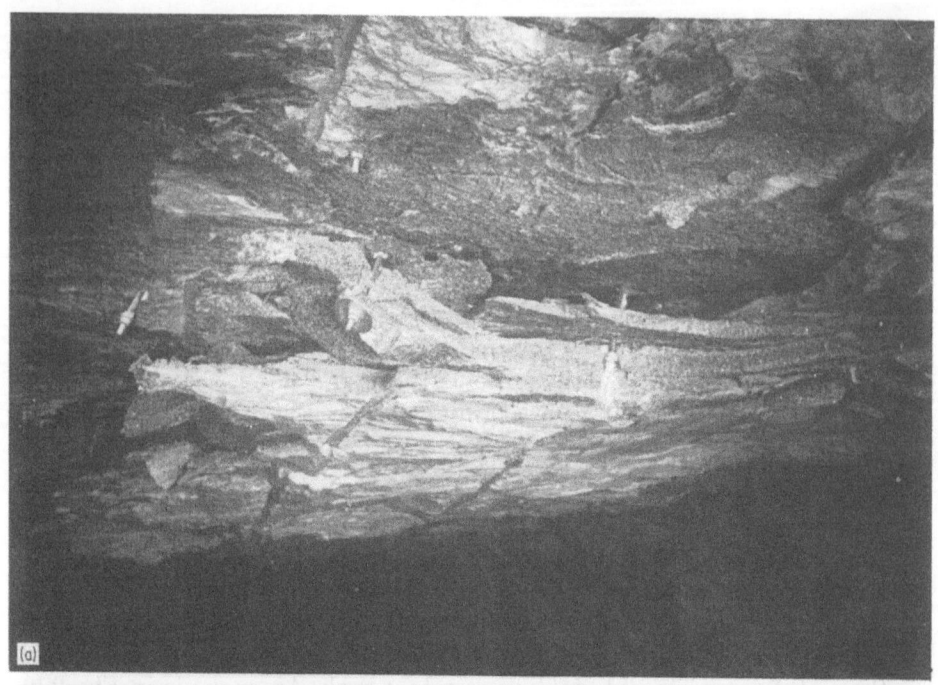

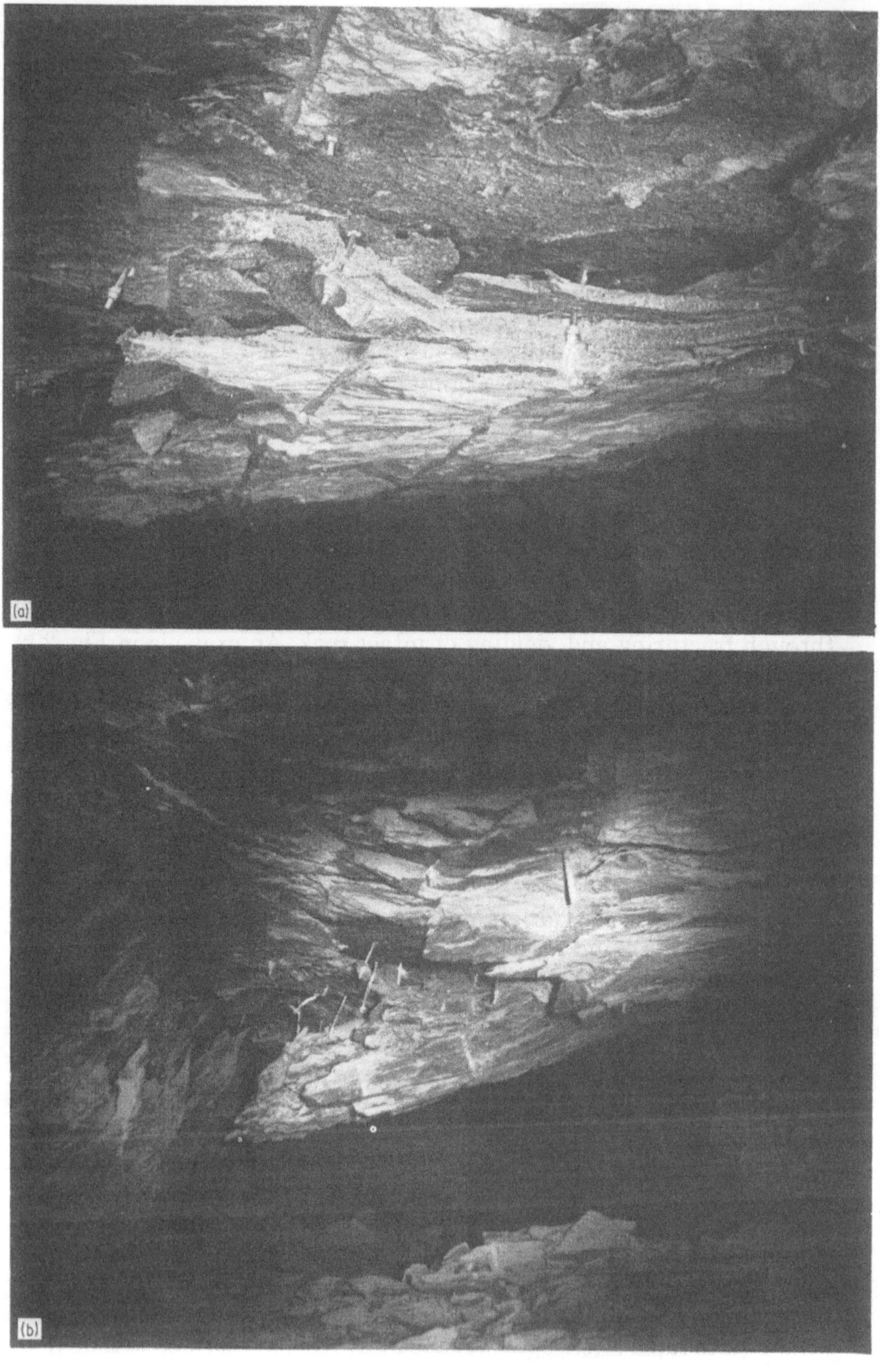

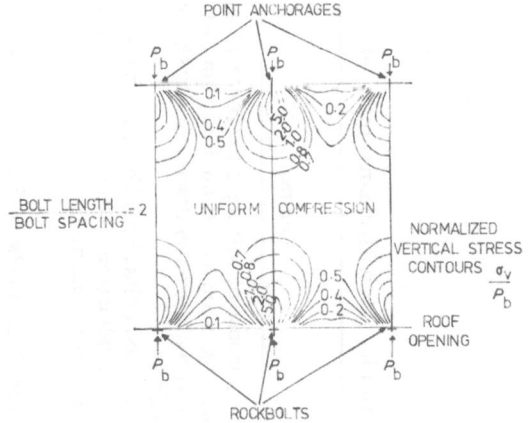

Fig. 2.20 Two-dimensional representation of stress distribution in an elastic roof beam induced by point loads with a spacing equal to half the beam thickness.

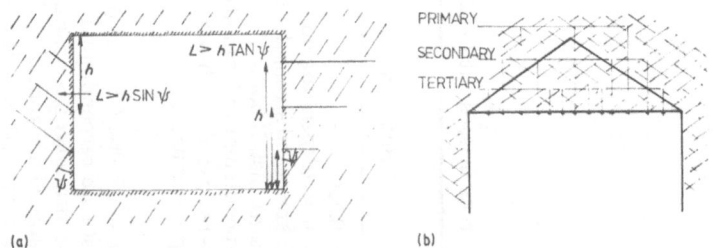

Fig. 2.21 (a) Determination of bolt lengths in excavation sidewalls; (b) primary, secondary and tertiary bolting patterns in a large span excavation.

$3\sigma_3$. There is, therefore, a requirement for support at the excavation periphery, and for a limited extent into the sidewall, although more careful design may be required where very high corner, edge stresses or large span bending occur.

The optimum ratio of bolt length to bolt spacing can be determined by simple analysis of stresses beneath an elastic half-space. Thus, by use of the Boussinesq distribution beneath a point force, it can be shown that a bolt length to bolt spacing ratio of 2 is required to create a zone of uniform

---

Fig. 2.19 Deterioration after: (a) 2 months; (b) 8 months of a bolted room in a potash mine formed in a roof of anhydritic marl with properties similar to those illustrated in Fig. 1.4. The mine was at a depth of 1100 m; rooms were 6 m wide by 3 m high; pillars 30 m × 30 m; strata temperatures were 46°C.

Table 2.7 Rockbolt parameter design rules for rock masses with < 2 and < 3 discontinuity sets with clean tight interfaces

Excavation span	Number of discontinuity sets	Bolt design	Comments
< 15 m	< 2 inclined at 0–45° to horizontal	$L = 0.3B$ $S = 0.5L$ (depending on thickness and strength of strata). Install bolts perpendicular to lamination where possible with wire mesh to prevent flaking	The purpose of bolting is to create a load-carrying beam over span. Fully bonded bolts create greater discontinuity shear stiffness. Tensioned bolts should be used in weak rock; sub-horizontal tensioned bolts where vertical discontinuities occur
	< 2 inclined at 45–90° to horizontal	For side bolts: $L > h \sin \psi$ (installed perpendicular to discontinuity) $L > h \tan \psi$ (installed horizontally). See Fig. 2.21(a) for h, ψ; L, bolt length; S, bolt spacing; B, excavation span	Roofbolting as above. Side bolts designed to prevent sliding along planar discontinuities. Spacing should be such that anchorage capacity is greater than sliding or toppling weight. Bolts should be tensioned sufficiently to prevent sliding

> 3 with clean tight interfaces		$L = 2S$ $S = 3-4 \times$ block dimension Install bolts perpendicular to excavation with wire mesh to prevent flaking	Bolts should be installed quickly after excavation to prevent loosening and retain tangential stresses. Prestress should be applied to create zone of radial confinement. Sidewall bolting where toe of wedge daylights in sidewall
> 15 m	< 2	$L_1 = 0.3B_1$ Primary bolting $S_1 = 0.5L_1$ $L_2 = 0.3S_1$ Secondary bolting $S_2 = 0.5L_2$ Install wire mesh to prevent spalling	Primary bolting conforms to smaller excavation design. Secondary (and tertiary) bolting supplements primary design (Fig. 2.21(b))
> 3 with clean tight interfaces		$L_1 = 0.3B_1$ Primary bolting $S_1 = 0.5L_1$ $S_2 = 3-4 \times$ block size. Secondary bolting $L_2 = 2S_2$	Primary bolting should have sufficient capacity to restrain major blocks. Decisions on block size for secondary bolting should be left to the section engineer

radial compression (Fig. 2.20). Bolt forces should be calculated to give a uniform compressive stress of the order of $30\,\text{kN}\,\text{m}^{-2}$ per metre of thickness of compression zone to support the body forces exerted by the compressed layer.

These arguments are the basis of most empirical design methods. Table 2.7 and Fig. 2.21 summarize the experimental and practical experience of Rabciewicz (1969) and Lang (1971).

This is a rather more complex approach than would normally be required for horizontally bedded Coal Measures strata.

3

TUNNELS AND DRIFTS

Tunnels and drifts, like rooms and pillars, considered in the previous chapter, are relatively simple structures which can be easily analysed. Because of this there may be a temptation to use over-simplified models to describe stress distribution.

The term *tunnel* used in coal mines may be taken to include major roadways driven in seam or as cross-measures drifts. *Access roadways*, which form an integral part of the longwall mining process and are subjected to high stresses, are considered separately in Chapter 6.

Over 90% of the tunnels and over 95% of the tunnels in rock in Britain are constructed in Coal Measures rocks during the development of underground coal mining operations. Construction of tunnels in Coal Measures rocks is affected by several factors, none of which is unique, but which combine to demand a particular approach to design and construction. The major factors are:

(a) The rocks of the Coal Measures occur in a rhythmic or cyclic sequence recurring around a coal seam–seatearth element at approximately 10 m intervals. The rocks in the sequence or cyclothem have been described in Chapter 1 and range from plastic clays to strong sandstones.

(b) Zones or layers of sheared argillaceous rocks (see Fig. 1.18) having negligible tensile strength and a shear resistance close to residual occur in each cyclothem, most frequently in the seatearths.

(c) Seatearths, some with significant montmorrillonite content and some of considerable thickness and falling, when saturated (Fig. 1.9), into Casagrande's CL classification (inorganic clays of low to medium plasticity) occur, usually in the inverts of development tunnels driven at the coal seam level.

(d) Tunnels are often subjected to high and variable stresses, partly because they are deep, but also because they may be affected by old workings, pillars, and tectonic disturbances.

The resultant effect of these factors is to cause yield, fracture or loosening of some of the rocks exposed in the periphery of a tunnel excavation. Depending on the relation between the strength of the peripheral rocks, the stresses redistributed by excavation, and the loosening of strata accompanying or following excavation, a yield zone or zone of deformation may be created around a tunnel. In strong rocks where brittle or strain-softening deformation behaviour occurs (Figs 1.1–1.3) strata can be supported relatively easily by mobilization of the residual strength of the deformed strata by low support pressures. In weaker rocks subjected to high stresses, where ductile or strain-hardening behaviour (Fig. 1.4) occurs – possibly over a period of time – much higher restraint may be required to support the strata.

3.1   Stress distribution around tunnels

Stresses around tunnels in the Coal Measures strata associated with coal mining are generally high. This is because the tunnels tend to be deep compared with similar tunnels in civil engineering, but also because they may be affected by the presence of mine workings, pillars, faults or steeply inclined strata.

The available evidence (Brown and Hoek, 1978) suggests that the in-situ stress field in horizontally layered, relatively undisturbed, Coal Measures strata at depths greater than about 500 m comprises a vertical principal stress equal to the pressure of overlying strata and two horizontal principal stresses of equal magnitude. At depths less than 500 m and in layers of particularly strong rock, it is probable that the horizontal principal stresses will exceed the vertical principal stress. This may be due to the residual effects of over-consolidation or to forces created by tectonic disturbance. These effects will probably be less pronounced in the Coal Measures strata which are relatively weak and cyclic. Of greater importance are high horizontal stresses encountered in the vicinity of geologically disturbed ground – particularly shear and thrust faults, where horizontal stresses up to five times the vertical stress have been reported. A possible explanation for this phenomenon is that in weak rocks, failure will take place in the Rankine passive state as overburden is reduced and when $\sigma_{hf}/\sigma_v = K_p = (1 + \sin \phi)/(1 - \sin \phi) = 3.5$. It is significant (see Fig. 1.23) that high horizontal stresses, invariably associated with thrust faulting at relatively shallow depths, are of this order of magnitude.

High vertical stresses can exist as residual stresses in pillars adjacent to worked longwall faces (see Fig. 2.6). These can have magnitudes more than five times the in-situ vertical stress, depending on the strength and confinement of rocks in the pillar. The stresses can be retained in pillars

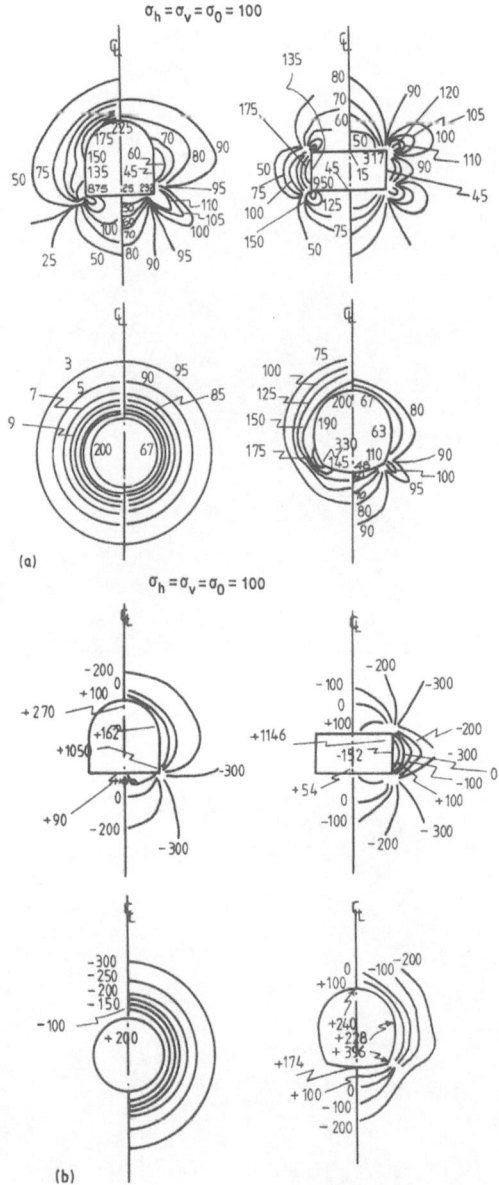

Fig. 3.1 Stress distribution around tunnels in a homogeneous stress field: (a) Distribution of deviatoric (LHS) and spherical (RHS) stresses (after Hoek and Brown, 1981) around four shapes of opening in a homogeneous stress field having an initial stress of 100 units at the tunnel axis. (b) Distribution of the McLintock–Walsh maximum tensile stress parameter $\sigma_{Tm} = (\sigma_1 - \sigma_3)(1 + \tan^2 \phi)^{\frac{1}{2}} - (\sigma_1 + \sigma_3) \tan \phi$ where $\phi = 40°$ and tensile stresses are positive.

for some considerable time following mining, and can extend for some considerable vertical distance. The mechanics of stress transfer in layered strata have been discussed by Gaziev and Erlikman (1971) and this is illustrated in Fig. 8.10. The effect of vertical stress transfer from pillars through 200–300 m of Coal Measures strata on tunnel stability has been illustrated by several authors (see for instance Billington and Jacomb-Hood, 1976).

However, if it is assumed that the vertical and horizontal stresses are equal, then several specific aspects of tunnel design in Coal Measures rocks can be illustrated by an examination of simple theoretical stress distributions around a tunnel in a continuum based on simple elastic stress analysis. Figure 3.1 illustrates the deviatoric and spherical stress distributions, and also the distribution of the McLintock–Walsh (McLintock and Walsh, 1962) tensile failure parameter around four tunnel shapes in a hydrostatic stress field in a homogeneous continuous material, calculated from base data obtained by Hoek and Brown (1981) from boundary element analysis.

The implications are obvious, since the circular and arch/inverted arch profiles lead to reduced deviatoric, and hence shear and tensile stresses. The rectangular, and particularly the arch shaped tunnel results in increased deviatoric stresses at the sharp corners of the tunnel periphery. Where the high deviatoric stresses occur in conjunction with the weakest rock, as they may do in Coal Measures strata where clays invariably occur in the floor of the tunnel, then disturbance of the floor is likely, and convergence and disturbance of the tunnel will occur.

This is demonstrated particularly in Fig. 3.2, which compares tunnel profiles from the Sea Drift at Dawdon Colliery, Durham, Britain (see Farmer and Glossop, 1983; Snowdon *et al.*, 1983). This shows tunnels in the same lithology 6 years after construction in 1975. Although the circular profile is undisturbed, and inspection showed virtually no contact between carcass and the rings, the arch profile shows signs of floor heave. The depth of this tunnel was 476 m below sea level and 449 m below sea-bed level. It is interesting to note that at the Selby project, Yorkshire (discussed later in this chapter) the extensions of the main tunnels, started in 1981, have been driven with a full face tunnelling machine and a

---

Fig. 3.2 Photographs taken in October 1981, six years after completion of tunnels 40 m apart at chainage 990 m in (a) the Sea Drift and (b) the East Heading at Dawdon Colliery, Co. Durham, Britain. The tunnels at this point were both in slightly disturbed siltstone having a strength of about $60\,MN\,m^{-2}$ at a depth of 449 m below sea level. The 3.65 m diameter machine driven tunnel was completely undisturbed with virtually no contact between the surrounding strata and the rings.

partial face machine (Hayward, 1982). The tunnel driven with the latter machine has a flat base and no invert support. It is reported to be showing up to one metre of floor heave at a similar depth to the Dawdon tunnel, while the circular profile tunnel is substantially undamaged.

3.2 Rock deformation and support

Many of the approaches to tunnel support design assume an instantaneous stress distribution around a tunnel which determines lining reactions. This is a simplification of a complex relation between the time and position of support installation, the stiffness of the support and the load-bearing characteristics of the support. This has been discussed by Ward (1978) amongst others.

If support is erected at the tunnel face, then the initial support load will be the setting load, since the tunnel face will act as a dome and forces redistributed by excavation will be carried by the face as well as the floor, roof and side abutments. As the face advances, the support provided by it is reduced and the tunnel section will act in the fashion of an arch under conditions of plane strain. With associated relaxation of the strata, the loads acting on the support will, therefore, change considerably during the first few radii of tunnel advance.

Provided the radial deformation, u_a, of the tunnel is known, then various idealized solutions can be obtained to determine the support pressure exerted on a lining installed at the face. Daemen and Fairhurst (1972) used a rather over-simplified analysis by De la Cruz and Goodman

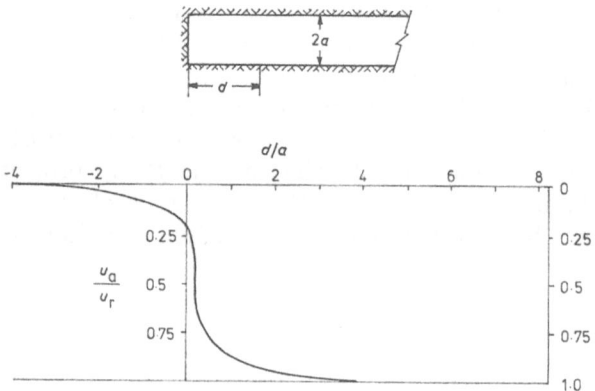

Fig. 3.3 Distribution of radial displacement u_a as a proportion of plane strain radial displacement u_r at the boundary of a circular tunnel in an isotropic stress field and relative to the tunnel face position (Ward, 1978).

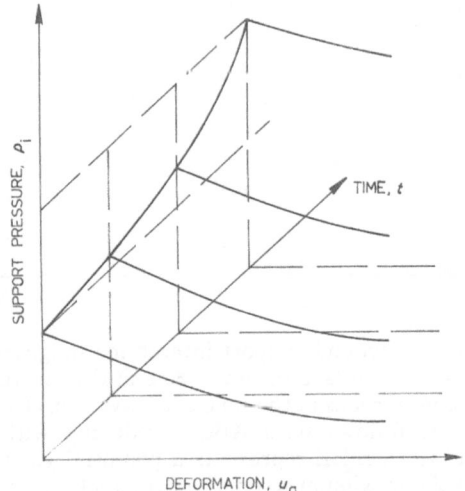

Fig. 3.4 An example of a characteristic support surface in p_i, u_a, t space, based on the concept that lining support pressure increases with time but reduces with radial deformation before support installation (see Daemen and Fairhurst, 1972; Muir Wood, 1979).

(1970) to obtain a solution for the support pressure, p_i, generated on a circular tunnel lining placed in a tunnel formed in an elastic material in an isotropic geostatic stress field (σ_0):

$$\frac{p_i}{\sigma_0} = \frac{K_s(1-F)}{K_s + 2G_r} \tag{3.1}$$

where K_s is a measure of the lining stiffness given approximately by $K_s = tE_s/a$ where t is the lining thickness, a the tunnel radius and E_s the modulus of the lining; G_r is the modulus of rigidity of the rock and F is the ratio between the actual radial displacement, u_a, and the potential fully developed elastic displacement $\sigma_0 a/2G_r$ of the tunnel.

It is less easy, however, to relate u_a to the position of the tunnel face. Ward (1978) synthesizing various approaches suggests a relation of the form illustrated in Fig. 3.3, in which the influence of the face reduces to 5 % of the face effect at a distance of 2.25 radii in front of and behind the face. It is immediately obvious that this is a less than totally satisfactory approach to support pressure and convergence. Factors which are not taken into account include time dependent deformation of the rock and various operational factors such as the method of excavation and speed of excavation.

It is therefore perhaps better to accept the theoretical justification and

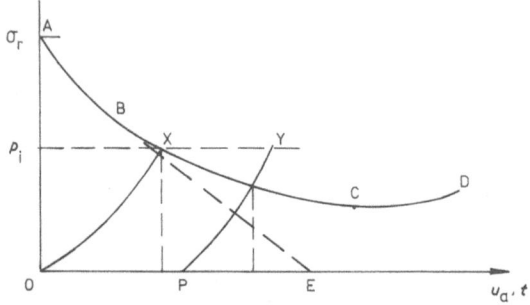

Fig. 3.5 An example of ground support interaction diagrams used in the New Austrian Tunnelling Method (see Braun, 1980 and Brown, 1981). σ_r is the radial pressure in the peripheral rocks at the time of excavation. Over an advance of 1–5 tunnel diameters this follows path ABCD, reducing with increasing radial deformation, u_a and time, t. BE represents a potential yield path. Minimum σ_r value C, represents the maximum strength of the rock. If further deformation is allowed the rock will loosen (CD). Support should therefore be applied as near to C as ground conditions allow. If the support resistance is p_i the delayed support characteristic PY represents a safer support than the initially applied support characteristic OX, although a combination of both may be necessary for safe support. Detailed knowledge of ground and support reactions is required to apply this method in practice.

develop an empirical approach along the lines of the characteristic support surfaces suggested by Daemen and Fairhurst (Fig. 3.4) or the ground support interaction diagrams of the New Austrian Tunnelling Method (Fig. 3.5). These are not used widely as a basis for support in coal mines, which is a pity since they describe very accurately the process of tunnel deformation and support. The basis of the New Austrian Tunnelling Method is described in the caption to Fig. 3.5.

In weak strata, strongly jointed strata, or strata which include weak layers, such as the Coal Measures, the stresses induced around an unsupported opening may exceed the strength of the weak layers. This will lead to failure of these layers on a similar yield path to BE in Fig. 3.5. Provided failure is of a strain-softening type, the volume of the adjacent ground will be increased, leading to closure of the opening, and redistribution of forces away from the immediate abutment, roof and floor, into the ground further away from the opening.

The process of failure and force redistributions will continue until the strata around the opening have reached a state of equilibrium. For this state of equilibrium to be reached the support system must be capable of accepting closure due to the expansion of failed strata, while maintaining sufficient resistance to mobilize any residual strength of the fractured rocks at the excavation periphery.

A hypothetical example of this mechanism can be illustrated from the peak and residual failure envelopes in Fig. 1.7, reproduced in Fig. 3.6 for a tunnel assumed to be in Coal Measures mudstone. Assuming a hydrostatic geostatic stress $\sigma_h = \sigma_v = \sigma_0$ the initial stress conditions at a depth of (say) 1600 m will be represented by point A assuming a unit weight of $25\,\mathrm{kN\,m^{-3}}$. If the rock behaves elastically as the excavation is constructed, the stress path at the excavation circumference will be given by $\sigma_1 + \sigma_3 = 2\sigma_0$ and will follow the excavation stress path AB. If and when this path intersects the peak strength curve at B, the rock will fail following a stress path BC to the excavation boundary where there will be zero (radial) confining pressure and a tangential stress equal to the residual strength of the rock. Inside the excavation sidewall, confining

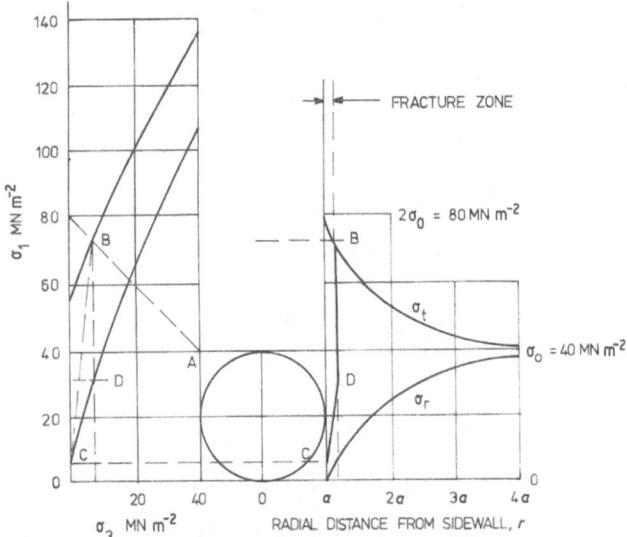

Fig. 3.6 Excavation stress paths at the surface of a circular section tunnel in the mudstone of Fig. 1.3 in an isotropic stress field. The stress distribution (radial, σ_r, tangential σ_t) in the sidewall is also illustrated together with a simple method for computing the extent of the fracture zone and the stress distribution in the fracture zone. The stress distributions in the sidewall can be computed from thick cylinder solutions for no internal pressure:

$$\sigma_r = \sigma_0\,(1 - a^2/r^2)$$
$$\sigma_t = \sigma_0\,(1 + a^2/r^2)$$

and for support pressure p_i

$$\sigma_r = \sigma_0 + (p_i - \sigma_0)a^2/r^2$$
$$\sigma_t = \sigma_0 - (p_i - \sigma_0)a^2/r^2$$

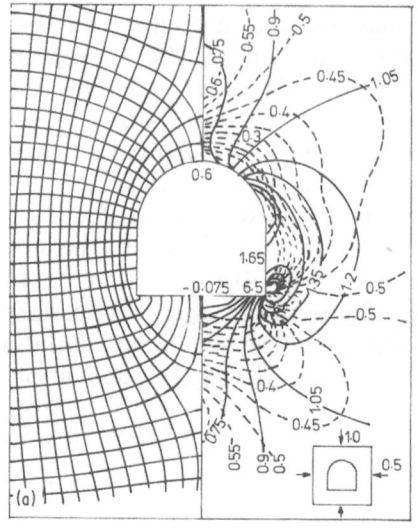

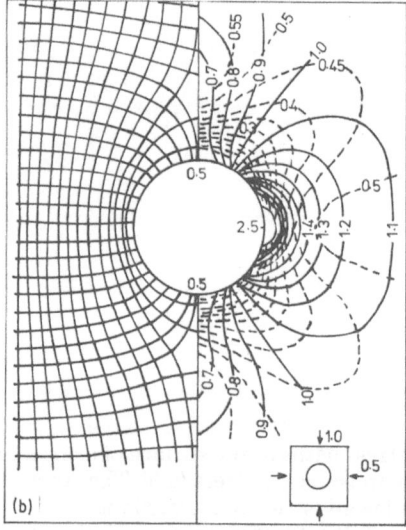

Fig. 3.7 Principal stress trajectories (LHS) and contours (RHS) of the ratios of major (solid line) and minor (dotted line) principal stress to vertical applied stress for (a) an arch shaped opening and (b) a circular opening in a non-uniform ($\sigma_v = 2\sigma_h$) stress field (from Hoek and Brown, 1980, using Bray and Hocking's two dimensional boundary element analysis). Note that this and similar data in Fig. 2.11 can be used to estimate the extent of a fracture zone around an opening by substituting for σ_1 and σ_3 in a failure envelope of the type $\sigma_{1f} = \sigma_{cf} + k_p\sigma_3$ (eq. 1.1).

stresses and tangential stresses will increase until they reach a magnitude D, where further failure will be suppressed. This stress will mark the boundary of a zone of fractured rock. The width of this *fracture zone*, which will be independent of the residual strength envelope, but dependent on the peak strength and the rock strength can be estimated by calculation or graphically as illustrated in Fig. 3.6. If the excavation is supported by some form of lining offering a resistance p_i, then the fractured zone will be reduced in proportion to the support resistance. This can be computed from the equations under Fig. 3.6. The same technique can also be used with the stress distribution data obtained by finite or boundary element analysis for an irregular opening in a non-homogeneous stress field (Fig. 3.7)). If the hypothetical distribution of stresses around the opening is known, then from peak and residual strength envelopes, the extent of the fracture zone can be estimated. This approach can also be extended through yield zone analysis in which the interaction of a fracture zone and a surrounding continuum of unfractured rock is considered. Various phenomenological models (see Table 3.1) can be used to describe the properties of the intact and fractured rock and to obtain a mathematical solution for the extent of the fracture zone. Since some of these are based on plastic analysis, the term yield zone is confusingly used.

The *fracture zone* analogy is quite valid in weak rocks at depth, where redistributed stresses are high compared with rock strength. It does, however, demand several rather over-simplified assumptions, namely that the strata are isotropic, homogeneous and originally elastic, that the opening is circular and deforms under plane strain conditions, that body forces approximate to zero and that the stress field is hydrostatic. A departure from most of these on their own would under certain conditions invalidate the approach. Thus body forces will tend to deform the fracture zone; a K_0 value less 0.14 will extend the fracture zone to infinity. There is also the difficulty of estimating rock mass properties, a daunting prospect to which some debatable simplifications such as Wilson's (1980) *f*-factor and Hoek and Brown's (1981) *rock mass strength criterion* have been applied. The analogy also demands a choice of failure criteria. The five most common assumptions are listed in Table 3.1.

The original form proposed by Fenner (1938) assumes the rock to be a Coulomb material with the ultimate strength of the rock mass controlled by the Coulomb criterion. If the rock is loaded beyond failure it maintains its ultimate strength values, so that the failure criteria of the failed and intact rocks are identical. Various adaptations of this approach have been proposed. Wilson (1977, 1980) has suggested that the failed rock should be assumed to act like a granular material, without cohesion, or as a Coulomb material, with a residual cohesion. Hobbs (1970) carried out tests

Table 3.1 Methods of fracture zone analysis

$\sigma_0 = \sigma_z = \sigma_h = \sigma_v = $ geostatic stress
$\gamma = $ Rock unit weight
$z = $ Depth from surface

$k_p = \dfrac{1 + \sin\phi}{1 - \sin\phi} = $ triaxial stress factor

$\sigma_c = $ Unconfined compressive strength
$\sigma_r = $ Residual strength of broken rock
$\sigma_{Tf} = $ Tensile strength

Analysis	Fenner (1938)	Wilson (1977)	Wilson (1980)
Failure criterion – fracture zone	$\sigma_{1f} = \sigma_{cf} + k_p\sigma_3$	$\sigma_{1f} = k_p\sigma_3$	$\sigma_{1f} = \sigma_r + k_p\sigma_3$
Failure criterion – solid rock	$\sigma_{1f} = \sigma_{cf} + k_p\sigma_3$	$\sigma_{1f} = \sigma_{cf} + k_p\sigma_3$	$\sigma_{1f} = \sigma_{cf} + k_p\sigma_3$
$\dfrac{R}{a}$	$\left\{ \dfrac{2[\sigma_0(k_p-1)+\sigma_{cf}]}{[p_i(k_p-1)+\sigma_{cf}](k_p+1)} \right\}^{1/(k_p-1)}$	$\left[\dfrac{2\sigma_0 - \sigma_{cf}}{p_i(k_p+1)} \right]^{1/(k_p-1)}$	$\left\{ \dfrac{2\sigma_0 - \sigma_{cf} + \dfrac{\sigma_r(k_p+1)}{(k_p-1)}}{\left[p_i + \dfrac{\sigma_r}{(k_p-1)}\right](k_p+1)} \right\}^{1/(k_p-1)}$

Analysis	Hobbs (1970)	Ladanyi (1974)
Failure criterion – fracture zone	$\sigma_{1f} = B\sigma_3^b + \sigma_3$	$\sigma_{1f} = \sigma_{cf} + k_p\sigma_3$
Failure criterion – solid rock	$\sigma_{1f} = \sigma_{cf} + k_p\sigma_3$	$\sigma_{1f} = \sigma_3 + \dfrac{\sigma_{cf}}{m+1}\left[2\left(1+\dfrac{n\sigma_3}{\sigma_{cf}}\right)^{\frac{1}{2}} + m - 1\right]$

$$\dfrac{R}{a}$$

Hobbs (1970):

$$\exp\left\{\dfrac{\left[\dfrac{2\sigma_0 - \sigma_c}{1+k_p}\right]^{(1-b)} - p_i^{(1-b)}}{B(1-b)}\right\}$$

where B, b define curvature of failure envelope for broken rock

Ladanyi (1974):

$$\left[\dfrac{\sigma_0 + H - m\sigma_{cf}}{p_i + H}\right]^{1/(k_p-1)}$$

where $n = \dfrac{\sigma_{cf}}{-\sigma_{Tf}}$

$m = (n+1)^{\frac{1}{2}}$

$H = \dfrac{\sigma_{cf}}{k_p - 1}$

$M = \dfrac{\left[1 + \dfrac{n\sigma_0}{\sigma_{cf}} - \dfrac{1}{4}(m-1)^2\right]^{\frac{1}{2}}}{m+1}$

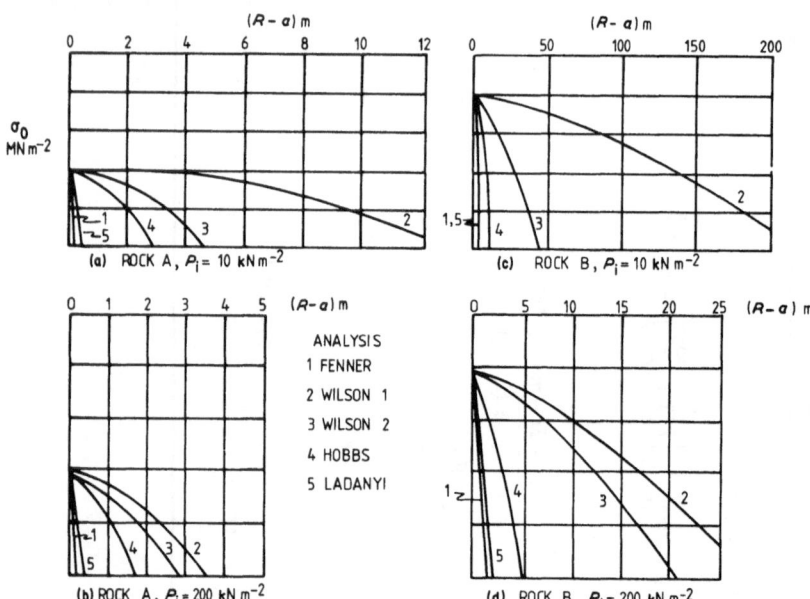

Fig. 3.8 Comparison of fracture zone widths from two support pressures (10 and 200 kN m^{-2}) and two hypothetical rock types. Rock A has $\sigma_{cf} = 30$ MN m^{-2}, $k_p = 4$, $\sigma_{Tf} = 10$ MN m^{-2}, $B = 4.82$, $b = 0.709$; Rock B has $\sigma_{cf} = 10$ MN m^{-2}, $k_p = 2.5$, $\sigma_{Tf} = 1$ MN m^{-2}, $B = 4.07$, $b = 0.74$.

on broken rock (mudstone from Bilsthorpe Colliery, Nottinghamshire, Britain) and proposed a curved failure criterion for the failed material. Ladanyi (1974) has proposed a curved failure envelope for the intact rock. Hoek and Brown (1981) have proposed curved failure envelopes based on the Griffith failure criterion.

In Fig. 3.8 the extent of the fracture zone at different depths in each case in Table 3.1 is compared for two typical support pressures and two hypothetical rock types. Rock A is Hobbs' Bilsthorpe silty mudstone with an assumed σ_{Tf} value; Rock B is the carnallite marl of Fig. 1.4 with estimates of σ_{Tf}, B and b. The plots clearly demonstrate the importance of the chosen failure criterion in estimating fracture zone width. It is interesting to compare the relatively narrow fracture zones predicted by Fenner and Ladanyi's method with the rather larger estimates from Wilson's approaches.

Wilson, when using his approach in design, reduces rock strength by a factor (f) between 3 and 5 to allow for reduced strength of the massive rock. This is a much greater reduction than Ladanyi's 60% allowance for long-term strength.

Hobbs' description of the shear resistance of broken rock by a curved envelope gives an intermediate fracture zone estimate. Nevertheless, the majority of rock in the Coal Measures cyclothem should show no tendency to yield or fracture at the stresses existing under mining conditions. The critical yield stress σ_{crit} is given by $2\sigma_{cf} - (1 + K_p)p_i$, or, in the Ladanyi case, by $(\sigma_{cf} - p_i)/M$ and this is likely to represent short-term stability. Long-term stability will lie somewhere between a residual and peak strength criterion depending on the structure and environmental conditions of the exposed rocks.

Closure can be calculated or estimated. Ladanyi (1974) proposes a method for calculating dilation of the fracture zone, due to fracturing of rock and opening of discontinuities, by means of average plastic dilation from the associated flow rule of the theory of plasticity applied to a limited portion of the post-failure strains. A more simple approach can be developed from the measured volumetric expansion of laboratory specimens. The average confining stress in the fracture zone can be estimated at about $0.25\sigma_0$. If an estimate of the bulking factor β of broken rock at this confining pressure is obtained from tests as in Figs 1.1–1.4 then the displacement can be estimated from:

$$u_t = a - [(1 - \beta)R^2 + \beta a^2]^{\frac{1}{2}} \tag{3.2}$$

using R, a values calculated from Table 3.1.

It must be stressed that formation of a classic yield zone requiring yield of the intact rock – requires a combination of high stresses and low rock strength. Except in the case of rocks such as seatearths and weak mudstones this is unlikely to occur in coal mine tunnels at depths less than 1000 m, and even then the extent of the zone will be very limited.

Depending on the mechanics of rock deformation, the fracture zone will take a finite time to reach a state of equilibrium (see the representations in Figs 3.3–3.5). Ladanyi (1974) and Muir Wood (1979) point out that very little is known about this process, and it undoubtedly represents a major area for future research. It will depend on the structure – particularly jointing and bedding – of the rock mass and of the broken rock in the failed zone where friction, interlocking and over-turning will affect the overall stress gradient in the fracture zone. Ladanyi concludes that long-term strength in intact rocks will be between 70 and 90% of short-term strength and as low as 60% of short-term strength in saturated rocks. In the case of both intact massive and broken rocks a curved failure envelope to allow for inter-block distortion at low confining pressures is recommended.

In the much more common case where deformation is caused by *loosening*, a combination of bedding plane separation and parting along joint surfaces affected by gravity, and the changed physical and hydrological environment resulting from tunelling, the controlling yield zone factor may be the residual strength envelope of the rock. This can be illustrated for the hypothetical case of a tunnel in silty sandstone in Fig. 1.7. There the initial stress conditions at a depth of 400 m may be represented by point G. Then the excavation stress path will be given by GH and fracture of the intact rock (peak envelope) will not occur. However, loosening of the massive rock along natural discontinuities imposed by blasting, effectively reducing the long-term strength of the rock to residual, may eventually lead to formation of a yield zone bounded by stress, J.

At any time before equilibrium is reached, it is possible to prevent further ground deformation by increasing the internal support pressure to mobilize high residual strength at a higher confining pressure. However, the smaller the volume of ground which is allowed to fail, the greater will be the support resistance, p_i, required to prevent further failure. This is essentially the argument presented for the New Austrian Tunnelling Method in Fig. 3.5.

3.3   Tunnel supports

The basic requirement of a tunnel support system is that it should control the movements of strata surrounding the tunnel in such a way that the function of the tunnel is not impaired. This can have various interpretations depending on the use of the tunnel, its life and the stresses to which it will be subjected. In the case of the main, semi-permanent transport and ventilation tunnels in a coal mine system, a high degree of roadway stability and resistance to deformation may be required. The ultimate requirement is in the spine roadways at the Selby project, Yorkshire, Britain (Forrest, 1978), where a maximum diametral deformation of 1 mm has been specified to accommodate cable belts.

In the case of secondary, and particularly access, roadways in weak rock, subject to high stresses and deformations, it is more realistic to design the support system to accommodate rather than to resist ground movement. In this case supports should have sufficient ductility to allow large deformations prior to failure while retaining their main function of supporting the roof.

The type of support system is therefore ultimately determined by the degree of deformation tolerable. Colliery arches (Table 3.2) will allow the largest deformation, while segmental linings may control and limit deformation in exceptional cases. In normal cases they would not be considered where deformation is likely to be significant. The recent

Table 3.2 Allowable deformation (expressed as percentage of vertical height) of typical support systems

Segmental lining	$< 1\%$
Rockbolts	$< 2\%$
Rockbolts/shotcrete	3%
Shotcrete	4%
Square/rectangular set	12%
Rigid arch	25%
Yielding arch	40%

installation of concrete segmental linings in British mines (Bloor, 1982) illustrates this point.

Typical examples of coal mine support performance are illustrated in Fig. 3.9 where four case histories of deformation above the crown of arch supported tunnels in Coal Measures strata are summarized. Anchors attached to extensometer wires in the crown show that deformation is not deep seated and that it is rapidly stabilized as the residual strength of the strata is mobilized by the supporting arch.

The contrasting case of weak unsupported floors referred to earlier requires a different explanation. In this case failure will be of a strain-hardening or ductile/plastic type and 'flow' of the deforming rock will exert pressures from the rock mass on the rocks at the tunnel floor surface. This type of deformation is dominant in seatearths and where these occur in the floor of arched roadways quite large deformations can occur. This type of deformation is illustrated in Fig. 3.10. Data comparable to that in Fig. 3.9 are illustrated for floor heave in Fig. 3.11. This is an access roadway profile, but the deformation characteristics can be used to illustrate the general case. These demonstrate as in Fig. 3.9 rapid deformation of roof strata – quickly reaching a stable state but accompanied by continuing deformation of floor strata albeit at reducing rate, resulting from plastic flow of the weaker seatearths.

In this case it is not sufficient to mobilize some residual strength. In order to reduce deformation, a very high support resistance, combined with reduction of stresses in the tunnel section is required. Although it is probably preferable to construct a large enough tunnel section to accommodate deformation, there are several ways in which deformation can be minimized:

(a) By construction of a circular or invert arch tunnel to reduce peripheral stresses as illustrated in Fig. 3.1 with the potential to achieve the results illustrated in Fig. 3.2

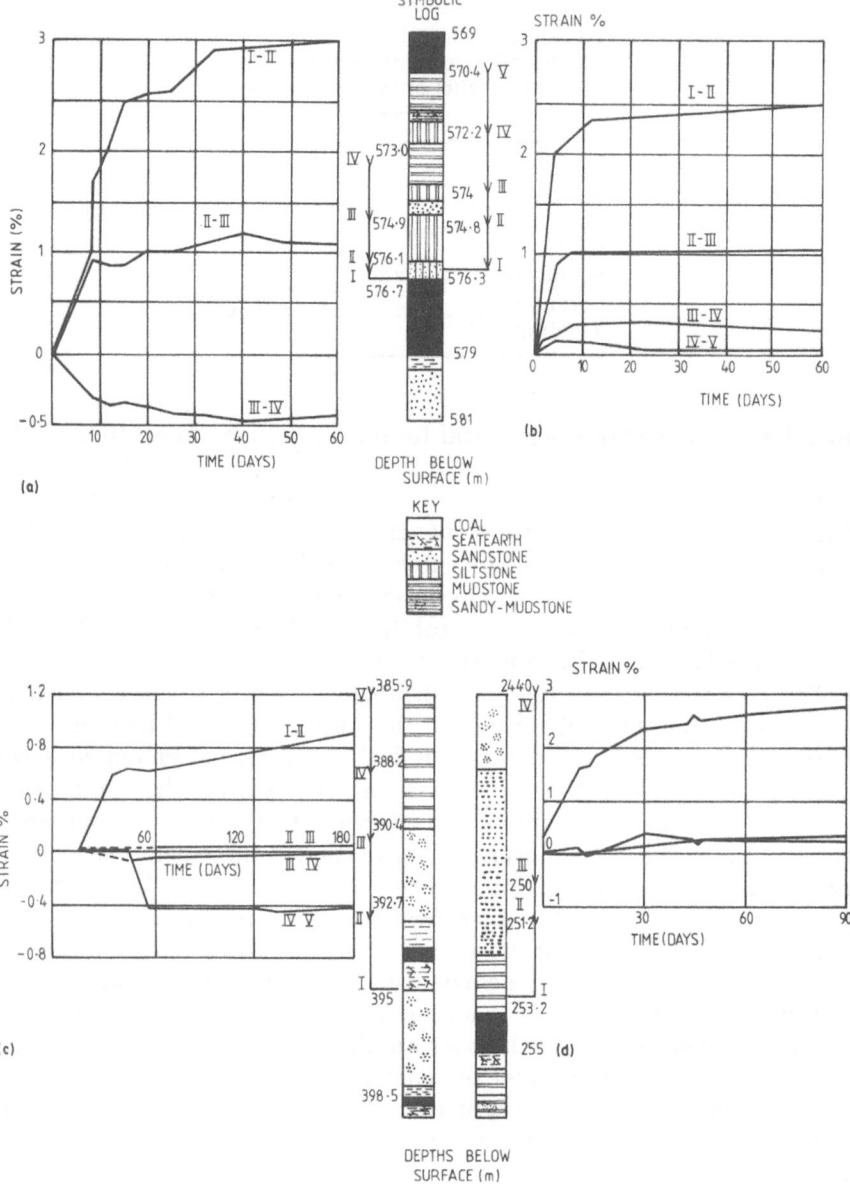

Fig. 3.9 Strains (extension positive) computed from anchor extensometer measurements in the roof of tunnels in Coal Measures strata: (a, b) Maudlin seam, Wearmouth Colliery, Tyne and Wear, Britain (Singh, 1968); (c) above Harvey seam, Woodhorn Colliery, Northumberland, Britain; (d) Low Main seam, Bedlington Colliery, Northumberland, Britain (Kirmani, 1972).

Fig. 3.10   Floor heave in the Deep Hard seam at Cotgrave Colliery, Nottinghamshire, Britain, at a depth of 570 m (photograph by L.J. Thomas).

(b) By reinforcing floor beams using piles or rockbolts to increase resistance to heave or
(c) By attempting to equalize stresses around the tunnel by reducing
point contact

The *comparative* performance of support systems in a shallow tunnel in
weak rock has been illustrated by Ward *et al.* (1976). This case history also
examines the effect of the tunnel profile on lining stability. The Kielder
Aqueduct experimental tunnel was driven in a fissile shale, very similar to
Coal Measures shales, in order to provide performance data for potential
tenderers for the 32 km of tunnel included in the Kielder Water Scheme.
The main experimental tunnel (Fig. 3.12) was horizontal, 100 m long and
3.3 m in diameter at an average depth of 90 m. It was driven in a lower
carboniferous shale, know locally as the Four Fathom mudstone. Part of
the tunnel was formed by drilling and blasting, and part by machine

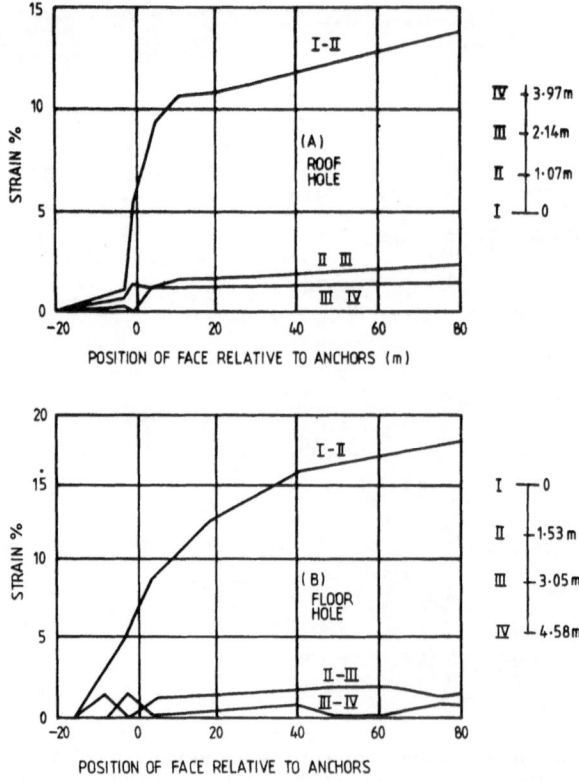

Fig. 3.11 Roof and floor strains related to face advance in an arch roadway adjacent to an advancing longwall face in the Low Main seam at Easington Colliery, Durham, Britain (Dudley, 1977). See Fig. 8.14 for the strata sequence.

cutting with a Dosco Roadheader. Support systems used were:

(a) Circular H-section steel ribs
(b) Rockbolts
(c) Sprayed concrete – circular arch
(d) Sprayed concrete and rockbolts
(e) Circular steel liner plates
(f) Unsupported

Details of support systems and their location are listed in Table 3.3. Properties of the rock were:

Unit weight (kN m^{-3})   24.3–26.6

Unconfined compressive
strength σ_{cf} (MN m^{-2})   34–54

Deformation modulus (MN m^{-2})
(tangent modulus at $0.5\sigma_{cf}$)   4800–14 200

RQD (rock quality designation)   0–8 %

Discontinuities (Number/m)
    Blasting   24–32
    Machine cut   17–26

The final machine cut length of the tunnel, which was unsupported, was intercepted centrally by a borehole from the surface (Fig. 3.13) containing magnet extensometers 0.3 (A), 1 (B), 2 (C) and 3 (D) m from the roof of the tunnel and 1 (F) and 2 (G) m from the floor. Deformation started immediately behind the tunnel face and continued rapidly as the sides and shoulders of the tunnel collapsed. After 2.5 years the unsupported section has collapsed to a height of 2.5 m above the unsupported section.

Deformation of the supported sections was monitored using anchor extensometers installed in the tunnel roof after excavation. Figure 3.14 summarizes displacement at a distance 0.3 m vertically above the tunnel

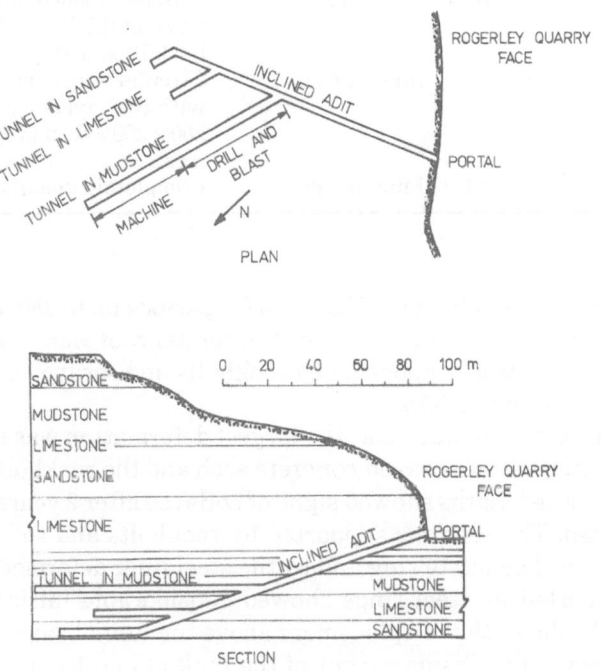

Fig. 3.12 Section and plan of the Kielder Aqueduct experimental tunnel at Rogerley Quarry, Co. Durham (after Ward *et al.*, 1976).

Table 3.3 Support systems – Kielder experimental tunnel (Ward *et al.*, 1976)

Section	Distance (m)	Construction method	Support system
1	0–19	Drill and blast	$127 \times 114 \times 20$ lb ft^{-1} $(0.265\,\text{kN m}^{-1})$ circular *H*-section steel ribs at 1 m centres backed with steel bars, galvanized sheeting, rock and wood packing
2	19–30	Drill and blast	Rows of 7 resin bonded untensioned 1.8 m long roof bolts in the roof at 0.9 m centres and linked with $50 \times 50 \times 3.2$ mm diameter steel mesh
3	30–44	Drill and blast	One layer sprayed concrete in the roof plus rockbolting as in (2) linked with $200 \times 200 \times 6.4$ mm diameter steel mesh followed by second layer sprayed concrete
4	44–61	Drill and blast	As (3) but without rockbolts
5	44–76	Machine cut	As (3) but rows of 5 rockbolts
6	76–85	Machine cut	Circular welded mild steel liner 12.7 mm thick, 11×0.7 m rings backfilled with p.f.a.–cement grout
7	85–95	Machine cut	Circular ring of sprayed concrete with two layers with intermediate $200 \times 200 \times 6.4$ mm diameter steel mesh
8	95–104	Machine cut	Completely unsupported

roof in each section listed in Table 3.3 for periods up to 280 days. Displacements at distances up to 3 m above the tunnel roof were measured in the machine cut sections supported by rockbolts and sprayed concrete. These are illustrated in Fig. 3.15.

It is interesting to note that the largest deformation was allowed by the partial systems – the sprayed concrete arch and the rockbolted roof and in fact both these lengths showed signs of collapse after 2 years and had to be resupported. The sections supported by rockbolts and sprayed concrete, by the sprayed concrete ring and by the steel liner deformed only slightly. That supported by steel rings showed considerable initial deformation. Figure 3.15 shows that displacement above the tunnel varies with the type of support system. Displacement of the rock at depths greater than 0.3 m above the steel ribs is much smaller than at the tunnel surface. However, in the case of the rockbolted section, deformation 3 m above the roof is approximately half the deformation 0.3 m above the tunnel. This illus-

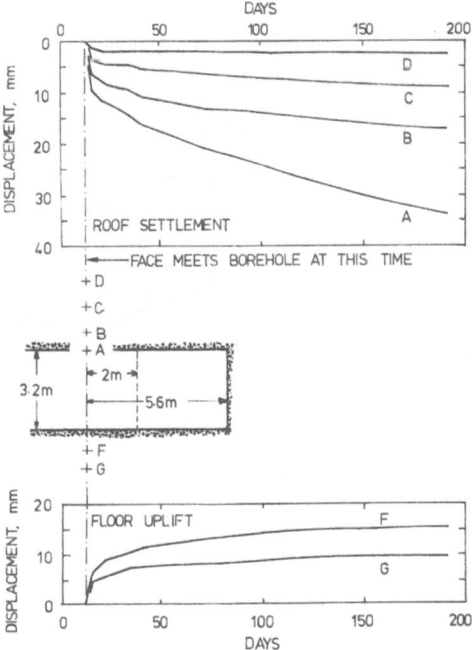

Fig. 3.13 Vertical roof and floor movement in the unsupported part of the Kielder Aqueduct experimental tunnel up to 190 days after excavation (after Ward *et al.*, 1976).

trates the positive roof support given by the steel ribs and the relative instability of the rockbolts. The section supported by sprayed concrete and rockbolts is supported rapidly and displacement above the tunnel is limited.

Figure 3.16 shows how the thrust on the steel rings builds up at the same rate as the displacement or dilation reduces. The variation between forces on the upper sides of the ring can be accounted for by eccentric loading – depending on the position of lagging and pads. Changes in loading result from small rock falls around the ring. Average diametral deformation was 5 mm. The mean thrust at 300 days of 28 kN is equivalent to 0.6 m of overburden pressure. This compares with 1200 kN m^{-1} equivalent to 30 m of overburden in the case of the sprayed concrete ring which gave virtually immediate support to the roof. In the case of steel liners support loads were equivalent to 28 m of overburden for a liner erected at the face, reducing to 12 m for a liner erected 3 m behind the face.

A major conclusion is that the loading of a support system depends on the amount of dilation of rock surrounding the tunnel which is allowed to

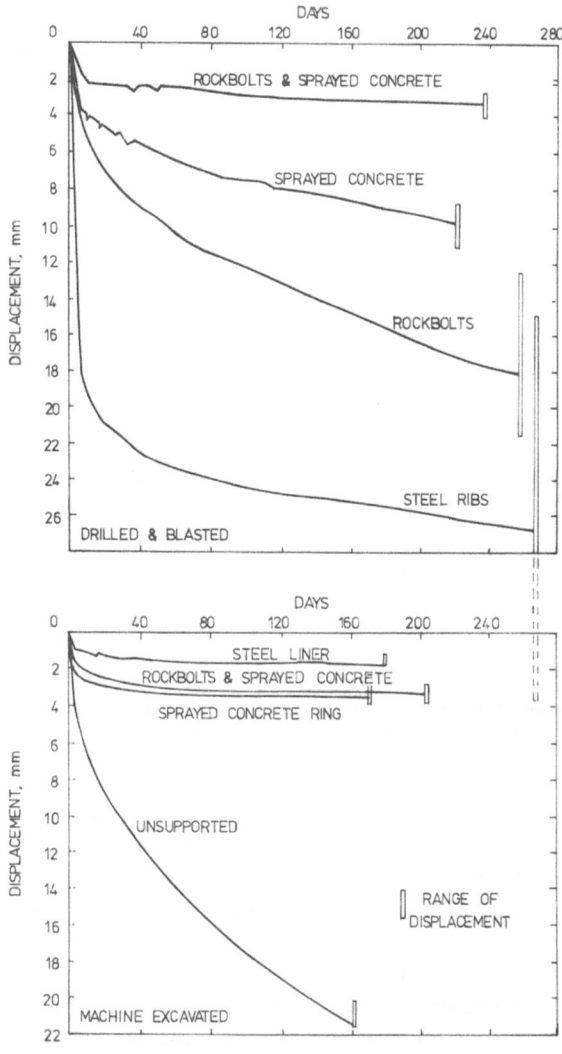

Fig. 3.14 Typical vertical displacements in different sections of the Kielder Aqueduct experimental tunnel (see Table 3.4) at 0.3 m above the crown of the tunnel (after Ward *et al.*, 1976).

take place before the supports are fully loaded, and on the subsequent stiffness of the support system. The greatest support loads occur when there is least dilation and when rapid support is given to the rock immediately after excavation. This supports again the basic philosophy of the New Austrian Tunnelling Method in Fig. 3.5.

The Kielder Aqueduct experiments while not directly relevant to coal mining are useful in that they illustrate certain fundamental aspects of support of tunnels in rock. While the results are applicable to mining conditions, the depth of the tunnel is shallower than those commonly encountered in many coal mines. Where depths are greater and rocks weaker, larger deformations may render many of the support systems invalid. In that case colliery arches may provide a more flexible type of

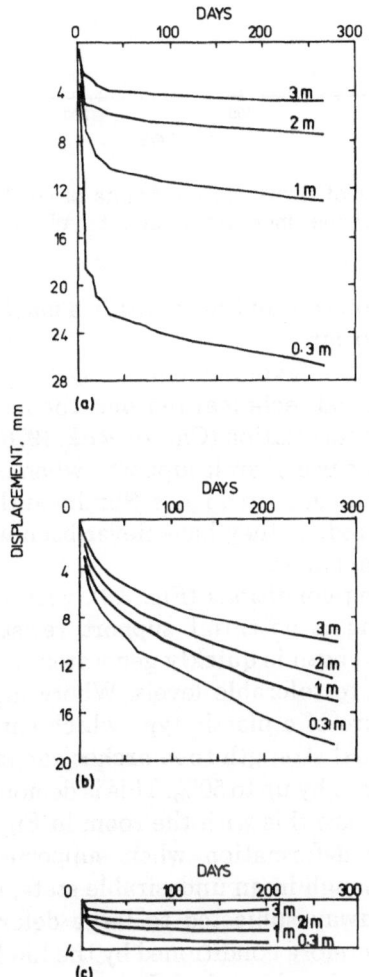

Fig. 3.15 Typical displacement-time curves for different roof support types in the rock 0.3 m, 1 m, 2 m, 3 m above the crown of the Kielder Aqueduct experimental tunnel (after Ward *et al.*, 1976).

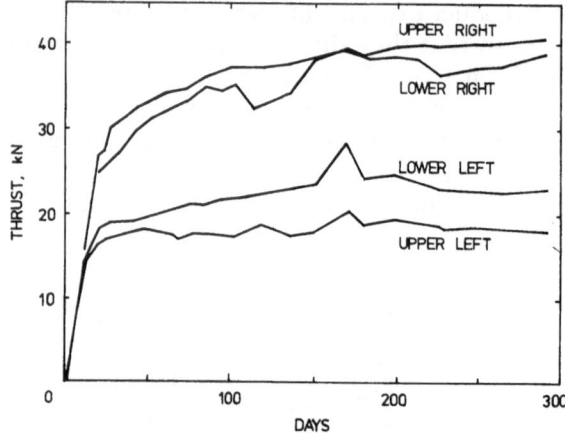

Fig. 3.16   Development of thrust on quadrants of steel in rings at the Kielder
Aqueduct experimental tunnel (after Ward *et al.*, 1976).

support, although shotcrete and rockbolts in a machine-cut tunnel appear
a valid long-term alternative.

Colliery arches are the favoured support system in British coal mines.
Although there are good technical reasons, the main reasons are safety
and tradition. A close correlation (Carver *et al.*, 1976) can be demonstrated
between safety and the use of arch supports, when compared with timber,
straight girders and unsupported roofs. Significantly roof bolts and shotcrete are not considered, as they have never been widely used for tunnel
support in British coal mines.

In most coal mining conditions (Fig. 3.9), with the use of fairly light
section arches giving a potential support resistance of $0.1\,\mathrm{MN\,m^{-2}}$,
sufficient residual strength is quickly generated in Coal Measures rocks
to limit deformation to tolerable levels. Where high stresses cause deformation and fracture of a plastic type which cannot be controlled by
mobilization of residual strength then arches can still maintain a stable
roadway when deformed by up to 50%. This is demonstrated in Fig. 3.17. It
is interesting to compare this with the room in Fig. 2.19 which has been
subjected to similar deformation when supported by rockbolts. The
arched roadway, although in an undesirable state, is demonstrably safer
than the bolted roadway subjected to large deformation. Selection of
support systems is therefore conditioned by the likely deformation which
will be encountered, and rockbolt based systems are usually only suitable
in relatively strong sandstone roof beds, of the type found in the United
States, where deformation is relatively small.

The performance of arches in roadways in coal mines can be improved either by improving the cross-sectional shape of the arch, or by increasing its flexibility and yield characteristics. The major advantage of the traditional H-section arch is its cheapness. The traditional arch does, however, have disadvantages and there has been some research aimed at alternatives. Billington and Jacomb-Hood (1976) point out that the performance of an arch support depends largely on the strength of the arch section in compression, and that this is determined by the ratio between the major and minor axis moments of inertia of the section and by the torsional stiffness. In the case of H-section arches the moment of inertia ratio ranges from 3 for square sections to well over 8 for the more elongated sections.

Any open section arches, including H-section, may be susceptible to torsional failure, and particularly post-yield plastic failure. The alternative is to use a section with similar moments of inertia about both axes. To date the most successful types of arch section have been the TH and

Fig. 3.17   An arch supported roadway subjected to major deformation at a depth of 900 m in the Parkgate seam at Hickleton Colliery, Yorkshire (photograph by L.J. Thomas). Note that despite the deformation the roadway can still be safely used.

Usspurweiss. The TH arch has relatively low torsional stiffness, but the Usspurweiss section has both high torsional stiffness and similar moments of inertia about both axes.

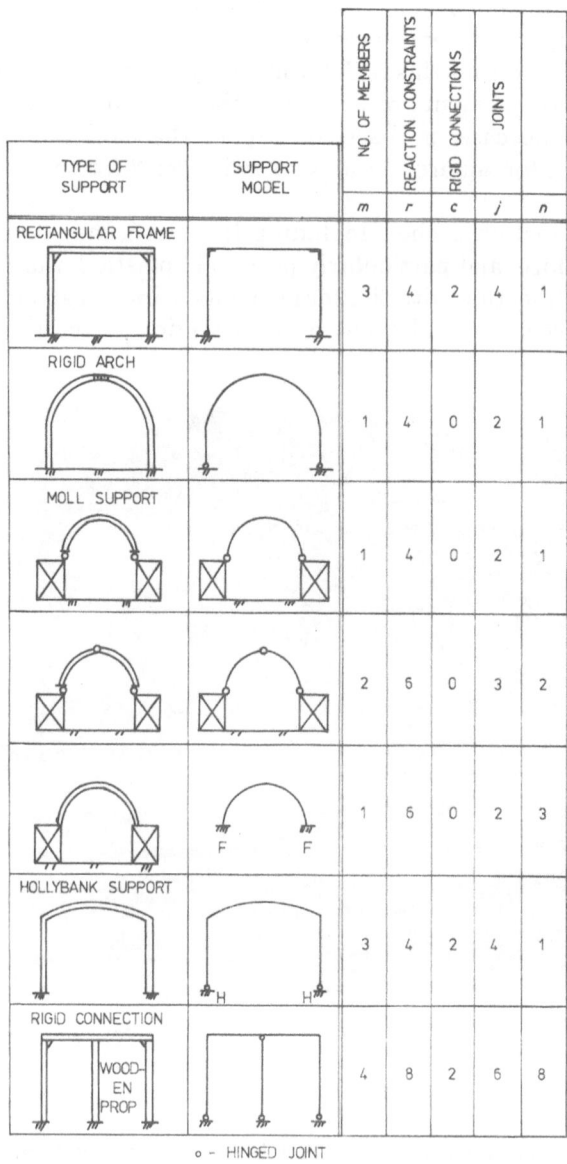

TYPE OF SUPPORT	SUPPORT MODEL	NO. OF MEMBERS	REACTION CONSTRAINTS	RIGID CONNECTIONS	JOINTS	
		m	r	c	j	n
RECTANGULAR FRAME		3	4	2	4	1
RIGID ARCH		1	4	0	2	1
MOLL SUPPORT		1	4	0	2	1
		2	6	0	3	2
		1	6	0	2	3
HOLLYBANK SUPPORT		3	4	2	4	1
RIGID CONNECTION		4	8	2	6	8

o - HINGED JOINT

Fig. 3.18 Statical indeterminacy of some typical coal mine roadway supports (after Arioglu, 1976).

Billington and Jacomb-Hood (1976) suggest that an improved alternative, costing the same as an H-section arch, but with the advantage of increased minor axis moment and high torsional rigidity can be obtained from a box-shaped rolled hollow section (RHS). Since the basic cost per unit weight of RHS arches is 50 % greater than RSJ arches, an equivalent saving in weight is required for RHS arches of comparable cost to RSJs. This can be relatively easily obtained.

The basic requirement for stability of any structure is that the external and internal forces acting on it are in static equilibrium. If the structure is arranged so that the internal and external forces can be determined for any loading conditions simply from the conditions of equilibrium then the structure is said to be *statically determinate*. If the structure is arranged so that it contains a greater number of forces or moments than are required to satisfy the equations of equilibrium then the structure is said to be *statically indeterminate*. The majority of supports in coal mining practice are statically indeterminate structures. The degree (n) of statical indeterminacy is usually given by the formula:

$$n = m + r + c - 2j$$

where m is the number of members, r the number of reaction constraints at supports, j the number of joints and c the number of rigid connections capable of transmitting moments and horizontal and vertical forces.

Figure 3.18 gives values of n for some typical and less-typical roadway supports. If $n < 0$, the structure is unstable, if $n > 0$, the structure will usually be stable depending on the force distribution. There are advantages in indeterminacy. Stresses in indeterminate structures are generally smaller than in determinate structures and they also possess the capacity to redistribute internal forces should the structure become overstressed, or subject to point or axisymmetric loading. This reduces buckling. Arioglu (1976) has examined the reaction of typical roadway support systems for combinations of vertical and horizontal forces, using elastic and plastic analysis. His main conclusions are that under the loading conditions existing in most mines, bending moments can be reduced and buckling best resisted using articulated arches mounted on chocks within parabolic axes, or on yielding legs. However, a better method of reducing buckling which also increases the bearing capacity of the supports is to reduce point loads on arches from the surrounding rock. The National Coal Board has developed a PVC bag which can be placed behind an arch and grouted with cement grout at $100\,\mathrm{kN\,m^{-2}}$ pressure to ensure even contact. A quilted bag to form a lagging and completely seal the roadway between the arches is also being developed. This aims, as well as reducing point loads, to preload the arch at an even setting pressure, making the overall arch support more effective.

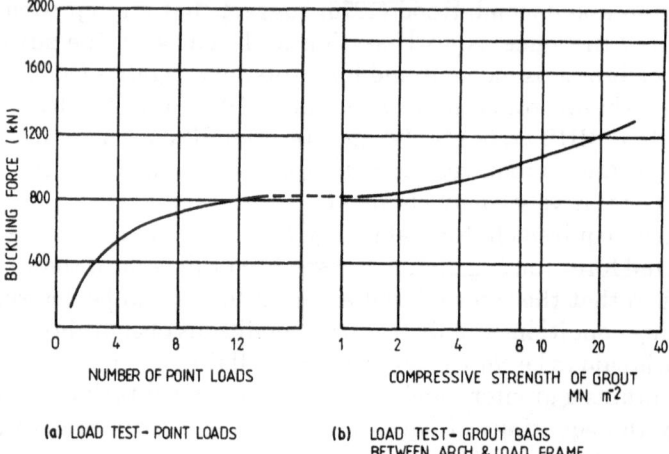

Fig. 3.19   Effect of (a) point loading and (b) evenly distributed load from grout bags containing cement grouts of various strengths, on the buckling resistance of 4.8 m × 3.6 m RSJ 150 mm × 125 mm section arches (work carried out at the National Coal Board Mining Research and Development Establishment with financial aid from ECSC Contract No. 6220-AB/8/801).

Figure 3.19 shows the effect of these bags and subsequent equalization of stress on the buckling resistance of a 150 mm by 125 mm section RSJ arch. The resistance of the arch increases as the number of contact points increases. When a limiting number of contact points is replaced by grout bags, the resistance continues to increase, in this case in proportion to the set grout strength.

3.4   Pressures on tunnel supports

Apart from assumptions of tunnel support to estimate the extent of fracture zones (Table 3.1 and Fig. 3.8) and the estimation of buckling forces on arches, no attempt has been made to estimate support loading in coal mine tunnels. This is sensible since most supports used are not designed to mobilize the full pressure which may be applied from the deforming tunnel carcass (see Figs 3.3 to 3.5) but to accommodate the deformation. This is the only sensible approach to support design, unless massive reinforced concrete segmental linings as in the Campine coalfield in Belgium (Stassen and van Duyse, 1977) or cast-iron segments as in the Selby coalfield, Britain (Forrest, 1978 and Section 3.5) are to be used at almost unlimited expense.

Computation of support pressures is possible using the fracture zone approaches in Table 3.1, but the simplifying assumptions are likely to lead to large scale inaccuracies – not least because of the impossibility of predicting or estimating relaxations with time and face/support position. It is simpler – if an estimate of pressure is required – to consider the empirical methods outlined in Chapter 2.

The simplest of the empirical methods is Terzaghi's (1946) classification system. This can be used (see Table 2.4) to estimate roof loading on arches and square set supports, and to estimate stand-up time. The different categories of rock in Terzaghi's classification represent different frictional resistance, with category 6 approaching the theoretical value. It is important that loosening is allowed for in the final load.

A wholly empirical extension of Terzaghi's method which is used extensively in United States coal mines, and was the precursor of many other systems (see Barton *et al.*, 1974 and Table 2.6) is Whickham, Tiedman and Skinner's (1972) rock structure rating (RSR). It was developed partly for coal mines under contract to the United States Bureau of Mines.

The principle of these systems is that a number of parameters are selected which are considered important in determining rock mass behaviour and which can easily be estimated by a geologist. These parameters are allocated a range of numbers and for a particular rock a numerical assessment is made within the designated range. These numbers are then summed or multiplied to give an overall index which can be related to selected behavioural characteristics of the rock.

The RSR system is based on a study of support performance in 33 tunnels in various strata. Three parameters, *A*, *B* and *C* defining *general geology*, joint pattern related to drive directions and *groundwater and joint condition* are allocated values along the lines proposed in Table 3.4.

The rating is then obtained from:

$$\text{RSR} = (A + B) + C$$

It is interesting to note that, as with Terzaghi's method, rock strength is not considered important. The rating numbers are related directly to rock loads for tunnels of various diameter and these are used as a basis for proposing support systems, as in Fig. 3.20.

3.5   Stresses on cast-iron segmental lining

Most tunnels constructed in coal mines are by their nature temporary or short-life structures, and a relatively high degree of deformation can be tolerated without seriously affecting efficiency. This is not the case in

Table 3.4 Rock structure rating system parameters (after Whickham et al., 1972)

(a) Rock structure rating parameter A. General area geology. Maximum value 30

	Basic rock type				Geological structure			
	Hard	Medium	Soft	Decomposed				
Igneous	1	2	3	4				
Metamorphic	1	2	3	4				
Sedimentary	2	2	3	4				
					Massive	Slightly faulted or folded	Moderately faulted or folded	Intensely faulted or folded
Type 1					30	22	15	9
Type 2					37	20	13	8
Type 3					24	18	12	7
Type 4					19	15	10	6

(b) Rock structure rating parameter B. Joint pattern. Direction of drive. Maximum value 45

	Strike normal to axis					Strike parallel to axis		
Direction of drive:	With dip			Against dip		Both		
Dip of prominent joints:*	Both Flat	Dipping	Vertical	Dipping	Vertical	Flat	Dipping	Vertical
1. Very closely jointed	9	11	13	10	12	9	9	7
2. Closely jointed	13	16	19	15	17	14	14	11
3. Moderately jointed	23	24	28	19	22	23	23	19
4. Moderate to blocky	30	32	36	25	28	30	28	24
5. Blocky to massive	36	38	40	33	35	36	34	28
6. Massive	40	43	45	37	40	40	38	34

*Notes: flat = 0–20°; dipping = 20–50°; vertical = 50–90°.

(c) Rock structure rating parameter C. Groundwater. Joint condition. Maximum value 25

	Sum of parameters $A + B$					
	13–44			45–75		
	Joint condition**			Joint condition**		
Anticipated water inflow ($l\,min^{-1}$ per 100 m of tunnel)	Good	Fair	Poor	Good	Fair	Poor
None	22	18	12	25	22	18
Slight ($< 250\,l\,min^{-1}$)	19	15	9	23	19	14
Moderate (250–$1500\,l\,min^{-1}$)	15	11	7	21	16	12
Heavy ($> 1500\,l\,min^{-1}$)	10	8	6	18	14	10

(d) Rock pressure on tunnel arches of various diameters corresponding to RSR $[= (A + B) + C]$ values

| *Rock pressure*
(kNm^{-2}) | Tunnel diameter (m) | | | | |
	3	4.5	6	7.5	9
50	49.9	57.8	62.5	65.5	67.7
100	32.7	43.0	49.9	54.5	57.8
150	21.6	32.6	40.2	45.7	49.8
200	13.8	24.7	32.7	38.5	43.1
250	.	18.6	26.6	32.6	37.4
300	.	.	21.6	27.7	32.6
350	.	.	17.4	23.4	28.4
400	.	.	.	19.8	24.7
450	.	.	.	.	21.5
500	.	.	.	.	18.6

** Notes: good = tight or cemented; fair = slightly weathered or altered; poor = severely weathered, altered or open.

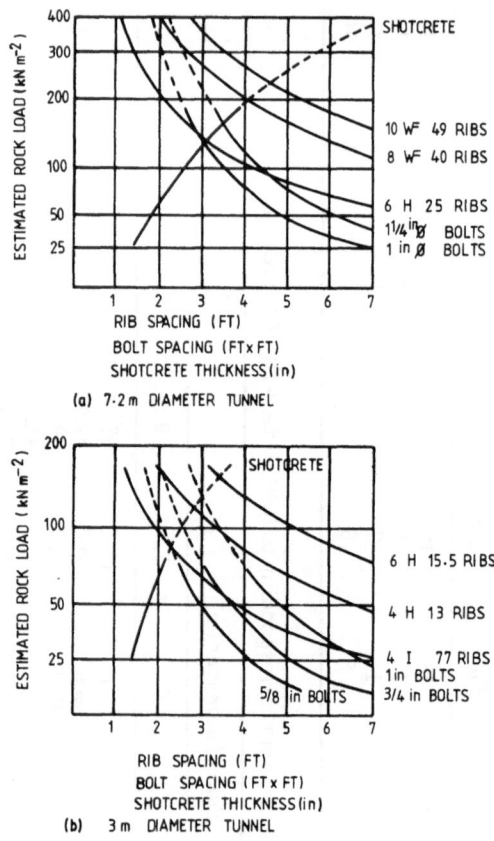

Fig. 3.20 Support requirement charts produced by Whickham *et al.* (1972). Note that imperial units are retained because of the complexity of relations (1 ft = 0.3 m, 1 in = 25.4 mm).

major access roadways where highspeed bulk, material or manpower transport systems are used. These create an increasing demand for deformation-free roadways. Major surface drifts are a particular example of such roadways, and in the case of the Gascoigne Wood Colliery drifts sunk as main access drifts for the Selby Coalfield, the need to eliminate movement led to the choice of spheroidal graphite bolted cast-iron segmental lining. The use of this very expensive type of lining – designed to resist the full hydrostatic and geostatic stresses at the depth of construction – requires much more detailed information on total and hydrostatic stresses imposed on the lining by the surrounding rock.

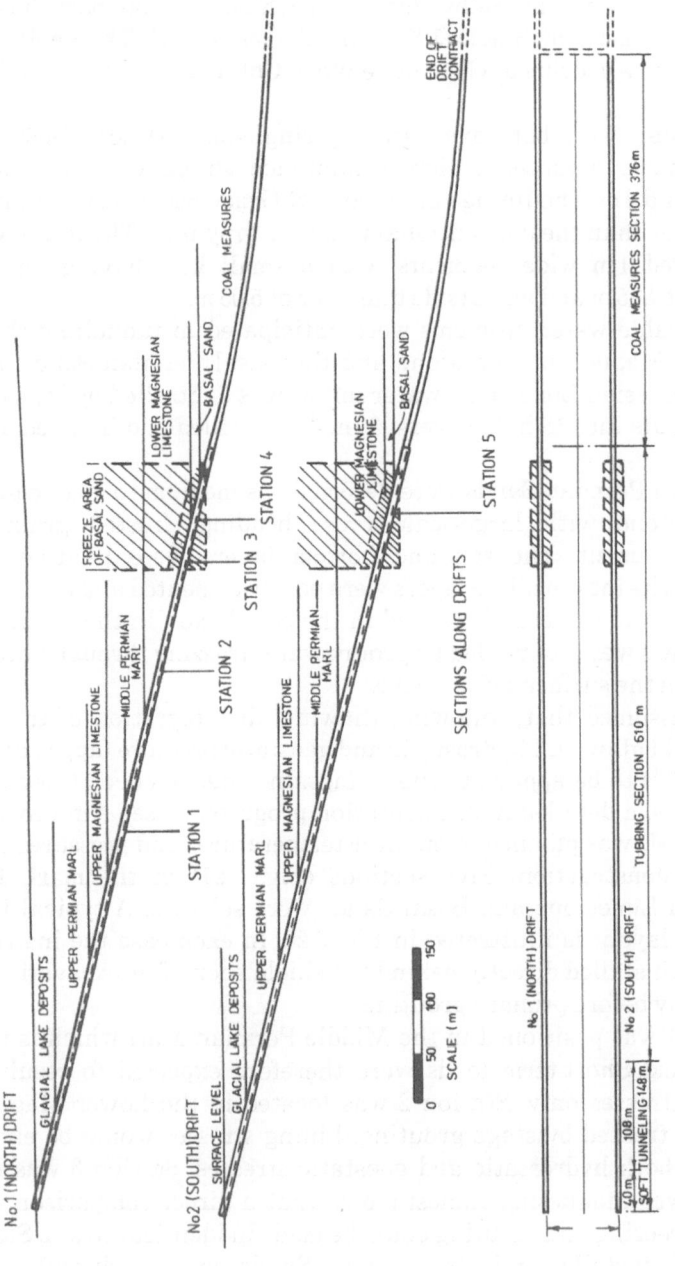

Fig. 3.21 Section through and plan of the twin drifts at Gascoigne Wood Colliery, Yorkshire, showing the positions selected for instrumentation and the main stratigraphical boundaries.

Detailed description of the Gascoigne Wood drift design and construction, together with the Selby Project of which it forms part, has been given by Forrest and Black (1979) and Forrest (1978). The geology and layout of the twin drifts at Gascoigne Wood Colliery are illustrated in Fig. 3.21.

The tunnels were both excavated by single-head Dosco SB600 boom-type machines mounted inside a Laurence shield with an external diameter of 5.2 m. The lining comprised SG Grade 600/3 cast iron, lighter and stronger than the conventional Grade 12 grey iron. The lining was in seven bolted 1 m wide segments with a small key, having an inside diameter of 4.75 m and an outside diameter of 5.05 m.

Considerable water problems were anticipated in tunnelling through the Lower Magnesian Limestone and the Basal Permian Sands. In the Lower Magnesian Limestone, water inflow was controlled by injection of cement grouts into 18 hole covers, 30 m long, drilled and injected at 12 m intervals.

The Basal Permian Sands were described as medium to coarse grained soft sandstone with large-scale cross bedding, frosted grains and carbonate cement. The top and bottom layers were relatively well cemented. The intermediate layers were poorly cemented and the material disintegrated to structureless sand in the core-boxes. In the Basal Sands water inflows were controlled by groundwater freezing through boreholes drilled from the surface 170 m above.

It was assumed that, following thawing, disintegration of the Basal Sands would allow full hydrostatic and geostatic pressure (approximately $4.25 \, \text{MN m}^{-1}$) to be applied to the linings. In order to check these design assumptions, a detailed instrumentation programme (see Altounyan and Farmer, 1981) was planned to monitor temperatures and pressures during and after construction. Five sections (Fig. 3.21) in the marl, Lower Magnesian Limestone and Basal Sands were selected. A typical instrumentation layout is illustrated in Fig. 3.22. In each case the instrumentation was installed directly behind the shield soon after excavation, and immediately before primary grouting.

Station 1 was positioned in the Middle Permian Marl which is not an aquifer rock. Short-term loads were therefore expected to result from geostatic stresses only. Station 2 was located in the Lower Magnesian Limestone treated by stage grouting. Lining stresses would be expected here from both hydrostatic and geostatic stresses. Station 3 was in the frozen Lower Magnesian Limestone so that a direct comparison of the effects of freezing and grouting could be made in identical strata. Stations 4 and 5 were installed in the frozen Basal Sands, one in each drift.

As access to each station was possible throughout the observation period, the instrumentation was not constrained by uniformity of trans-

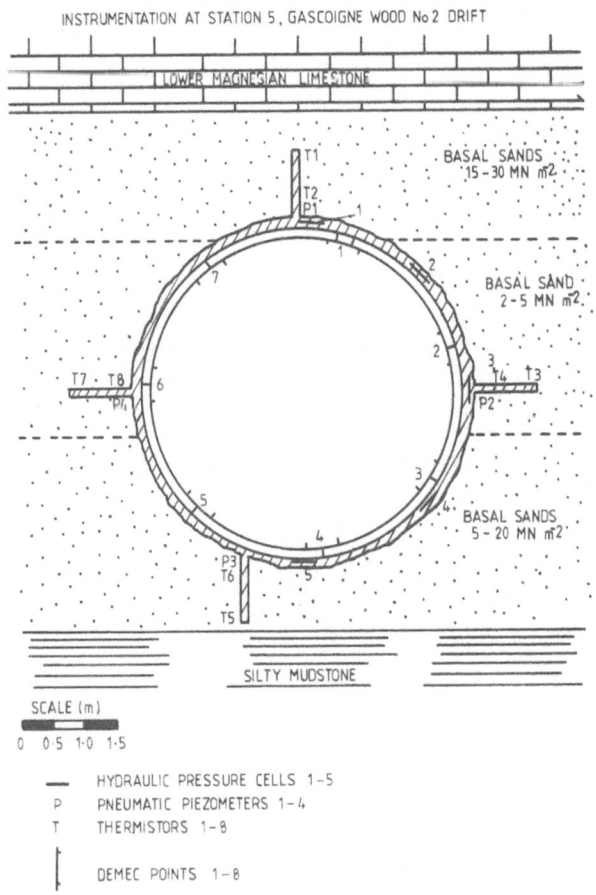

INSTRUMENTATION AT STATION 5, GASCOIGNE WOOD No 2 DRIFT

Fig. 3.22 Section through an instrumented section of Gascoigne Wood No. 2 Drift, Station 5 (after Altounyan and Farmer, 1981).

ducer output, and the choice of instrument for each type of measurement was governed by reliability, simplicity and cost. The important parameters to be monitored were *total stress* on the lining, *deformation, water pressure* and *position of the icewall* in frozen ground. Details of the instrumentation are given in Altounyan and Farmer (1981).

Figure 3.23 shows total radial stress and piezometric pressure measurements at Station 1 in the Permian Marl and Station 2 in the unfrozen, grouted Magnesian Limestone, respectively. Total radial stresses reached a constant level between 200 and 300 days. There were no piezometric pressures at Station 1. Piezometric pressures reached a constant level

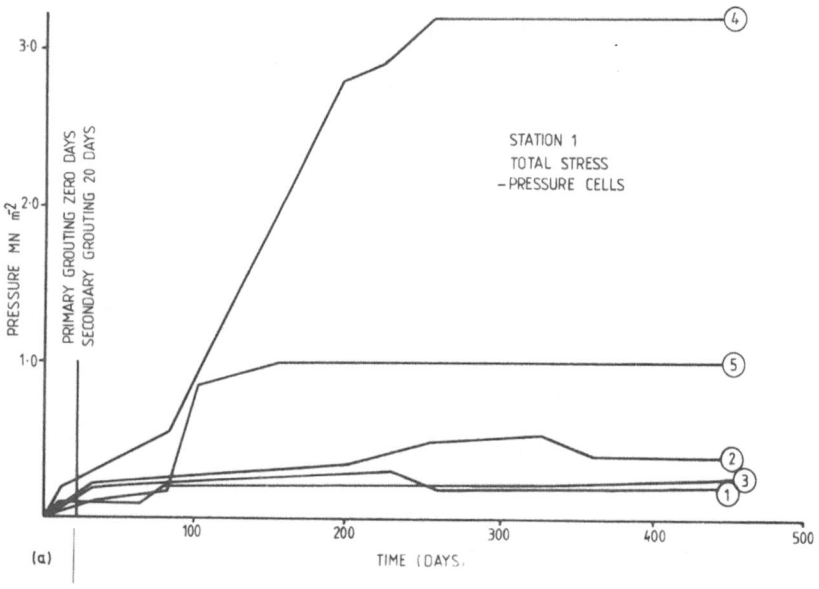

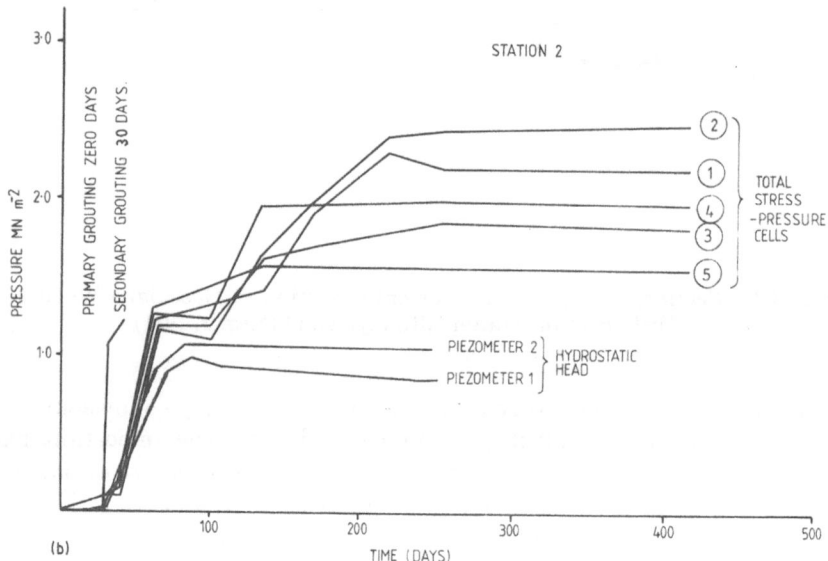

Fig. 3.23 Development of (a) total stresses at Station 1 in the Middle Permian Marl and (b) total and piezometric stresses in the Lower Magnesian Limestone, with time at Gascoigne Wood No. 1 Drift (after Altounyan and Farmer, 1981).

after about 100 days at Station 2. The depth at Station 1 was 98 m (equivalent to a vertical total geostatic stress of 2.40 MN m^{-2} and vertical hydrostatic pressure of 0.96 MN m^{-2} assuming a rock unit weight of 25 kN m^{-3} and a surface piezometric level) and the depth at Station 2 was 134.5 m (equivalent to a vertical total geostatic stress of 3.30 MN m^{-2} and a vertical hydrostatic pressure of 1.32 MN m^{-2}).

The radial lining stresses built up evenly at Station 1, not unduly affected by secondary contact grouting after 20 days, to a relatively low level of 0.3 MN m^{-2} at the crown cells (Nos 1, 2, 3). The pressure at the invert cell (No. 5) rose to about 1 MN m^{-2} and the invert shoulder cell to 3 MN m^{-2}, quite rapidly after 100 days. These anomalous results are probably due to eccentric loading from the bolted lining string. The tunnel had been allowed to dip approximately 50 mm below axis level at this point and was subsequently realigned.

The measurements at Station 2 illustrate a characteristic pressure build-up on a bolted lining. Before secondary contact grouting at 22 days there was a minor build-up in total geostatic stress. The absence of any significant hydrostatic pressure indicates the leakage of water through the grout and lining. Secondary contact grouting is accompanied by a rapid and equal increase in total and hydrostatic stress up to 100 days as the secondary grouting at this and other rings reduces water flow around the tunnel. It is, however, significant that even after 400 days there was sufficient flow into the tunnel to reduce piezometric pressures to about 1 MN m^{-2} – considerably below the estimated piezometric head.

The radial geostatic total stresses continued to rise after this time, reaching a maximum pressure of about 2.0–2.2 MN m^{-2} above the crown (cell Nos 1 and 2) at about 250 days. It is interesting to note here the components of stress distribution on the lining plotted in polar form (plotted in Figure 3.25(b)). The relation between crown (0.9 MN m^{-2}) axis (0.5 MN m^{-2}) and invert (0.25 MN m^{-2}) effective stresses is very close to that observed by Attewell, Farmer and Wickson (1976) and Attewell and Farmer (1977) in a similar investigation on the Tyne Syphon Tunnel.

In Fig. 3.24, the development of radial total stresses at Stations 3 and 5 with time is plotted, together with the average radial thickness of the thaw zone, estimated from temperatures measured in the radial boreholes.

At Station 3 there is considerable increase in stress at four of the cells (cell 3 appears defective) before breach of the icewall. This appears to be related partly to refreezing and subsequent increase in thickness of the radial thaw zone. Immediately before the breach of the icewall there is very rapid fluctuation in total stress, and break of the icewall (leading to rapid increase in piezometric head at piezometers 1 and 4 and a slow increase over two weeks at piezometers 2 and 3) is not accompanied by an increase in total stress equal to the increase in piezometric stress. The

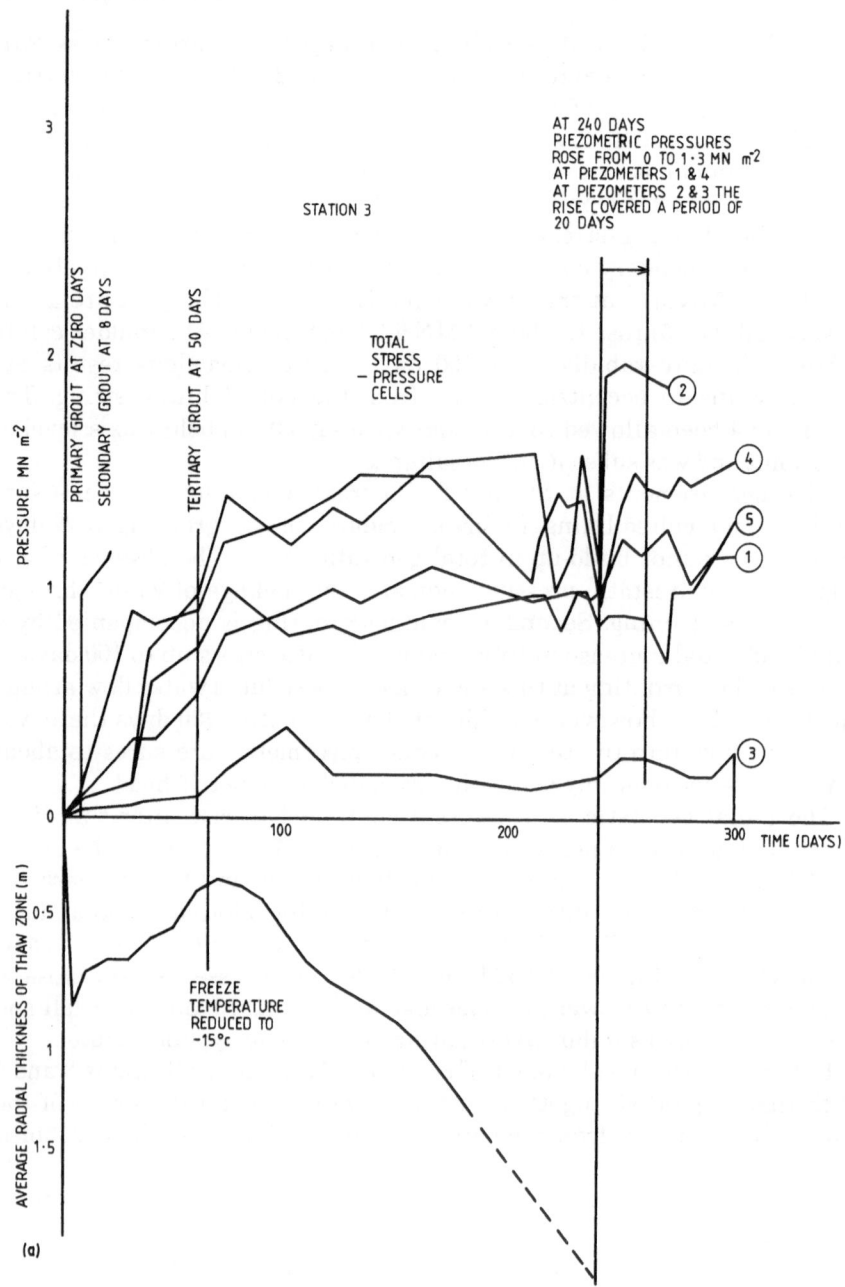

AT 240 DAYS
PIEZOMETRIC PRESSURES
ROSE FROM 0 TO 1·3 MN m⁻²
AT PIEZOMETERS 1 & 4
AT PIEZOMETERS 2 & 3 THE
RISE COVERED A PERIOD OF
20 DAYS

STATION 3

TOTAL
STRESS
— PRESSURE
CELLS

PRESSURE MN m⁻²

PRIMARY GROUT AT ZERO DAYS

SECONDARY GROUT AT 8 DAYS

TERTIARY GROUT AT 50 DAYS

3

2

1

0   100   200   300   TIME (DAYS)

AVERAGE RADIAL THICKNESS OF THAW ZONE (m)

0·5

1

1·5

FREEZE
TEMPERATURE
REDUCED TO
−15°c

(a)

116

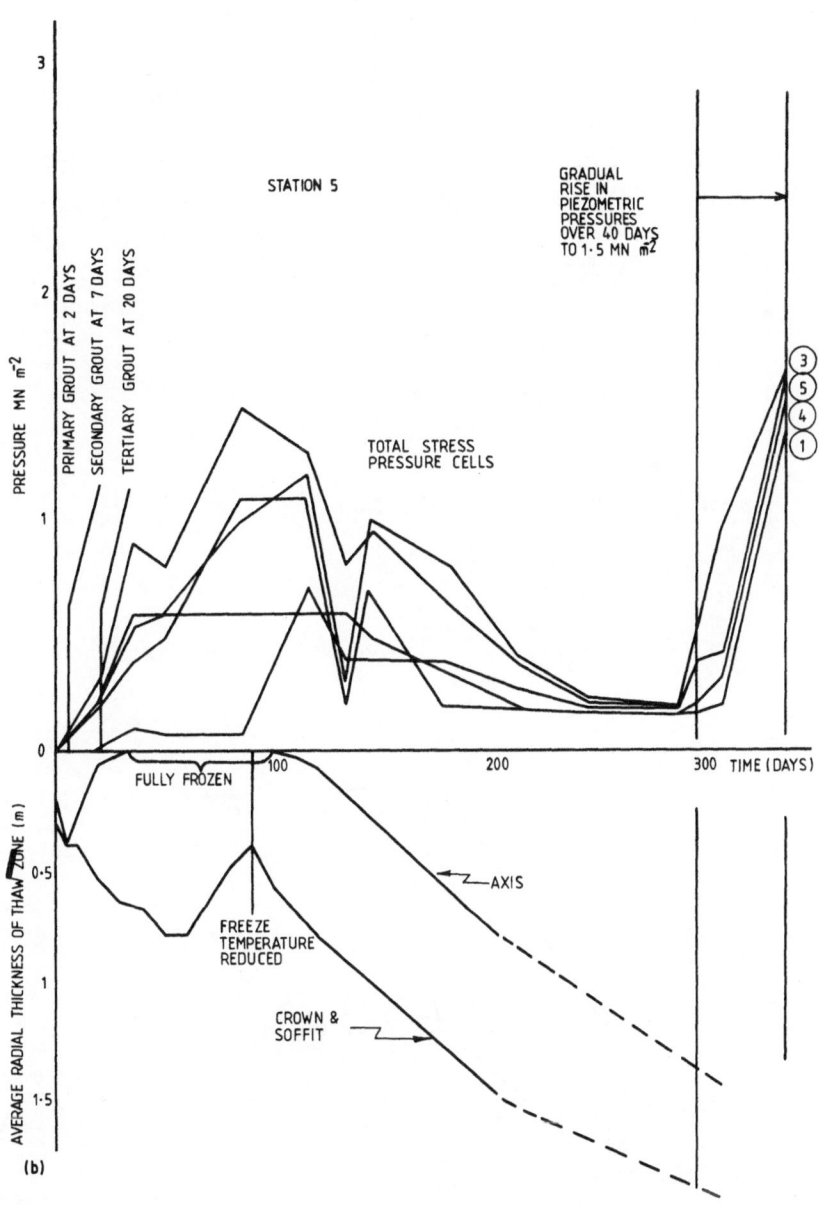

Fig. 3.24 Development of radial total stresses and the extent of the thaw zone at the tunnel periphery with time at (a) Station 3, No. 1 Drift and (b) Station 5, No. 2 Drift at Gascoigne Wood (after Altounyan and Farmer, 1981).

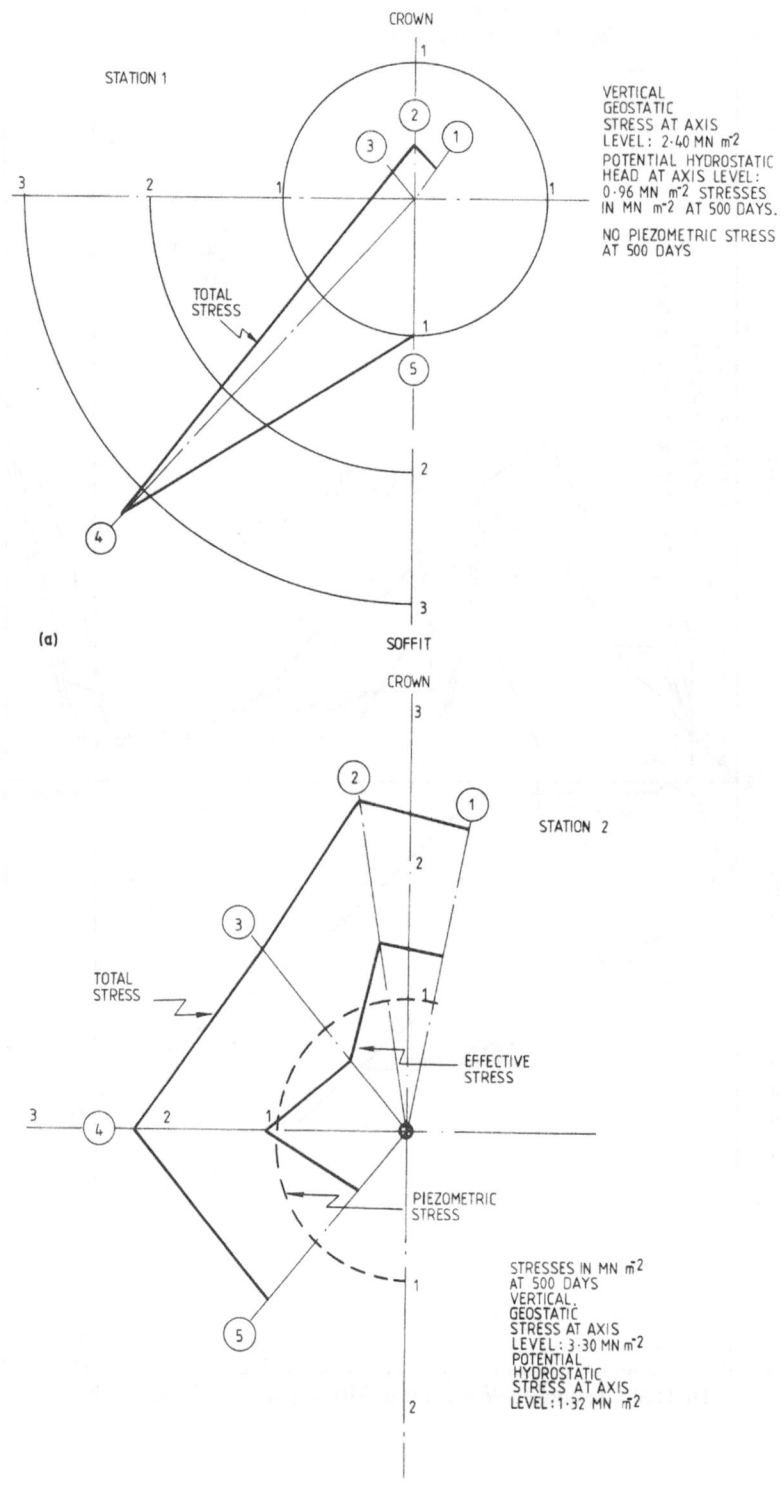

CROWN

STATION 1

VERTICAL
GEOSTATIC
STRESS AT AXIS
LEVEL: 2·40 MN m⁻²
POTENTIAL HYDROSTATIC
HEAD AT AXIS LEVEL:
0·96 MN m⁻² STRESSES
IN MN m⁻² AT 500 DAYS.

NO PIEZOMETRIC STRESS
AT 500 DAYS

TOTAL
STRESS

(a)

SOFFIT

CROWN

STATION 2

TOTAL
STRESS

EFFECTIVE
STRESS

PIEZOMETRIC
STRESS

STRESSES IN MN m⁻²
AT 500 DAYS
VERTICAL,
GEOSTATIC
STRESS AT AXIS
LEVEL: 3·30 MN m⁻²
POTENTIAL
HYDROSTATIC
STRESS AT AXIS
LEVEL: 1·32 MN m⁻²

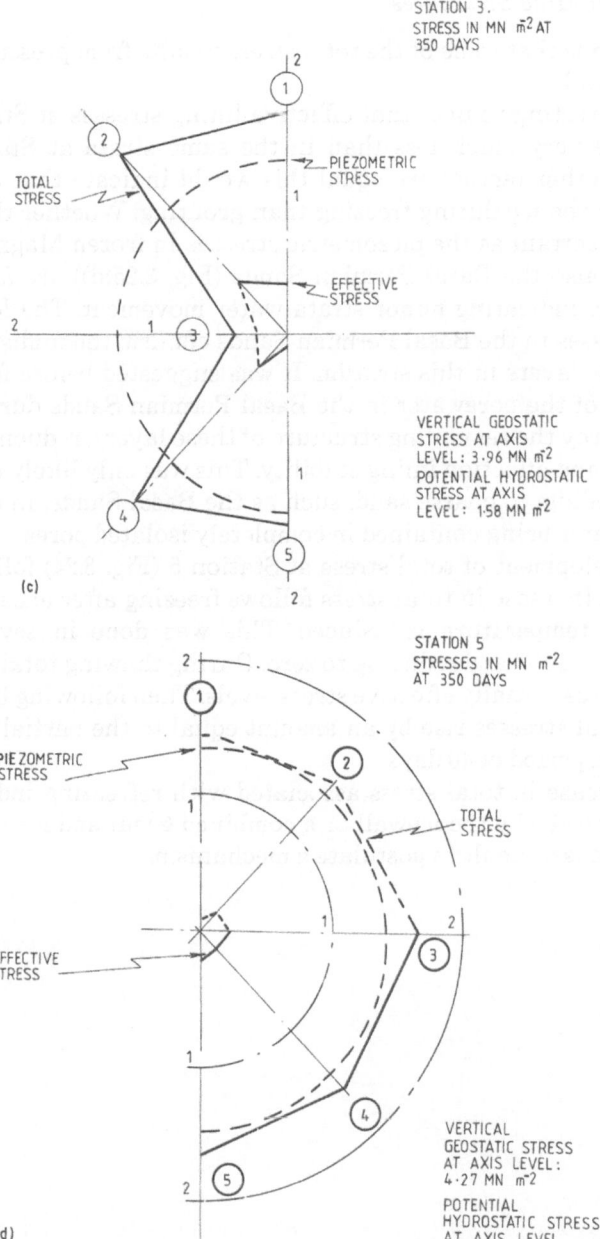

STATION 3.
STRESS IN MN $\overline{m}^2$ AT
350 DAYS

2
1

2

PIEZOMETRIC
STRESS

TOTAL
STRESS

1

EFFECTIVE
STRESS

2   1   3

VERTICAL GEOSTATIC
STRESS AT AXIS
LEVEL: 3·96 MN m^2
POTENTIAL HYDROSTATIC
STRESS AT AXIS
LEVEL: 1·58 MN m^2

1

4

5

(c)

2

STATION 5
STRESSES IN MN m^{-2}
AT 350 DAYS

2

1

2

PIEZOMETRIC
STRESS

1

TOTAL
STRESS

1   2

3

EFFECTIVE
STRESS

1

VERTICAL
GEOSTATIC STRESS
AT AXIS LEVEL:
4·27 MN m^2

4

POTENTIAL
HYDROSTATIC STRESS
AT AXIS LEVEL
1·71 MN m^2

(d)

2   5

Fig. 3.25 Total, effective and piezometric stresses at (a) Station 1 – 500 days; (b) Station 2 – 500 days; (c) Station 3 – 350 days; (d) Station 5 – 350 days (after Altounyan and Farmer, 1981).

119

implication is that some of the total stress results from pressures induced by the icewall.

It is interesting to note that effective lining stresses at Station 3 (Fig. 3.25(c)) are very much less than in the same strata at Station 2 (Fig. 3.25(b)). If other factors are equal this would indicate that there is less strata disturbance during freezing than grouting. Whether this will continue is uncertain as the piezometric stresses in frozen Magnesian Limestone and also the Basal Permian Sands (Fig. 3.25(d)) are *less than full hydrostatic*, indicating minor strata water movement. The low effective lining stresses in the Basal Permian Sands confirm the undisturbed state of the loose layers in this stratum. It was suggested before freezing that expansion of the porewater in the Basal Permian Sands during freezing would destroy the remaining structure of these layers, reducing frictional resistance and affecting lining stability. This was only likely to happen in a dense medium to coarse sand, such as the Basal Sands, in the unlikely event of water being contained in completely isolated pores.

The development of total stress at Station 5 (Fig. 3.24) follows a quite clear path. Increase in total stress follows freezing after excavation until the freeze temperature is reduced. This was done in several stages, starting at $-15°C$ and reducing to zero. During thawing total stresses fall to what are essentially effective stress levels. Then following breach of the icewall total stresses rise by an amount equal to the partial hydrostatic head over a period of 40 days.

The increase in total stress associated with refreezing indicates some pressure exerted by the icewall or a combined grout and icewall pressure for which it is difficult to postulate a mechanism.

4

SHAFTS AND INSETS

Shafts and insets are dissimilar types of structure. The former are invariably circular and have a vertical axis; the latter are arch shaped and have a horizontal axis. Vertical shafts are inclined normally to most Coal Measures strata, and often pass through overlying aquifer rocks. Insets are large structures often formed at considerable depths in weak but dry Coal Measures rocks.

Shafts and insets have in common the feature that they are the main access to the mine, continually in use over design periods of 50–100 years. They must remain dry and safe with sufficient stability to allow rapid passage of winding and transport equipment. In the case of shafts this predicates the use of a water-resistant lining – usually plain concrete, although cast-iron tubbing may be considered in extreme circumstances. The design of this lining, and treatment of saturated strata during its construction, constitutes the major problem in shaft design.

The case for concrete lining (invariably reinforced) of insets is less easily made, although often accepted in default. The design approach is essentially the same as for tunnels, with emphasis on the appearance of the finished structure. Barton *et al.*'s (1974) concept of *excavation support ratio* might be used to justify some of the more extreme examples of reinforced concrete construction.

The sinking of new shafts in the Selby coalfield, Britain (see Forrest, 1978) has provided a unique opportunity to examine the design and performance of shafts and insets and their linings.

4.1 Shaft lining design

Shaft lining design is determined by assumptions made about the stresses acting on the lining and about the behaviour of the lining material. In the next section, measured data on rock-lining interaction are discussed. It is useful, however, to outline the conventional assumptions.

Since the shaft is vertical it is usually assumed that radial stresses from body forces exerted through relaxation of circumferential rocks are not a significant design factor. This is justified by the inclination of the shaft and also in the case of concrete linings by the time delay of 3 months during which the concrete is in a plastic state – and a much longer period when it is able to creep under load. This would be more than long enough to allow stabilization of redistributed stresses. In any case the conventional methods of shaft construction (see Tuffs, 1982) in which lining is cast in 6 m lengths approximately 1.5–2 diameters above the shaft bottom allow sufficient relaxation (see Fig. 3.3) to prevent build-up of 'elastic' stresses.

The major part of the radial lining stress is invariably caused by *hydrostatic stresses* where the shaft passes through aquifer rocks. In such cases it is conventional to assume a radial lining stress equal to the pressure of a static head of water beneath the piezometric surface. Any minor error resulting from this assumption will be compensated by the generous safety factors used in concrete design.

Design methods for concrete shaft linings have been summarized by Auld (1979, 1982) and Bell (1982). The traditional design method utilizes the Lamé elastic analysis of a thick cylinder to give an equation for the lining thickness t in terms of the maximum tangential lining stress σ_{tmax} at the inside of a lining cylinder of radius r_i and the radial stress σ_r on the outside of the cylinder:

$$\frac{\sigma_{tmax}}{F} = \frac{2\sigma_r(r_i+t)^2}{t(t+2r_i)} \tag{4.1}$$

where F is a safety factor.

This can be expressed in British Standard (1972) concrete design by substituting $\sigma_{tmax}/F = 0.67\,Cu/2.1$ where Cu is the cube strength of concrete.

This is an extremely conservative approach to design, worth examining in detail because it has considerable similarities with the fracture zone approach to estimating tunnel deformation in Table 3.1. It is conservative because only the inside wall of the shaft is postulated to reach an ultimate stress ($\sigma_{tmax}, 0$) at which the concrete will fracture. The remainder of the lining section (Fig. 4.1) is stressed below its failure level. The current practice for concrete design (BS CP 110: Part 1, 1972) uses a *limit state design* approach based on the ultimate limit state for strength on a serviceability limit state for deformation. This is a much more useful approach, in which it is assumed that under a radial stress σ_{ri} a highly stressed zone occurs at the inner surface of the lining and that as the concrete approaches failure, plastic yield takes place at the inside face and spreads across the lining thickness. The ultimate limit state occurs

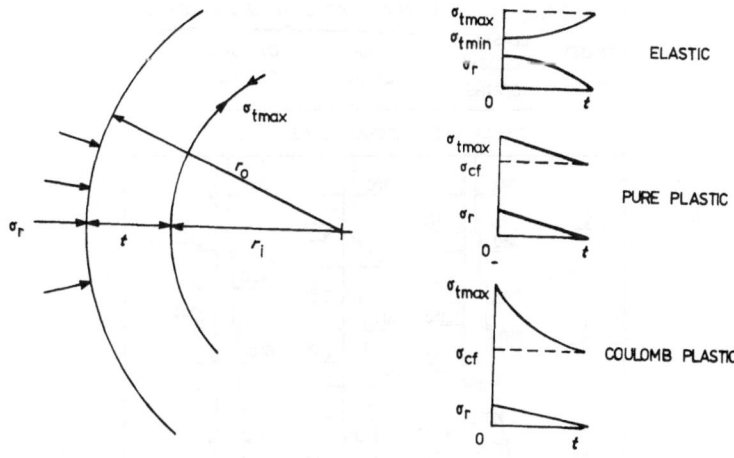

Fig. 4.1 Stresses in a thick cylinder for various design assumptions (after Auld, 1982).

when failure spreads through the whole lining. Design depends on the ultimate load failure mechanism.

Two very simple mechanisms (Fig. 4.1) are proposed by Auld (1982) – *pure plasticity* in which a constant shearing stress equal to $(\sigma_{tmax} - \sigma_r)/2$ (or twice the unconfined compressive strength σ_{cf} of the concrete) is assumed to exist through the lining at the ultimate state, and *Coulomb plasticity* in which the strength increases with confinement. The equivalent equations using the same nomenclature are, for pure plasticity:

$$\sigma_r = \sigma_{cf} \log_e \frac{(r_i + t)}{r_i} \tag{4.2}$$

and for Coulomb plasticity

$$\sigma_r = \frac{\sigma_{cf}}{(A-1)} \left[\left(\frac{r_i + t}{r_i} \right)^{A-1} - 1 \right] \tag{4.3}$$

where $A = \tan^2 (45 + \sigma/2)$.

The equations are derived, and partial factors of safety defined by Auld (1982), to give the lining thicknesses for the three cases in Fig. 4.2. While relatively simple, the approach illustrates the considerable differences in design obtained by the three assumptions. The Coulomb plasticity assumption is not recommended by Auld since backwall grouting pressures would raise σ_r beyond the ultimate limit state.

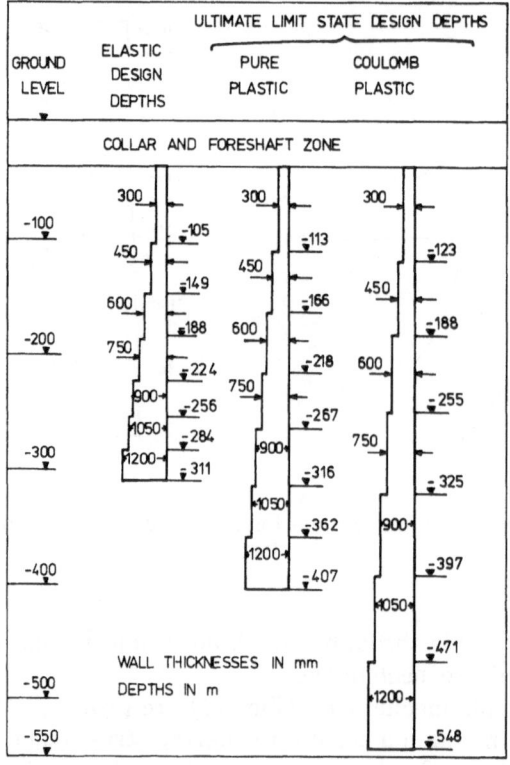

Fig. 4.2 Plain concrete (BS Grade 45) lining thicknesses calculated from the assumptions in Fig. 4.1 for a 7.315 m diameter shaft in saturated strata having a piezometric level at ground surface level. A factor of safety of 2.1, based on concrete cube strengths, was used in the calculations (after Auld, 1982).

4.2   Shaft lining behaviour

During the sinking of ten shafts at five sites for the Selby Project, Britain, most types of aquifer rock and ground condition were encountered, and most methods of groundwater control during construction were used. Figure 4.3 summarizes the geology and gives lining thicknesses and concrete strength – which were based, it can be seen by comparison with Fig. 4.2, on elastic analysis. The shafts are arranged in Fig. 4.3 in west–east order. The geology comprises Permo-Trias strata unconformably overlying the Carboniferous Coal Measures, dipping at 1 in 40 to the east (see Fig. 3.21 for comparison). The general succession comprises 20 m of glacial clays and silts overlying the Bunter Sandstone which is 200–300 m

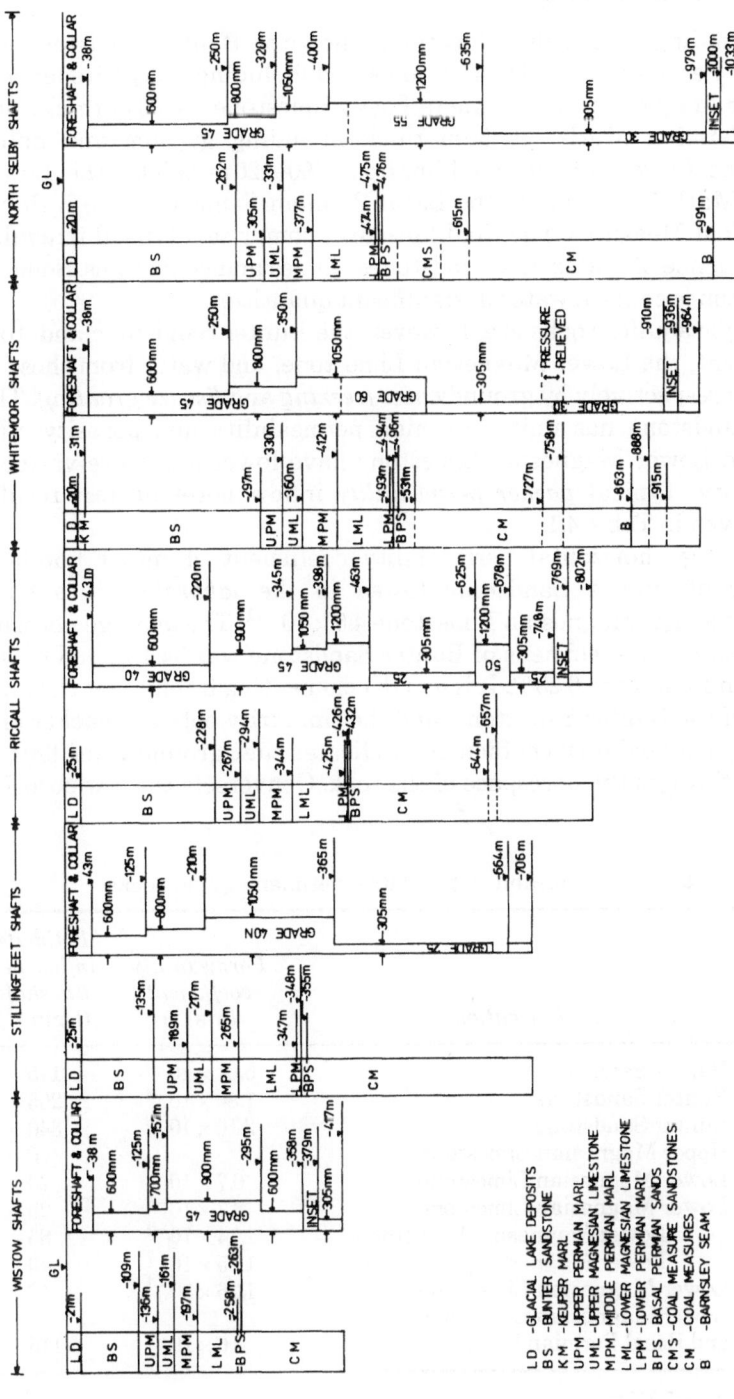

Fig. 4.3 Shaft strata sections for the mines of the Selby coalfield, Yorkshire, together with lining thicknesses and concrete strengths (after Bell, 1982).

thick depending on the shaft location. Below the Bunter Sandstone are the Upper Permian Marls (30–45 m thick and including anhydrite layers), the Upper Magnesian Limestone (a flaggy limestone 20–30 m thick), the Middle Permian Marls (30–50 m thick including gypsum and halite layers), the Lower Magnesian Limestone (60–120 m thick), the Lower Permian Marl (1–2 m thick), the Basal Permian Sands (up to 9 m thick) and the Coal Measures. The Coal Measures comprise a typical irregular cyclic sequence of seatearths, sandstones, shales and coal seams. Some of the sandstones contain water in significant quantities.

The major aquifer rocks are, however, the Bunter Sandstone and, to a lesser extent, the Lower Magnesian Limestone, and water from these is controlled respectively by groundwater *freezing* and *fissure grouting*. The Bunter Sandstone has uniformly high permeability and porosity. The Upper and Lower Magnesian Limestone have lower and more variable permeability. Typical *packer permeability* inflow borehole test results were as given in Table 4.1.

The average horizontal permeability coefficient of intact borehole specimens of Bunter Sandstone tested in the *laboratory* was $8.1 \times 10^{-6} \, \mathrm{m \, s^{-1}}$ and of Magnesian Limestone 4.5×10^{-11}. The average porosity of intact borehole specimens of Bunter Sandstone was 34.4%, and of the Magnesian Limestone 0.23%. It is worth commenting briefly that the high porosity of the Bunter Sandstone, and the similarity between packer and laboratory permeability coefficients, indicates that groundwater flow is primarily through the porespace of the rock. Conversely the low porosity

Table 4.1   Permeability test data – Permian aquifer rocks

Depth (m)	Formation	Permeability coefficient $(\mathrm{m \, s^{-1}})$	Estimate inflow per 10 m shaft* $(\mathrm{l \, min^{-1}})$
45–52	Bunter Sandstone	5.01×10^{-6}	175
131–144	Bunter Sandstone	1.80×10^{-6}	205
201–211	Bunter Sandstone	2.10×10^{-6}	340
283–301	Upper Magnesian Limestone	–	0
344–358	Lower Magnesian Limestone	6.7×10^{-6}	20
356–372	Lower Magnesian Limestone	8.7×10^{-6}	25
372–390	Lower Magnesian Limestone	1.14×10^{-7}	35
387–405	Lower Magnesian Limestone	1.15×10^{-7}	40
405–424	Lower Magnesian Limestone	1.26×10^{-7}	40
424–432	Lower Magnesian Limestone and Basal Permian Sands	3.6×10^{-7}	115

* Shaft diameter 7.315 m.

of the Magnesian Limestone and the low laboratory permeability coefficient (four orders of magnitude less than the packer permeability) indicates that groundwater flow is primarily through the fissures in the rock. Relations between groundwater flow, pore size and porosity and fissure frequency and size (see Attewell and Farmer, 1976) can be used to estimate the average pore passage diameter in Bunter Sandstone as 0.005 mm and the average fissure width in the Magnesian Limestone as 0.05 mm. The former is effectively inpenetrable by most grouting materials; the latter can be penetrated by cement particles in weak suspension using the ram pump of the *cementation* method. For this reason, groundwater freezing was chosen to prevent water flow from the Bunter Sandstone during shaft sinking, and cementation grouting to prevent flow from the Magnesian Limestone. Neither method will be

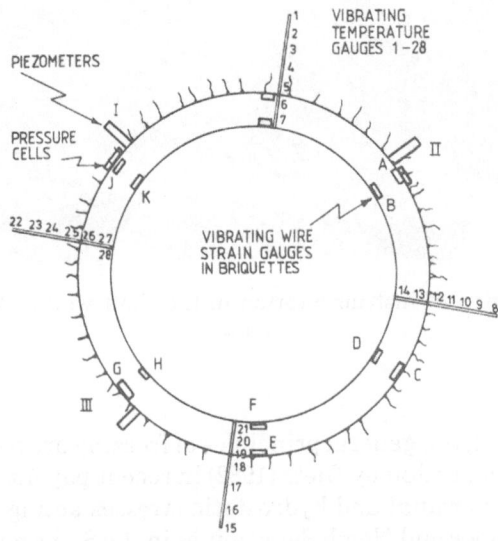

Fig. 4.4 General arrangement of instrumentation at a depth of 232 m at Whitemoor No. 2 shaft in frozen Bunter sandstone. The basic elements comprised:

(a) 300 mm diameter Soil Instruments Ltd pressure cells with high lateral stiffness ($> 200\,\mathrm{GN\,m^{-2}}$) for radial stress;
(b) 38 mm diameter by 240 mm long porous pot piezometers with deformable diaphragm installed in boreholes and plugged with bentonite;
(c) 140 mm long vibrating wire strain gauges installed free or precast into briquettes in the concrete lining;
(d) 65 mm vibrating wire temperature gauges installed in the lining and the shaft wall.

All transducer output was vibrating wire strain gauge which allowed multiplexing of signals at source to allow monitoring through a single shaft cable.

Fig. 4.5   Installation of instrumentation in the shaft wall at Whitemoor No. 2 shaft.

described in detail; the general principles of freezing are outlined by Klein (1982) and of cementation by Dietz (1982) in recent papers.

Measurement of radial and hydrostatic stresses acting on the finished lining at Whitemoor and North Selby mines in the Selby coalfield (Fig. 4.3) was carried out as part of a general shaft instrumentation programme (Altounyan, Bell, Farmer and Happer, 1982; Altounyan, Shelton and Wang Hao, 1983) which also included measurements of concrete strains and hydration temperatures. Measurements were made in the frozen Bunter Sandstone at depths of 232 m at Whitemoor No. 2 shaft and 234 m at North Selby No. 2 shaft. Average geotechnical properties of the Bunter Sandstone were:

Uniaxial compressive strength (saturated)	$9.8\,\mathrm{MN\,m^{-2}}$
Uniaxial compressive strength (saturated frozen)	$36.2\,\mathrm{MN\,m^{-2}}$
Uniaxial deformation modulus (saturated)	$5.5\,\mathrm{GN\,m^{-2}}$

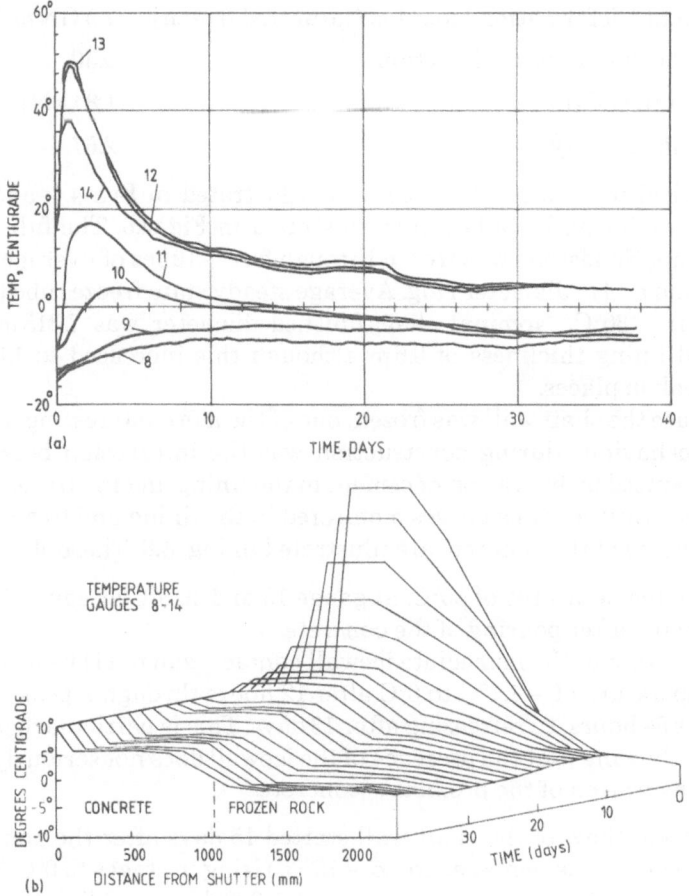

Fig. 4.6 (a) Change in temperature with time after pouring of concrete at gauges in array 8–14 (Fig. 4.4). Note the position of the gauges which are affected by overbreak:

Gauge number	Position
8	1170 mm into icewall
9	720 mm into icewall
10	320 mm into icewall
11	50 mm into icewall
12	290 mm from icewall
13	550 mm from icewall
14	930 mm from icewall and 115 mm from inner shaft wall

(b) An isometric projection of the temperature profile with time through the shaft wall at array 8–14.

Uniaxial deformation modulus (saturated frozen)	$7.5 \, \text{GN m}^{-2}$
Coefficient of internal friction	0.50
Dry unit weight	$1.84 \, \text{kN m}^{-3}$
Specific gravity	2.67

A typical instrumentation layout is illustrated in Fig. 4.4; and instrumentation during installation is illustrated in Fig. 4.5. The full depth of the Bunter Sandstone was frozen through freeze tubes of average spacing 687 mm on a 14 m diameter ring. Average steady state freeze tube temperature was $-30°C$. Nominal shaft finished diameter was 7.315 m with a nominal lining thickness of 0.6 m although this increased to 1.5 m with overbreak in places.

Because the shaft wall was frozen, one of the more interesting aspects of lining behaviour during construction was the interaction between the heat produced by hydration of cement in the lining and the frozen ground. The temperature – time curves monitored in the lining and frozen ground after placing of the concrete are illustrated in Fig. 4.6. These show:

(a) Peak temperatures of 50°C, at gauge 13 and 49.8°C at gauge 12, about 29 hours after pouring of the concrete.
(b) A thawing of the immediate icewall contact (gauge 11) from an initial temperature of $-1.1°C$ to 0°C after 12 hours through a peak of 16.9°C after 54 hours, to refreezing after 13 days. This is particularly interesting, showing that the presence of the icewall does not seriously inhibit the hydration of the proximate concrete.

Accelerated thaw of the shaft wall started 15 days after the pour with a reduction in freeze temperature to $-15°C$, rising in stages to 0°C. The first major breach of the icewall occurred after 209 days, and lining stresses – which exhibited a minor total stress component (compare with Fig. 3.24) after refreezing of the icewall at 15 days gradually rose to hydrostatic levels, at about 300 days. Specimen comparative values were as given in Table 4.2.

From data on total stress and hydrostatic pressures obtained from pressure cells and piezometers, the following points may be noted:

(a) There is a close correlation between piezometric and total stresses. The average piezometric stress for three gauges after 300 days was $2.21 \, \text{MN m}^{-2}$. This compares with a theoretical piezometric pressure of $2.18 \, \text{MN m}^{-2}$ if the groundwater level is assumed 10 m below ground surface. There was no noticeable leakage of groundwater through the lining. The average total stress was $2.16 \, \text{MN m}^{-2}$, indicating zero contribution to lining stresses from the backwall rock – even though this was a relatively weak rock. This confirms previous observations by

Table 4.2 Total and hydrostatic stresses on shaft lining, initially in frozen ground.

Time after concrete pour (days)	Total stress pressure cells (MN m^{-2})			Hydrostatic pressure – piezometers (MPa)		
	J	A	G	I	II	III
15	0.37	0.40	0.33	0	0	0
200	0.40	0.42	0.36	0.03	0.05	0.04
209	0.67	0.66	0.63	0.50	0.53	0.56
217	0.68	0.69	0.65	0.48	0.38	0.49
300	2.20	2.11	2.17	2.22	2.20	2.21

Altounyan and Farmer (1981) of the very low strata disturbance caused by groundwater freezing, around a tunnel in weak rock (see Fig. 3.25).

(b) The lining strains can be shown to exactly relate to the hydrostatic stresses in accordance with Lamé's theorem, and a computed deformation modulus of 33.6 GN m^{-2} compares with the design modulus range for 45 MN m^{-2} concrete of 27–38 GN m^{-2}.

Similar observations were obtained at North Selby No. 2 shaft at a depth of 234 m under similar circumstances (Table 4.3).

These results indicate:

(a) A small total stress component following refreezing of the icewall at 14 days
(b) An average hydrostatic pressure similar to the calculated piezometric pressure (2.35 MPa) at the instrumented level
(c) An average total stress similar to or lower than the hydrostatic stress

Once again there is similarity between the behaviour of this frozen ground and similar frozen ground around the tunnel illustrated in Figs 3.24 and 3.25.

Table 4.3 Total and hydrostatic stresses on shaft lining, initially in frozen ground.

Time after concrete pour (days)	Total stress – pressure cells MN m^{-2}			Hydrostatic pressure – piezometer (MPa)		
	13	14	15	16	17	18
14	0.42	0.45	0.34	0	0	0
220	0.73	0.75	0.66	0.50	0.57	0.57
310	2.52	1.56	2.08	2.36	2.40	2.43
385	2.64	1.94	2.09	2.20	2.43	2.43

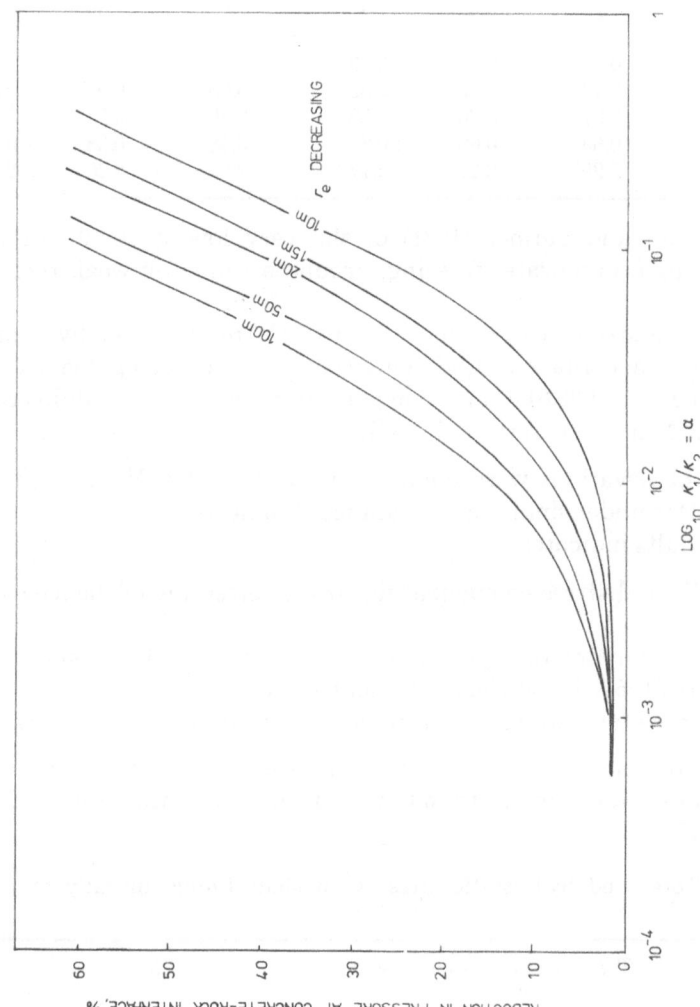

Fig. 4.7 Relation between reduction in radial lining pressure at a rock–concrete interface and the ratio of rock (K_1) to concrete (K_2) permeability coefficients for various assumed radii of influence (after Jennings, 1981).

The current specification for shaft lining *watertightness* is 4.5 l min^{-1} (1 gal min^{-1}) per 30 m of shaft lining – although ideally the lining should be virtually impermeable. Since the average permeability coefficient of good concrete is in the region of 10^{-12} m s^{-1}, this means that flow is mainly through joints or cracks in the concrete lining. It is sometimes interesting to speculate on the reduction in radial lining pressure which might result from flow through the lining. This can best be expressed (Fig. 4.7) by plotting the reduction in radial pressure against ratio of rock (K_2) to concrete (K_1) permeability coefficients. It can be seen that, for a significant reduction, a permeability match virtually equivalent to full drainage is required. This may of course be essential as a permanent or temporary construction expedient in the case of deep linings in saturated rock.

An example of this occurred at Riccall Mine (Fig. 4.3) where the Coal Measures contained one major aquifer rock – the Wooley Edge Sandstone – 12.5 m thick at a depth of 660 m. Laboratory samples gave the following average index data:

Compressive strength	19.7 MN m^{-2}
Tensile strength (indirect)	2.9 MN m^{-2}
Deformation modulus (secant at 50% strength)	8.3 GN m^{-2}
Poisson's ratio (at 50% strength)	0.24
Dry unit weight	21.0 kN m^{-3}
Saturated unit weight	22.5 kN m^{-3}
Porosity	0.13
Void ratio	0.15
Permeability	3.2×10^{-6} m s^{-1}
Potential water make	108 l min^{-1} per 10 m

Drainage was chosen as the preferred method of groundwater control because of the relatively low permeability of the sandstone which would have inhibited successful chemical grouting. The general dewatering layout is illustrated in Fig. 4.8. Water was collected from eighteen 25 m long holes drilled at an angle of 22° from a level about 8.5 m above the sandstone. Initially water flowed into the boreholes at a rate of 60 l min^{-1}, reducing to a rate of 18 l min^{-1} following excavation, the remaining water seeping into the excavation.

During lining construction, shaft wall surface water was kept from contact with the fresh concrete by a 'flexipane' sheeting. A secondary water collection system was installed behind this sheeting to prevent pressure build-up until the lining had developed sufficient strength. Grout seals were then installed below and the lining was extensively backwall grouted. Following this the dewatering holes were grouted.

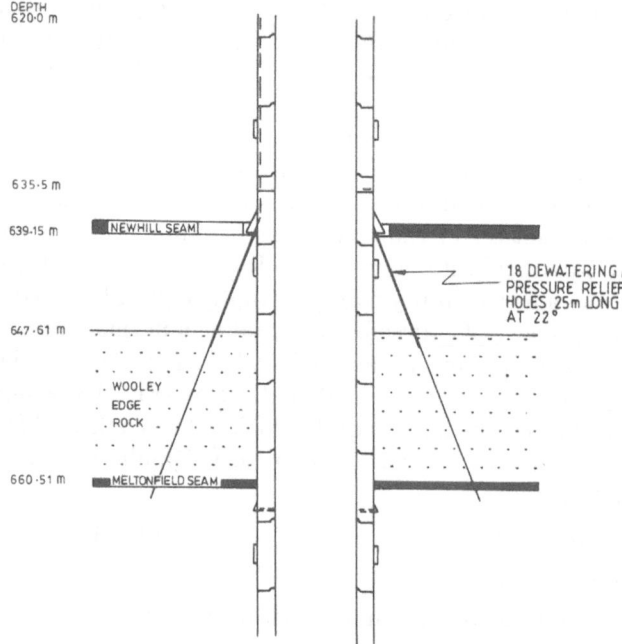

Fig. 4.8 Dewatering layouts used during lining construction in the Wooley Edge
sandstone at a depth of 660 m in the Coal Measures at Riccall Mine.

The effect of this grouting is illustrated in Fig. 4.9. Following grouting
at 42 days there was a general increase in pressure of 0.8 MPa which
reached a steady state when the grouting was completed, when a flow of
48 l min^{-1} was observed from the drain holes. Sealing of the grout holes
raised the pressure level to between 4.2 and 5.2 MPa, a magnitude equal to
between 62 and 80% of the full hydrostatic head. It is interesting to note
that at this point a total water flow of less than 6 l min^{-1} was estimated to
be leaking through the lining over the 12.5 m length of the shaft in the
Wooley Edge Sandstone.

During the succeeding 60 days, while continuous readings were not
taken, both piezometric and hydrostatic pressures increased to levels
close to hydrostatic pressure levels. Lining strains showed close correlation and there were no indications of any contribution to radial lining
stresses from the rock wall.

The break in slope of the curves in Fig. 4.9 cannot easily be explained.
The most likely reason for the differential recharge is that over the first 50

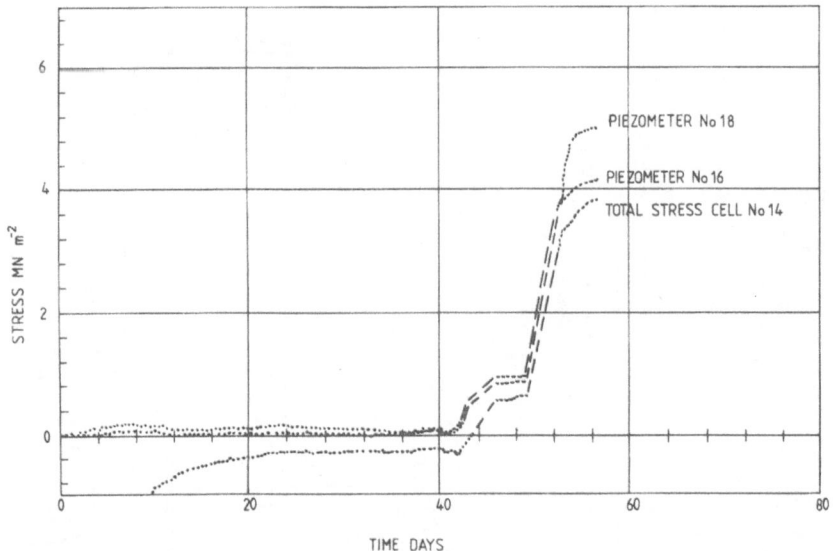

Fig. 4.9 Changes in piezometric and total stress up to 56 days at Riccall No. 2 shaft during grouting operations in the Wooley Edge sandstone.

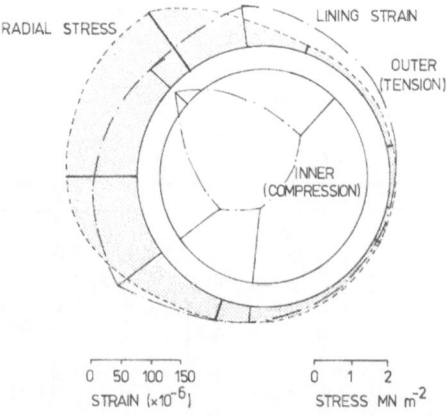

Fig. 4.10 Total radial stress and inner and outer lining strains around a shaft lining at a depth of 326 m in rock salt after 213 days at Riccall Mine (Gilbert, 1982). The stresses and outer ring strains are zeroed on the outside lining surface. The inner ring strains are zeroed on the inner surface.

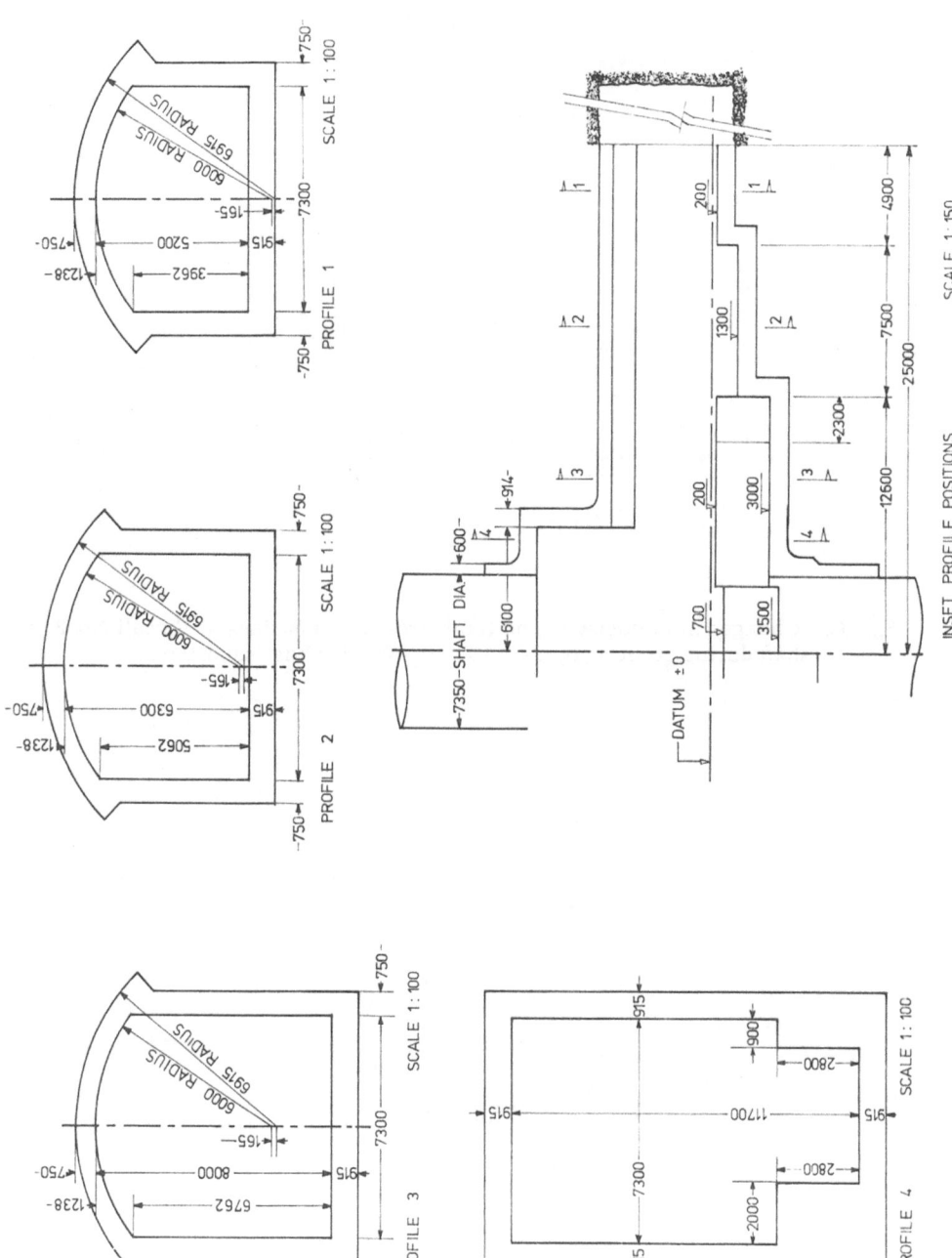

PROFILE 1   SCALE 1:100

PROFILE 2   SCALE 1:100

PROFILE 3   SCALE 1:100

PROFILE 4   SCALE 1:100

INSET PROFILE POSITIONS   SCALE 1:150

136

(a)

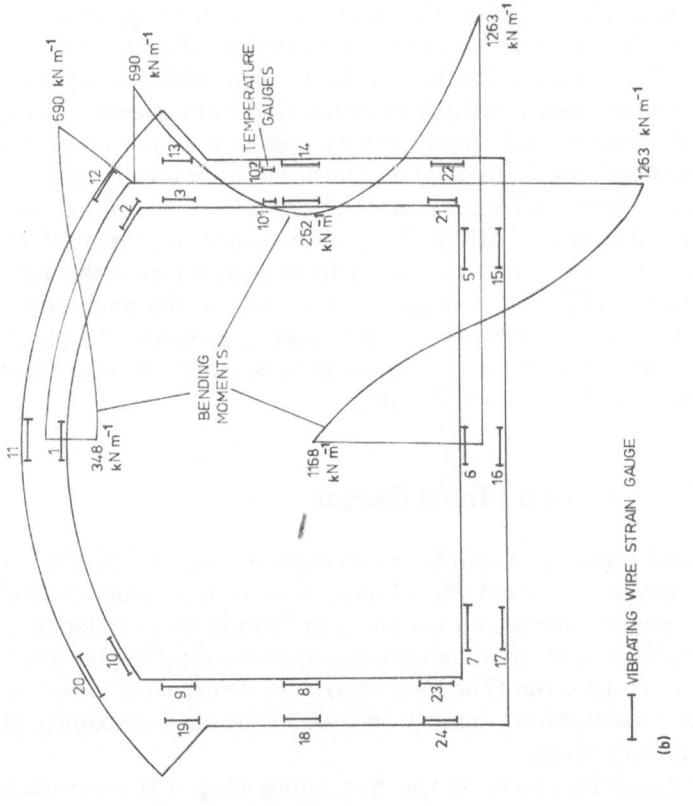

Fig. 4.11 Dimensions of reinforced concrete inset sections used as the basic design for the Selby Project mines with bending moments calculated for a continuous beam loaded with a uniform radial stress of 0.2 MN m^{-2} superimposed on profile 2.

137

days partial drawdown occurred through leakage into the adjacent shaft. Reduction of leakage in this shaft then allowed accelerated recharge. Since the outcrop of the Wooley Edge confined aquifer rock is about 20 miles to the west, it is likely that continual adjustments will occur for some time.

Although rock pressures can generally be ignored on shaft linings, where these pass through *evaporites*, there may be some contribution from the strata to radial stress. This is unlikely if the rock remains dry (see Gilbert and Farmer (1981)), but small quantities of water (see Varo and Passaris (1977)) can significantly affect the deformation of evaporites.

An example of this was found at Riccall Mine where the shafts passed through a layer of rock salt at a depth of 326 m. Hydraulic pressure cells, strain gauges, piezometers and anchor extensometers were placed in and around the lining (see Gilbert (1982)). The results were not totally successful but at 213 days (Fig. 4.10) after construction, part of the lining was subjected to a radial stress of $2\,\mathrm{MN\,m^{-2}}$, equal to about one-quarter of the estimated geostatic stress. This was found to have risen over the next three years to about $3\,\mathrm{MN\,m^{-2}}$ – roughly equivalent to the hydrostatic head, and yet there was apparently no piezometric pressure. To check this, cores were taken and the halite was found to be moist, but there was no evidence of water behind the shaft lining.

4.3   Inset design

There is a natural tendency towards overdesign in shaft insets. This is partly because they are the most immediately visible mine structure and partly because they are large and complex. In Britain this tendency is reinforced by the knowledge that major insets at Abernant (Shepherd and Wilson, 1960) and Wolstanton (Thomas, 1964) were deformed extensively during and after construction – mainly because of the close proximity of thrust faulting in both cases.

In the Selby coalfield, insets at the five mines (Fig. 4.3) were each designed to the same dimensions and to withstand the same ground pressures irrespective of depth. The inset structures were designed as a series of fully continuous plane sections (Fig. 4.11) restrained by an elastic surrounding medium. The concrete was designed to ultimate limit state principles as in British Standard CP 110, and reinforced to withstand bending moments calculated from an assumed $0.2\,\mathrm{MN\,m^{-2}}$ uniform load and 900 mm nominal thickness concrete.

The construction sequence for the insets is illustrated in Fig. 4.12. At each stage, excavation, temporary support and final lining was placed before continuation to the next stage. The sequence comprised:

Stage 1. This consisted of initial shaft widening at a height of 8.6 cm above the eventual 'rail level', to just above the roof of the inset forming the upper part of the inset box. Prior to the placement of the concrete pours 1a and 1b, the exposed rock was supported temporarily by hydraulic props, wood chocks and an array of rock anchors pre-sunk from a shaft crib 6.5 m above the inset roof.

Stage 2. On both sides of the inset this comprised the upper part of the inset with the floor on top of the Barnsley seam at Stillingfleet Mine (which will be discussed in greater detail). The sequence of excavation was 2h, 2g, 2c, 2b, 2a (south side); 2f, 2e, 2d (north side). The two 8.3 m wide by 4 m high and 19 m long headings were supported by 203 × 152 mm RSJ curved roof girders at 0.65 m centres. The additional excavations to develop the recessed shoulder abutments were normally completed by hand. Support in these areas was extended by 304 × 254 mm thrust blocks. The reinforced concrete lining was placed in three sections approximately 6.3 m long in each inset heading, in retreat fashion working towards the shaft.

Stage 3. This began with the drivage of a 4 m wide, 2 m deep excavation at the centre of each heading through the temporary floor. A 4 m by 3 m extension of the inset was then excavated a distance of 10 m to act as an absorption chamber, allowing a measure of stress relief to occur before the full stiffness of the concrete lining was mobilized. Following completion of the extension, the floor excavation was extended to the full width and depth of the inset in retreat fashion towards the shaft, undermining the short sidewall concrete of the roof wall. The reinforcement mats were completed and finally concreted, first in the floor and after a few days in the sidewall. This pattern was repeated in each of the sections of stage 3 moving towards the shaft. Finally, the two inset headings were linked across the inset box section, and shaft sinking (stage 4) resumed.

Two aspects of the inset design and construction are of particular interest – the first the extension heading, and the second, the recessed shoulder abutments of the roof arch. The former were designed to prevent transfer of high stresses to the inset lining during later extension of the heading. The latter were designed to take the full roof load, thus protecting the inset area during subsequent construction activity. It is probable that in fulfilling this design requirement the shoulder abutments radically alter the whole of the inset design. The reason for this is that the conventional design assumes a continuous beam, pinned by moment transmitting joints at the corners (Fig. 4.11). Beam theory assumes that no

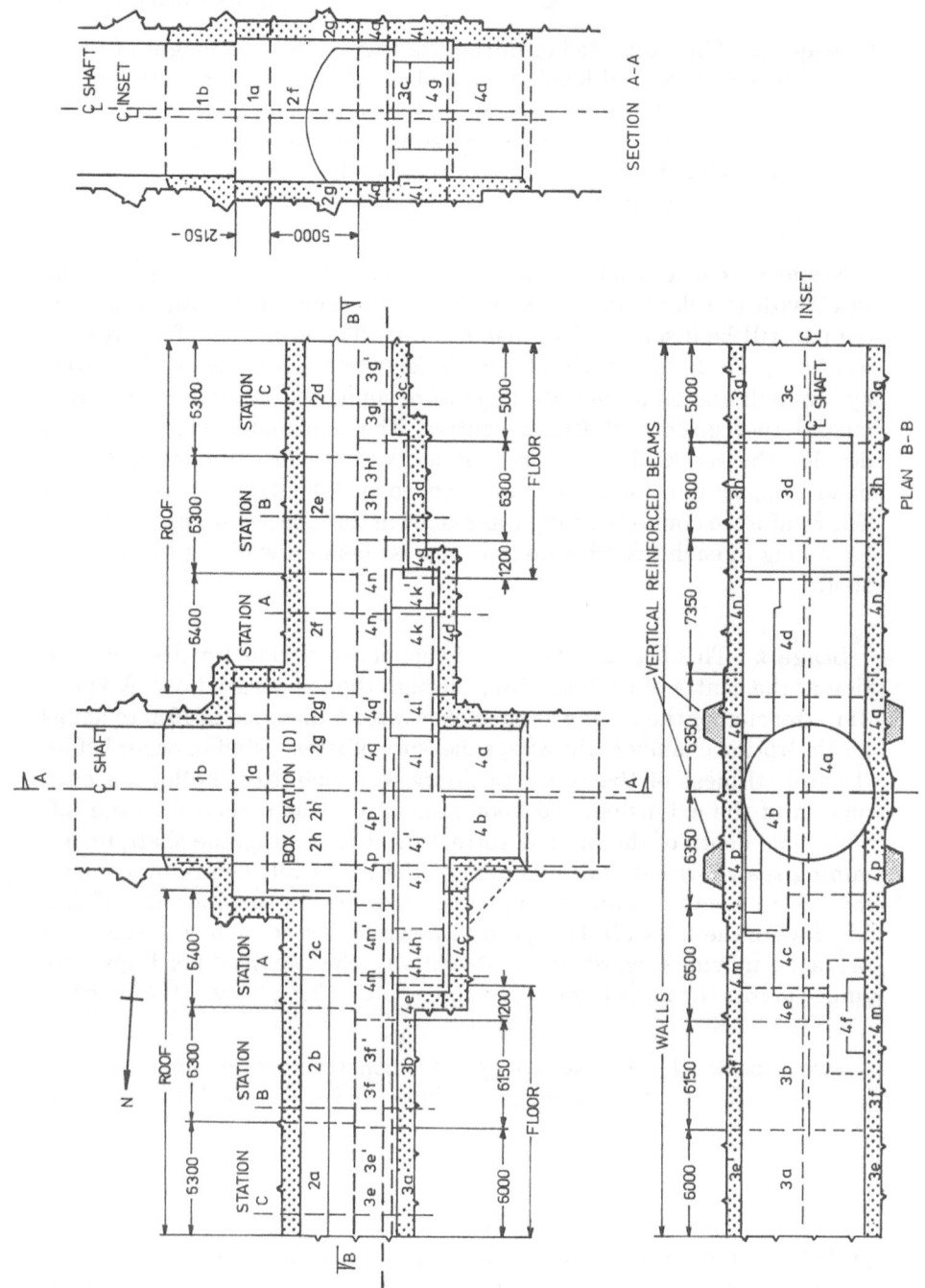

SECTION A-A

PLAN B-B

axial compression can exist in the beam, and radial stresses are calculated from the outer surface stresses less the mean stress. If the arch shoulders are keyed into the rock and the reactions to the radial stresses are taken through the shoulders, then the probability is that the curved part of the beam will be subjected to axial compression. In this case radial stresses can be related directly to outer surface strains.

Figure 4.13 illustrates the principal stress vectors generated in the concrete arch profile of Fig. 4.10 by the same uniform radial stress of $0.2\,\mathrm{MN\,m^{-2}}$ as in the original design calculations, using finite element (PAFEC 75) analysis of an arch restrained at the shoulders. Several interesting observations may be made:

(a) A quite uniform zone of compression is developed along the neutral axis of the arch.
(b) The reactions to the radial stresses are transmitted through the shoulders to the rock, rather than through the sidewall section which was also restrained.
(c) No tangential tensile stresses are developed in the structure.

There is a linear relation between tangential stresses and the uniform radial stress which is illustrated in Fig. 4.13. It is evident that the postulated stress levels would indicate an allowable radial stress very much higher than the $0.2\,\mathrm{MN\,m^{-2}}$ suggested as a design load. As in the case of shaft excavations, however, the magnitude of the *rock pressure* on the lining remains an unknown quantity. An observation programme was designed at Stillingfleet and Riccall Mines to examine the behaviour of the insets and inset linings during excavation and after construction of the lining, with the objective of improving design. This has been described in detail by Shelton and Farmer (1984). Some of the observations with particular emphasis on deformation during construction and lining reactions after construction are described below.

4.4   Inset behaviour during construction

During stage 2 excavation (Fig. 4.12), precise levelling and tape measurements were used to monitor peripheral deformation. Stage 2 was chosen when the excavation was 4 m high by about 9 m wide mainly for operational reasons, but also because this was the stage during which the

---

Fig. 4.12   Plan and elevation of the basic shaft inset design for the Selby Project showing excavation stages, concrete placement sequence and instrumentation stations at Stillingfleet Mine.

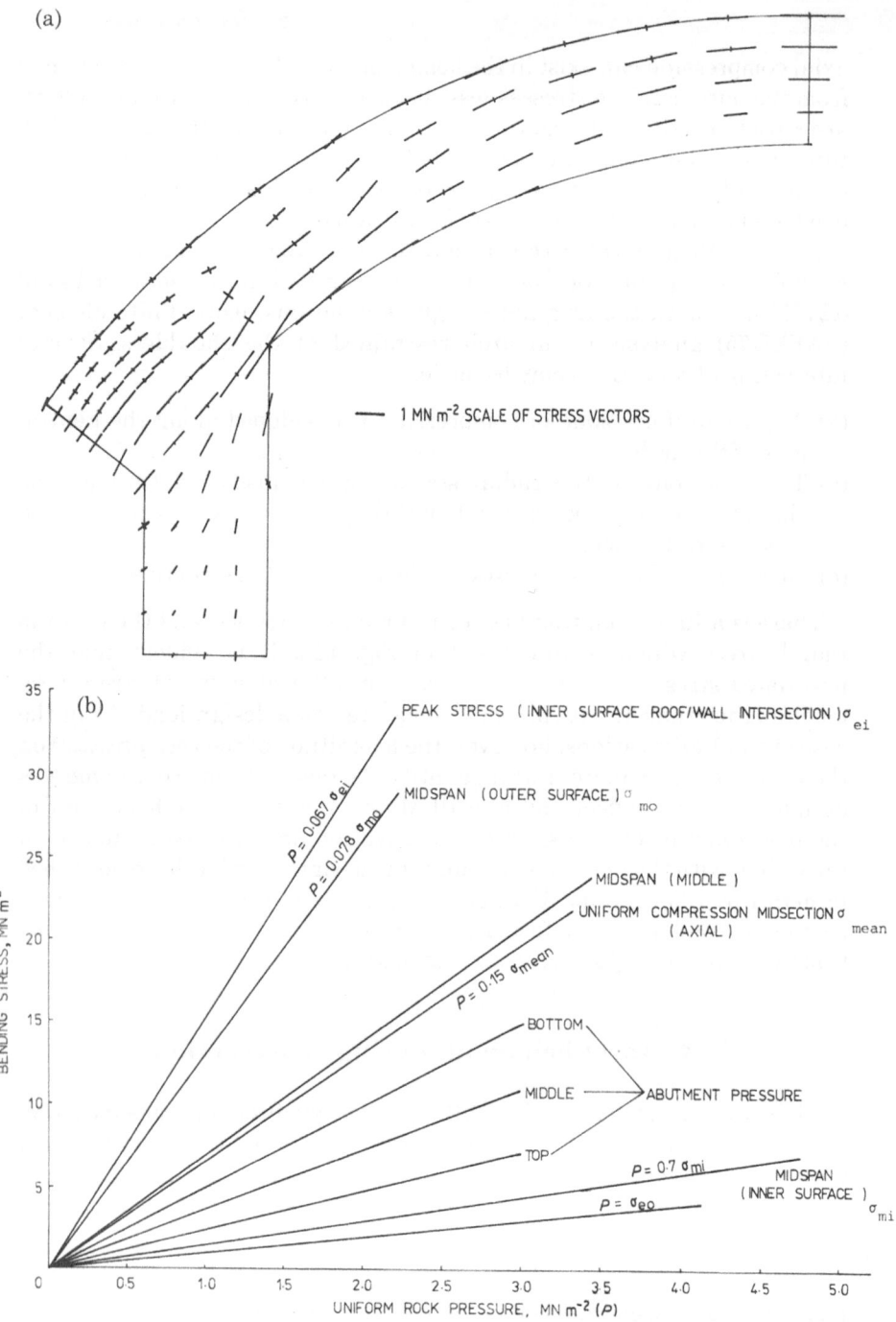

(a)

— 1 MN m⁻² SCALE OF STRESS VECTORS

(b)

PEAK STRESS (INNER SURFACE ROOF/WALL INTERSECTION) σ_{ei}

MIDSPAN (OUTER SURFACE) σ_{mo}

$P = 0.067\ \sigma_{ei}$

$P = 0.078\ \sigma_{mo}$

MIDSPAN (MIDDLE)

UNIFORM COMPRESSION MIDSECTION σ_{mean}
(AXIAL)

$P = 0.15\ \sigma_{mean}$

BOTTOM

MIDDLE → ABUTMENT PRESSURE

TOP

$P = 0.7\ \sigma_{mi}$

MIDSPAN
(INNER SURFACE) σ_{mi}

$P = \sigma_{eo}$

BENDING STRESS, MN m⁻²

35

30

25

20

15

10

5

0   0.5   1.0   1.5   2.0   2.5   3.0   3.5   4.0   4.5   5.0

UNIFORM ROCK PRESSURE, MN m⁻² (P)

excavation roof was exposed. Details of measurement procedure are given in Shelton and Farmer (1984). Three stations at Stillingfleet No. 1 shaft north and south insets were chosen for instrumentation, numbered A, B and C sequentially from the shaft in Fig. 4.12. In both cases A was 7.2 m from the shaft centre line and 18.8 m from the end of the inset heading; B was 13.2 m from the centre line and 11.8 m from the end of the heading and C was 19.2 m from the centre line and 5.8 m from the end of the heading. The time interval between excavation and lining was 31 days (A south), 14 days (B south), 6 days (C south), 29 days (A north), 20 days (B north) and 11 days (B north).

Graphs of vertical closure against time; horizontal closure against time and roof lowering and floor heave against time are illustrated in Fig. 4.14. A geological section at the inset level is included in Shelton and Farmer (1984). It is possible to outline from the vertical closure data three phases of deformation:

(a) Up to about 5 days or depending on the rate of advance an advance of about 5–10 m a closure rate of about 5 mm per day. This may be attributed to time related loosening of roof strata and squeezing of floor strata, inhibited by the supporting action of the proximate heading face, but accompanied by some deformation due to relaxation of the heading carcass as the face advances.

(b) Between about 5 and 15 days and a further 15 m advance a higher closure rate of about 12 mm per day. This may be attributed to time related deformation and relaxation.

(c) Beyond 15 days when excavation is completed, reducing time related deformation at a rate of about 3 mm per day as loosening and squeezing are mitigated by mobilization of residual strength in the surrounding strata.

The major observation, however, is that deformation is largely completed at sections A and B where the face has advanced a sufficient distance beyond the station to reduce face effects before the lining is installed.

It is useful at this point to refer to Figs 3.3–3.5, describing the basis of ground reaction curves around a tunnel excavation. There is considerable

---

Fig. 4.13 (a) Principal (compression) stress vectors generated by a uniform radial stress of 0.2 MN m^{-2} applied to the upper surface of a half inset roof profile. The only restraints applied to the structure were that the plane through the midspan was restricted to vertical movement and the shoulder and sidewall contact were held rigid. (b) Relation between a uniformly applied radial pressure and bending (or tangential) stresses generated at various locations within the concrete lining (after Shelton and Farmer, 1983).

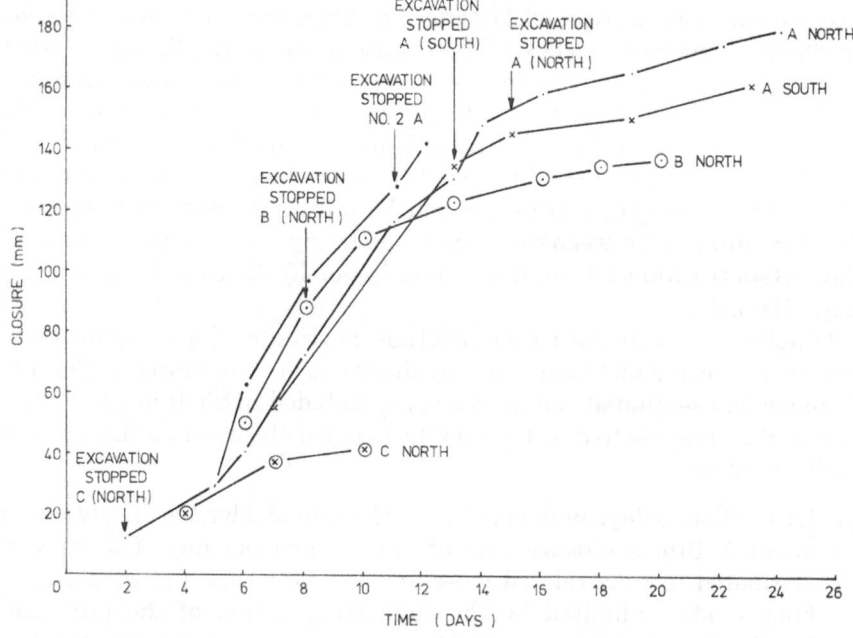

(a)

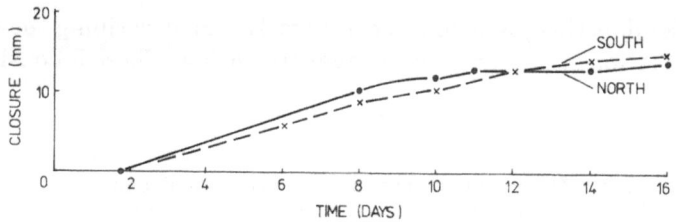

(b)

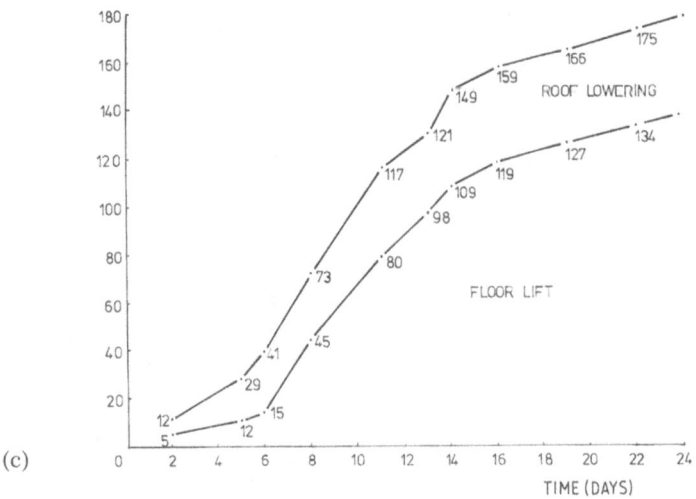

(c)

Fig. 4.14 Stillingfleet Mine No. 1 and No. 2 shaft inset. Relations between: (a) vertical closure and time at the centre line; (b) horizontal closure and time at the axis; (c) average roof lowering and floor heave with time at the centre line (after Shelton and Farmer, 1984). Note that these curves have been extrapolated to compensate for initial readings not possible immediately after blasting.

similarity between the displacement/distance curve of Fig. 3.3 and the related ground reaction curve of Fig. 3.5 and the closure/time curve of Fig. 4.14a. It can be argued therefore that the temporary supports installed during stage 2 excavation were able to mobilize the support potential of the rock at Sections A and B. At section C, the full *rock strength potential was not mobilized* and further extension of the inset might be expected to generate significant lining stress. This is discussed in the following section.

The concept of fracture around an opening has been discussed in the previous chapter (see Fig. 3.6). It is worth considering here since the relative roof, floor and sidewall deformations (Fig. 4.14) illustrate the concept well. The roof was mainly sandstone, the sidewalls coal and the floor, mudstones containing seatearth layers. In Fig. 4.15, peak and residual envelopes for the rock types involved are illustrated with a superimposed hypothetical excavation stress path.

Clearly the sandstone and mudstone are unlikely to fracture. The seatearth which demonstrates a degree of ductility is likely to fracture and certainly under high deviatoric stresses at the corners of the excavation (Fig. 3.1) will deform in a ductile manner. This will explain the high floor heave in Fig. 4.14(c). It is interesting to note that anchors in the floor

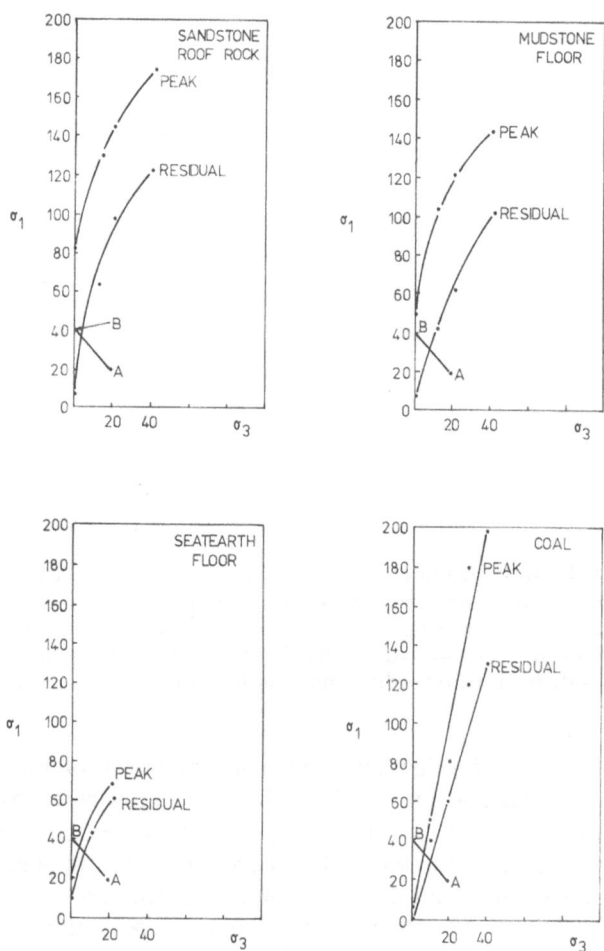

Fig. 4.15 Peak and residual strength envelopes in $MN\,m^{-2}$ plotted from servo-controlled triaxial test data on strata from Stillingfleet No. 1 shaft at inset level. Hypothetical stress paths A–B are superimposed for a point on the edge of a circular opening in a homogeneous stress field at a depth of 800 m.

showed that all but 14 % of this floor heave occurred above a depth of 3 m. The roof lowering on the other hand was much more deep seated and over 50 % of this occurred above a height of 3 m into the roof. The intact strength of the sandstone would indicate that this was due principally to loosening under the action of gravity or body forces, which are absent in the floor.

The limited sidewall deformation indicates the very high residual strength of the coal. The relatively low unconfined compressive strength resulting from dense cleating is rapidly compensated by residual strength, and continuing low levels of deformation (Fig. 4.14(b)) may be attributed to surface readjustment (possibly involving spalling) to roof and floor deformation.

Actual deformations were significantly less than predicted using a range of analytical techniques from yield zone analysis to finite element analysis. The difficulty lies partly in predicting the residual behaviour of apparently weak rocks such as coal and seatearths, but mainly in predicting the effect of excavation and support in mobilizing residual strengths. Most analytical techniques are based on plane strain analysis, whereas excavation is a gradual process where support from the face is replaced by artificial support. In addition action of body forces on discontinuities will tend to exacerbate roof deformation, while the action of the same forces will tend to inhibit floor deformation – even without invert support.

For this reason it is difficult to recommend analytical techniques for design in Coal Measures rocks. The empirical approaches summarized in Figs 3.3–3.5 combined with an understanding of deformation mechanisms based on Figs 1.1–1.5 and Fig. 4.15 and an observation programme during construction represent a more positive approach.

4.5   Lining behaviour after construction

Sections A, B and C in Stillingfleet No. 1 shaft south inset (Fig. 4.12) were instrumented to determine lining behaviour. Vibrating wire strain gauges were installed in pairs one close to the *inner* lining surface; the other close to the *outer* concrete/rock interface. Hydraulic pressure cells were installed at the rock/concrete interface to measure radial stress.

The concrete lining was cast in three sections from the end of each inset leading towards the shaft. Thus, of the instrumentation arrays, station C, was the first to be installed, followed by B and A respectively. The stations were chosen to correspond directly to the positions of the deformation measurement stations. A period of between 7 and 20 days was allowed to elapse before datum readings of the strain gauges were taken in order to allow the thick concrete sections to cool down to inset ambient temperature. Readings from the temperature gauges installed along with instrumentation showed this to be between 18°C and 22°C. The exothermic cement reaction raised the temperature to between 40°C and 60°C.

The results from measurements taken on the hydraulic pressure cells, positioned to measure radial stress, were disappointing, and have not been reproduced. The sensitivity of the read-out unit was poor at low

SECTION C
STILLINGFLEET
JUNE 1982

pressures between 0 and $0.5\,\mathrm{MN\,m}^{-2}$. The only conclusion, which in itself is significant, is that the radial pressures lie somewhere in this range.

Satisfactory measurements were obtained from the vibrating wire strain gauges installed in the concrete lining. Some gauges, principally from section B, were lost, probably due to cable damage during concrete placement. A specimen selection of strain data from station C is illustrated in Fig. 4.16, based on microprocessor printouts. Data at stations A and B followed similar paths, but increases in strain were lower with maximum strains of the order of 100 microstrains in the floor and sidewalls and 200 microstrains in the roof. Most measured strains were lower. Strains in the shaft box – also monitored – were negligible.

The increases in strain at station C can be related accurately to events occurring during the inset construction period. The initial event which occurred at 120 days prior to the placement of the floor and sidewall concrete was the drivage of the 10 m long overdrive. The second strain increase occurred uniformly through the inset after 650 days and coincided with the beginning of a second excavation phase beyond the overdrive, together with the construction of a junction in the south heading of the inset. Excavation activities in the north heading of the inset which occurred during these times did not have any appreciable effect on strains measured in the south side.

The pattern of strain distribution in the roof can be seen more clearly if the strains are plotted as *mean* and *differential* strains – respectively half the sum of, and the difference between the linear and outer strains. The value of mean strain can be taken to indicate the hoop strain; the differential strain the bending component within the concrete.

In Fig. 4.17, mean and differential inset roof strains are plotted – the malfunction of gauge 61 meant that an incomplete picture was given for station C. It can be seen that mean strains are constant and compressive after excavation of the overdrive; differential strains are tensile in midspan, compressive towards the shoulders. This pattern remains constant up to 500 days, and is indicative of a uniform of radial stress applied to the lining. Calculated from beam theory at station B and C it lies in the range 0.07–$0.11\,\mathrm{MN\,m}^{-1}$, and from finite element analysis about $0.25\,\mathrm{MN\,m}^{-1}$. After further excavation anomalous deviatoric stresses indicate a change in this loading pattern at 700 days. There are also significant increases in strains in the floor and sidewalls at station C.

Fig. 4.16 Variation of inner and outer lining strains (corrected for exact gauge location assuming linear variation through the concrete section) with time at Station C (Fig. 4.12). Stillingfleet Mine No. 1 inset (South) – after Shelton and Farmer (1984). Zero date is taken as the date of concrete pours. Compression is positive.

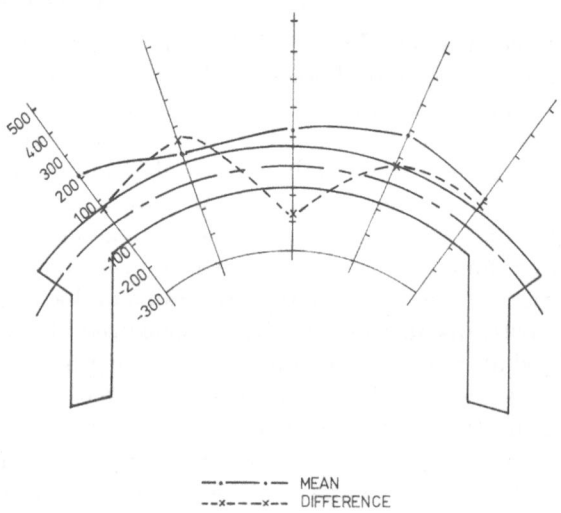

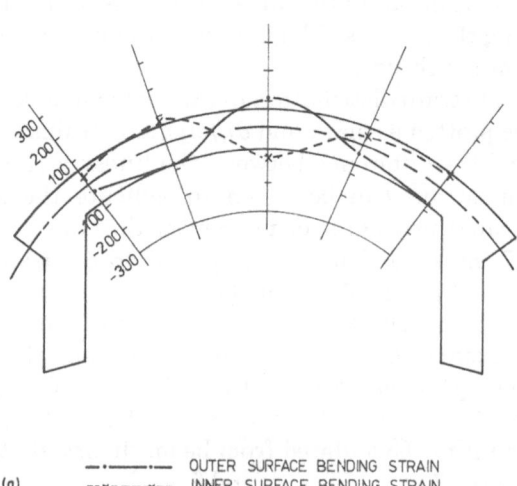

Fig. 4.17 Mean and differential roof strains at Stillingfleet Mine No. 1 inset (South): (a) 154 days at Station C – after excavation of the overdrive; (b) 500 days – following completion of concreting but before further excavation; (c) 700 days – following further excavation. In (a) inner and outer surface bending strains are also illustrated. Compression is positive.

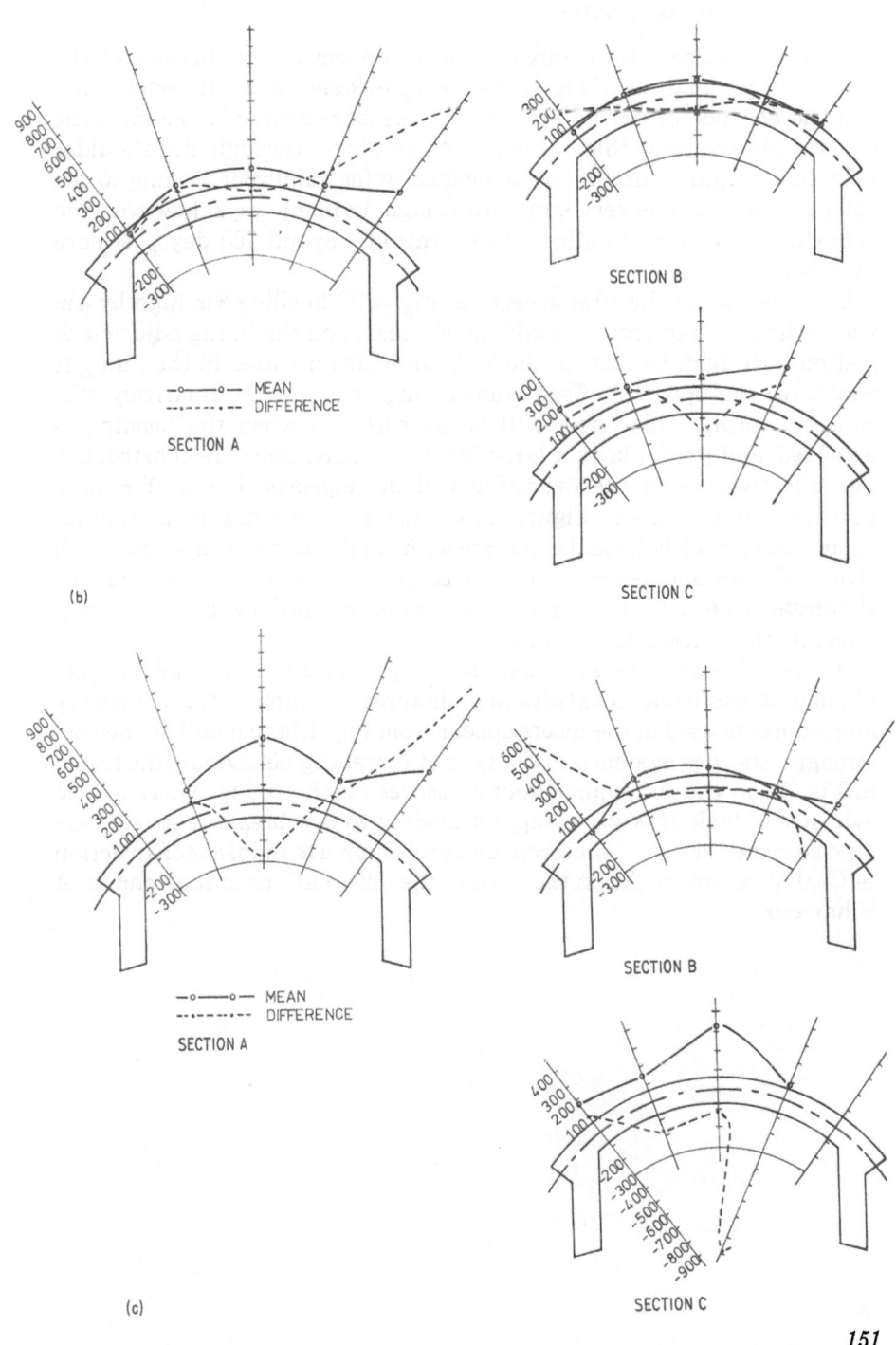

SECTION A

—o——o— MEAN
·—·—·— DIFFERENCE

SECTION B

SECTION C

(b)

SECTION A

—o——o— MEAN
·—·—·— DIFFERENCE

SECTION B

SECTION C

(c)

Until this stage, the strains generally confirm the predictions of the finite element analysis of Fig. 4.13 showing evidence of axial compression and the absence of tensile strains. The measurements also confirm the general observation, that transmission of stress through the shoulder abutments, significantly alters the capacity for axial roof loading to the extent where there is very large overdesign. In many ways, however, the behaviour after continuation of excavation, beyond 700 days, is more interesting.

Construction of the 10 m overdrive (Fig. 4.12) heading 3 m high by 4 m wide is designed to prevent build-up of stresses on the lining adjacent to station C. In fact, because of the 'soft' arch support used in the lining it may have the opposite effect, transferring stress to the relatively stiff concrete lining. This effect will be exacerbated when the heading is extended, and particularly if large junction excavations are constructed. The effectiveness of the overdrive will be improved if a stiffer more positive lining is used. Figure 3.14 compares deformation of tunnels supported by rock bolts and shotcrete with similar tunnels supported with steel arch ribs and demonstrates the early effectiveness of the rockbolt/shotcrete system. This positive recommendation has now been incorporated into the British inset design.

It is possible also to make a case for temporary and permanent support of shaft insets using rockbolts and shotcrete. Although the temporary arch supports used in the insets appear from Fig. 4.14 to mobilize the roof strength, there is evidence of the type of loosening behaviour illustrated in Fig. 3.5 in the subsequent roof pressures on the lining. There is also evidence of lack of positive support leading to overbreak. A strong case may be made for use of shotcrete and arch supports in inset construction in Coal Measures rocks on the basis of the observations of rock and inset behaviour.

5

STRATA DEFORMATION ABOVE LONGWALL EXCAVATIONS

In previous chapters the structures which have been considered are not radically different from those which can be found in any type of underground construction. They may be unique in so far as the rocks of the Coal Measures cyclothem and their associated geological structures are unique. But the simplifying assumptions used in design indicates that this uniqueness is of a low order.

Where coal mine structures *are* unique is in the longwall system, where coal is extracted from advancing or retreating faces, usually greater than 180 m wide and 1000 m long, so that the overlying strata cave into the void created by excavation. The resulting redistribution of stresses creates high stress concentrations in adjacent pillars (see Fig. 2.6) and particularly at the edges of the worked area, which can affect the stability of access roadways (see Chapter 6) and faces (see Chapter 7) bordering the longwall excavation.

There is considerable difficulty in modelling the deformation of strata above a longwall face. As has been demonstrated in Chapters 2 and 3 if an excavation is roughly equidimensional it is possible to analyse deformation using conventional continuum mechanics methods or, in the case of discontinuous rock, by a study of the kinetics of blocks able to move into the excavation. If the excavation has a high horizontal–vertical aspect ratio, the mechanics of deformation become so complex that no approach to numerical modelling can be satisfactorily used. This has been confirmed by the unsatisfactory nature of some of the best approaches to analytical and numerical modelling (useful summaries of some approaches are given by Salamon (1964), Litwiniszyn (1964), Voight and Pariseau (1970), Berry (1977) and Peng and Harthill (1981)).

For instance in the case of the strata overlying a longwall face in a coal mine, computation of tangential stresses from the common analogy of a horizontal ellipse in an infinite elastic plate and with a long major axis and with zero support, would show these to be compressive (see Coates, 1970). In practice deformation of the roof strata under gravity will induce tensile stresses which will rapidly reach failure levels. However, before this, the sagging of the roof layers partly from excavation induced stress and partly from self-weight stresses will gradually segregate the strata at weak horizons – specifically *shear zones* – into hinged sagging beams. Ward (1978) has observed the detailed breakdown process showing how with sagging (see Fig. 2.16(b)) the hinges will tend to be accentuated at vertical joints reducing the contact area at the joints and thus increasing the contact pressures and the tendency to collapse.

5.1   Strata deformation – case histories

Numerous attempts have been made to measure ground deformation immediately above caving longwall faces. These have involved direct measurements such as levelling and banding at the surface and at inter-mediate underground levels where access was feasible, and remote measurements through vertical boreholes drilled from the same locations. Remote measurements have usually been confined to vertical movements relative to the borehole mouth using anchor extensometer systems (viz. Hedley, 1969), although more sophisticated vertical and horizontal mag-netic extensometer and inclinometer systems of the type used in soft ground tunnel instrumentation (namely Attewell and Farmer, 1974) have recently been used with limited success.

Because of the large scale deformations associated with fracture zones above caving longwall faces, and because of the difficulty of access, the length of anchor wires involved and the cost and commitment of resources required, only limited success has been obtained in most measurement programmes. A detailed review of work carried out by and commissioned by the National Coal Board in Britain in the past has revealed only three case histories with sufficient detail to warrant consideration. These com-prise an investigation at Wearmouth Colliery described by Johnson (1963), Leigh (1963) and Potts (1964), an investigation at Markham Colliery described by Batchelor (1972) and by King, Whittaker and Batchelor (1972) and an investigation at Lynemouth Colliery described by Hodkin (1978) and Tubby and Farmer (1981).

The face and seam geometry and instrumentation layouts of each case are described in detail in the appropriate references, and are summarized briefly in Table 5.1. At *Wearmouth Colliery* vertical boreholes for anchor

Table 5.1   Case history details

Colliery Seam Face	Wearmouth, Co. Durham Harvey NW6	Markham, Derbyshire Threequarters L58	Lynemouth, Northumberland Brass Thill K4(K6)
Face geometry			
Depth from surface (m)	457	587	165 (205 m below MHWL)
Azimuth of face centre line	44°	114°	75°
Dip and dip direction	4°/225°	3°/90°	1°/95°
Seam extraction thickness (m)	1.1	0.9	1.6
Face width (m)	140	180	180
(rib–rib)	Advance	Advance	Retreat
Width–depth ratio	0.31	0.31	1.09
Type of instrument	Anchor wire	Anchor wire	Anchor wire
Instrumentation			
Height of instruments above seam roof (m)	(max) 69 (min) 23	71 5	77 7
No. of instrument levels	6	8	7
No. of boreholes	12	4	2
Reference	Leigh, 1963	Batchelor, 1972	Hodkin, 1978

extensometers were drilled into the floor and roof from a roadway constructed 30 years previously in the Hutton seam goaf area, 56 m above the proposed workings which were at a depth from the surface of 457 m in the Harvey seam. The roadway at an angle of 17° to the face centre line and boreholes containing anchors were located over the central 100 m of the face. Additional vertical movement data was obtained from boreholes drilled into the ribside and goaf from the face access roads in the Harvey seam.

At *Markham Colliery*, Derbyshire, Britain, vertical boreholes for anchor extensometers were drilled into the floor and roof from a roadway in the first Piper seam goaf area worked 5 years previously and 45 m above the proposed workings, which were at a depth of 587 m from the surface in the Threequarters seam. The roadway was at an angle of 65° to the face centre line and four boreholes containing anchors were drilled at positions over the centre line, 45 m from the centre line, directly over the ribside and 15 m over the solid on the northern side of the face. An additional single borehole was drilled into the roof over the face centre line.

At *Lynemouth Colliery*, Northumberland, Britain, access was obtained to pillar workings in the Main seam, 77 m above the proposed undersea Brass Thill seam workings. These were 165 m below sea bed and 205 m below MHWL. A total of five boreholes were bored into the intervening strata, which also contained some Yard seam pillar workings, 12 m below the Main seam. The instrumentation was installed over two faces, K4 and K6, at right angles to each other. Three boreholes were drilled over the face centre lines, one 45 m from the centre line and one 10 m over the solid ribside. The most important boreholes were above K4 face and their location is given in Fig. 2.9 with anchor positions in Fig. 1.25.

In the presentation of data from these cases, emphasis tended to be placed on settlement profiles above the faces and normal to the face centre line (Fig. 5.1) in a manner comparable with *Subsidence Engineers' Handbook* profiles (National Coal Board, 1975) and through the face centre line (Fig. 5.2). The effects of geology and the mechanics of deforma-

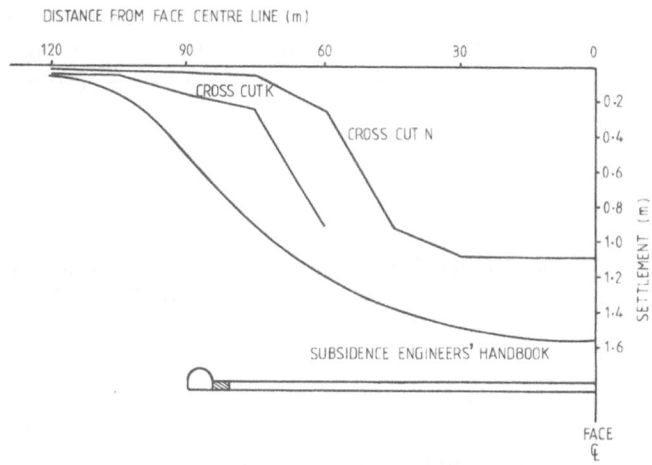

Fig. 5.1 Settlement semi-profiles measured at the Main seam level at Lynemouth Colliery (see Fig. 2.9 for location) above K4 retreating longwall face in the Brass Thill seam (after Hodkin, 1978).

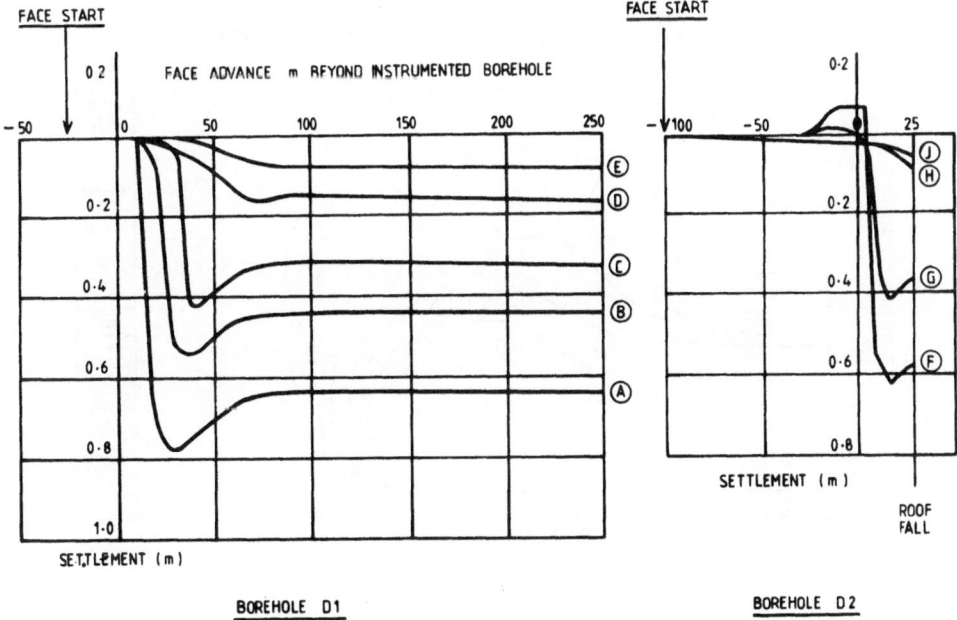

Fig. 5.2 Vertical settlements of borehole anchors relative to the mouth of boreholes D1 and D2 (see Fig. 2.9), bored from the Main seam to just above the centre line of K4 retreating longwall face in the Brass Thill seam at Lynemouth Colliery, Northumberland. The vertical position of the anchors is given together with the inter-seam Coal Measures sequence in Fig. 1.25 (after Hodkin, 1978).

tion were not seriously considered. In Figs 5.3 to 5.5 data from the three case histories recalculated by Farmer and Altounyan (1980) are presented as contours of percentage vertical strain plotted from computed values of vertical extension (positive) or compression (negative) between the vertical anchors. Figures 5.3(a), 5.4(a) and 5.5(a) show respectively strain distribution over the face centre-lines for the Wearmouth, Markham and Lynemouth cases. Figures 5.3(b), 5.4(b) and 5.5(b) show strain distribution normal to the face centre-line at 200 m behind the face for the same cases. The former are based largely on continuous measurements as the face passed a centre-line borehole; the latter on a degree of extrapolation between boreholes and settlement profiles across the face half-line.

Several comments may be made on the strain contours:

(a) There is evidence of yielding of the coal at the face line and ribsides, indicated by extended and relatively low compression strain zones ahead and to the sides of the face.

CONTOURS EXPRESSED AS PERCENTAGE STRAIN
TENSILE STRAINS POSITIVE

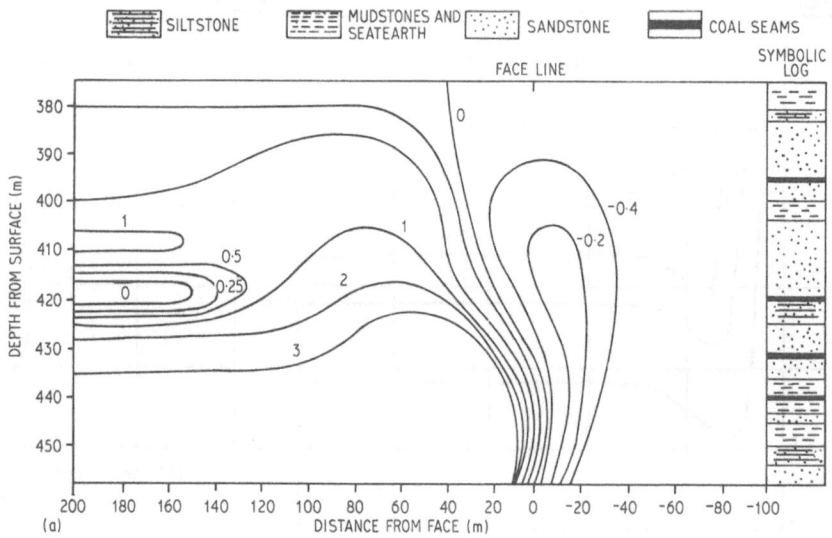

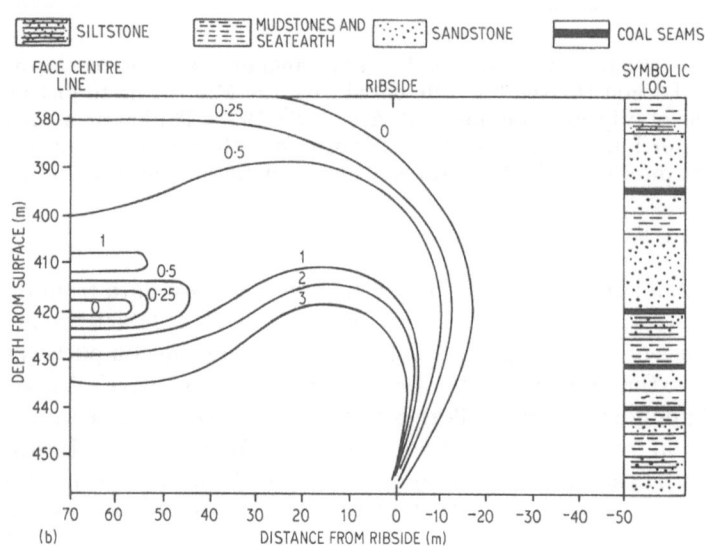

Fig. 5.3 Strain (%) contours computed from relative settlements between bore-hole anchors above NW6 face in the Harvey seam at Wearmouth Colliery, Durham, Britain: (a) parallel to and vertically above the face centre line; (b) normal to the face centre line and 200 m behind the face line (after Farmer and Altounyan, 1980).

158

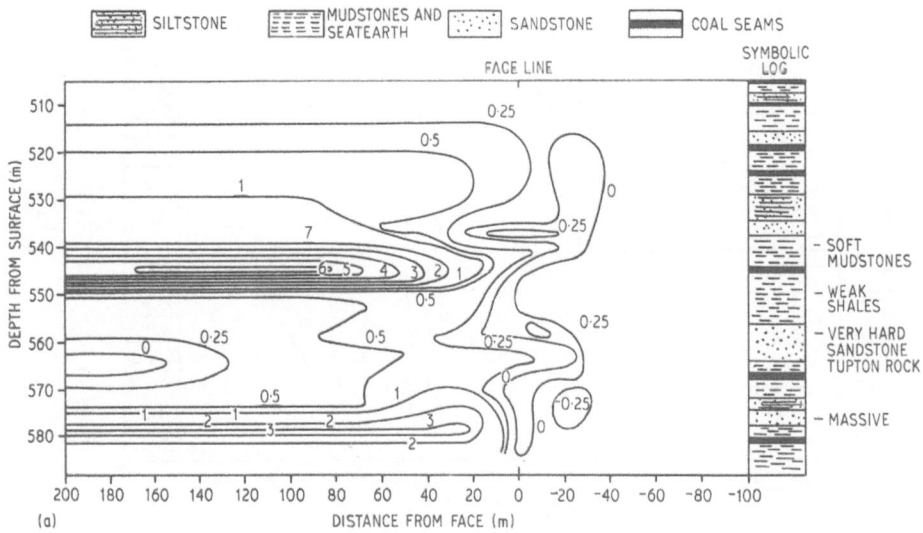

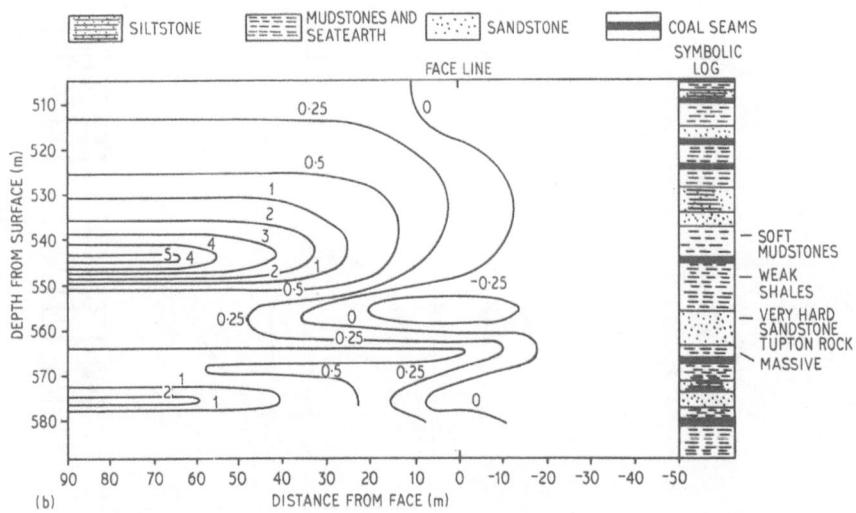

Fig. 5.4 Strain (%) contours computed from relative settlements between bore-hole anchors above L58 face in the Threequarters seam at Markham Colliery, Derbyshire, Britain: (a) parallel to and vertically above the face centre line; (b) normal to the face centre line and 200 m behind the face line (after Farmer and Altounyan, 1980).

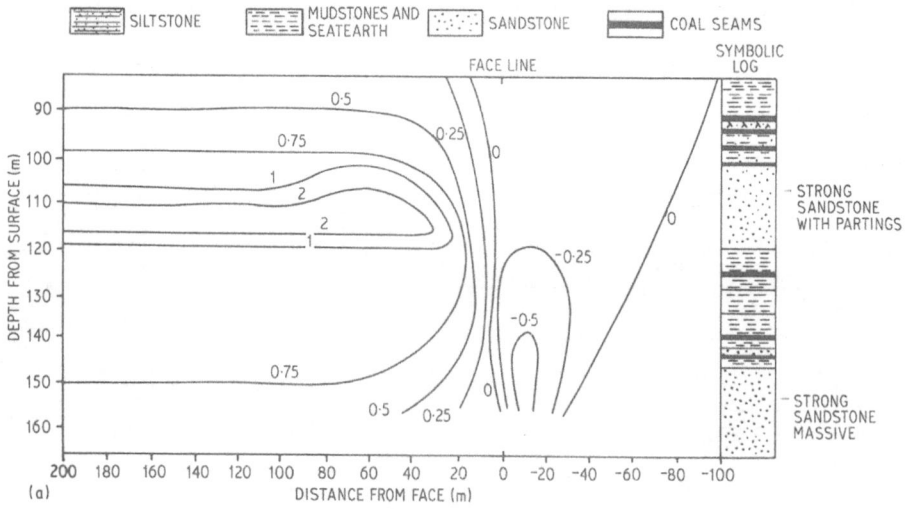

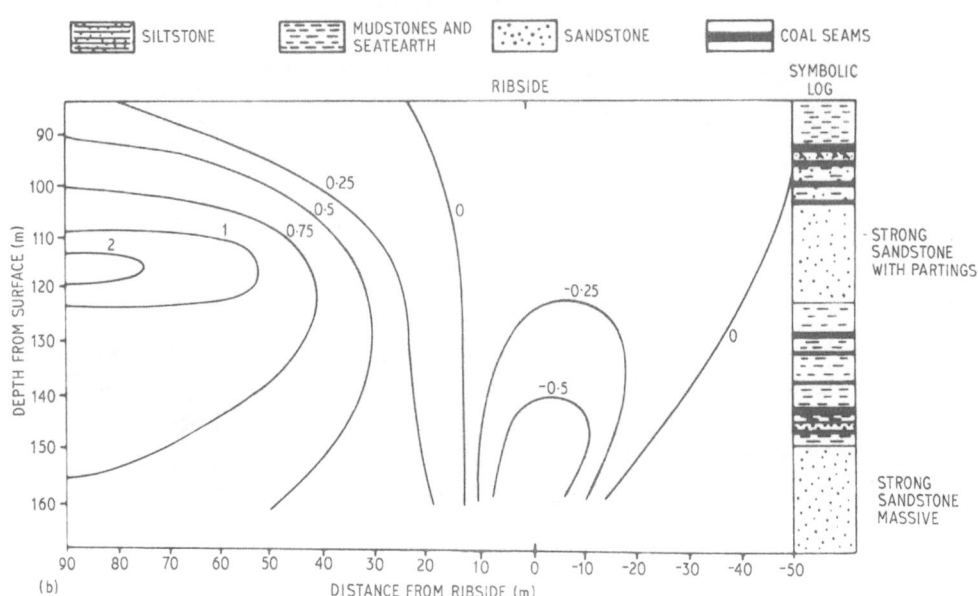

Fig. 5.5 Strain contours computed from relative settlements between borehole anchors above K4 face in the Brass Thill seam at Lynemouth Colliery: (a) parallel to and vertically above the face centre line; (b) normal to the face centre line and 200 m behind the face line (after Farmer and Altounyan, 1980).

(b) In the Markham (Fig. 5.4) and Lynemouth (Fig. 5.5) cases, tensile strains are confined to an approximately rectangular prism above the caving area. Compression strains exist in a vertical direction over the ribsides and extend slightly over the faces. At Wearmouth (Fig. 5.3) tensile strains extend some distance over the face and ribside abutments.

(c) Tensile strains tend to be more closely related to stratigraphy than to proximity to the face. At Lynemouth (Fig. 5.5) peak tensile strains (2 %) are associated with a layer of sandstone containing strong bedding plane shears. At Markham (Fig. 5.4) and Wearmouth (Fig. 5.5) very low strains are associated with sandstone beds, and much higher strains with weaker rock. The highest strains at Markham coincide with the relatively recent workings in the Piper seam which have been subjected to considerable dilation.

(d) With the exception of Markham (Fig. 5.4) where compression strain peaks of 0.25 % are associated with sandstone layers ahead of the face, compression strains appear less dependent on stratigraphy.

(e) Tensile strains – effectively dilation in a vertical direction – are not rapidly dissipated with passage of the face. Each of the contoured sections (Figs 5.3(b), 5.4(b) and 5.5(b)) normal to the face centre-line is taken at 200 m from the face. At Lynemouth and Wearmouth, where observations are available respectively 2 and 3 years later, there is no evidence of significant recompression. This corroborates the evidence of Wilson and Ashwin (1972), among others, who have observed long term pillar stresses in the ribside pillars (see Fig. 2.6) of longwall faces. It is also supported by the evidence of the *Subsidence Engineers' Handbook* (National Coal Board, 1975), which is considered further in the next section, that significant dilation occurs during settlement. The estimated 10 % volume loss between extraction level and surface for a 'critical width' subsidence trough would, however, represent a relatively conservative ground loss estimate based on the cases considered here.

(f) A particular feature in the Markham case (Fig. 5.4) is the low tensile strain in the sandstone beds 30–35 m above the seam and large dilations of 7 % in the mudstones immediately above. The large dilations are associated with mudstones around the previously worked out face in the first Piper seam. The low strains in the sandstone indicate that the sandstone beds have collapsed in single units, possibly exacerbating the dilation in the mudstones above. This particular sandstone known as the Tupton Rock is an extension of the Parkgate sandstone. Phillips (1944) describes how spectacular releases of strain energy associated with collapse of this sandstone were responsible for a rockburst at Barnborough Main in 1943.

Under the geostatic pressures existing at the depths considered in the case histories (200–600 m) vertical geostatic stresses in the range 5–15 MNnm^{-2} might be expected. Abutment stresses would be expected of several times this magnitude. The measured compression strains over the abutments are not compatible with high abutment stresses and indicate either a lack of resolution or a degree of yielding close to the seam level On the evidence of Figs 1.1–1.3, a linear compression strain of 0.8–1.5 would normally be needed at low confining pressures to cause shear fracture and significant dilation in Coal Measures rocks. This does not occur at Markham or Lynemouth. A much lower *tensile* strain would, however, be needed to cause fracture and dilation. Although very little experimental data exist, the ratio between tensile and compressive strengths of 1:10 would indicate that fracture in tension would occur at a linear strain of about 0.1 %. The indications are, therefore, that a *fractured* zone of considerable extent exists over the caving longwall workings in each of the cases considered.

5.2   Mechanics of strata deformation

The mechanics of deformation immediately above a caving excavation are well known. Removal of coal and face support is followed by collapse of the immediate roof layers subjected to a combination of bending and tensile stresses. The height to which they break is known as the *caving height* and is estimated by Kenny (1969) and Wilson (1975) to be between two and four times the extraction height. The broken rocks from the immediate roof partially support the superincumbent roof layers. However, the degree of support is insufficient to restore the initial geostatic stresses, with the result that deviatoric stresses in the rock mass are significantly increased – particularly by increases in the minor (tensile) principal stress above the caved zone. The resultant fracture and associated dilation may cause relatively small strains, but the rock strength is effectively reduced from a peak to a residual level with loss in load-bearing capacity and resultant stress redistribution. In addition sheared and strongly laminated rocks – of which there are many in the Coal Measures cyclothem – will be affected by bed separation, also increasing dilation in the massive rock.

The tensile strain contours in Figs 5.3–5.5, particularly those above 0.25 %, may be said to represent fractured rock which has deformed 'non-elastically'. The compression strain contours represent unfractured rock which, for modelling purposes, can be assumed to have deformed 'elastically'.

In all cases it is interesting to note that the zone of fracture defined by the extent of dilation extends for at least half the face width above the seam level, and probably to a higher level than this. The strength of rocks in the fractured zone will be reduced to some residual level with equivalent changes in deformability, and stresses will tend to be permanently redistributed to the abutments. In modelling ground deformation above a caving longwall face, therefore, a rectangular prism of fractured ground should be introduced above the caved area in order to simulate more closely the mechanics of ground deformation. Where stronger sandstone layers exist above the face, the potential for stored strain energy is illustrated by the low strain contours in these layers.

Although finite element and other models for simulation of fracture zones have been suggested (see for instance Daemen and Hood (1981)) it is difficult to model accurately the process of strata deformation above a longwall face. Under such circumstances it may be better to use empirical approaches such as that in the *Subsidence Engineers' Handbook* (National Coal Board, 1975) or to develop concepts of caveability (see for instance Fig. 2.17).

A useful method of illustrating the type and extent of deformation may be a two-dimensional physical model. A model study by Harwood (1980) and Singh (1981) in the author's laboratory proved valuable in this respect. It comprised a hundredth scale model of the strata above K4 face centre line at Lynemouth Colliery (Figs 1.25 and 5.5). It was made from plaster, mica-dust, sand, sawdust and coal dust in various combinations. Sixteen geomechanical and geometrical variables were simulated using the Buckingham Pi theorem. These included moduli of elasticity and rigidity, Poisson's ratio, uniaxial compressive, tensile and shear strengths, unit weight, discontinuity cohesion and frictional angle, face length and depth, seam thickness and rate of advance. Particular emphasis was placed on bedding plane and joint discontinuity surfaces in order to simulate block release during caving.

After extraction of the Brass Thill seam (in the model) below simulated pillar workings in the Main seam, which produced disappointingly low deformations with little relation to those actually measured it was decided to extract the Main seam. It should be said that legal requirements and local (National Coal Board, 1968) instructions precluded longwall working as close as this (85 m) to the sea bed. The photographs in Fig. 5.6 and the sketches in Fig. 5.7 show how the fracture zone above the face extended in the shape of a truncated wedge to the surface of the model.

It is particularly interesting to notice the way in which caving in these near horizontal strata is discontinuity controlled, with successive upward advances of the caved zone being halted at bedding plane partings until the collapse of the overlying bedding plane allows further upward extension. These extensions can be detected in practice as minor tensile seismic

Fig. 5.6 Development of fracture above a 0.2 m wide plane stress model of an extending face extraction in Coal Measures strata (after Harwood, 1980). The model is 1/100th scale and the positions represent advances (from top to bottom) of 24 m, 34 m and 46 m in a 2 m thick seam.

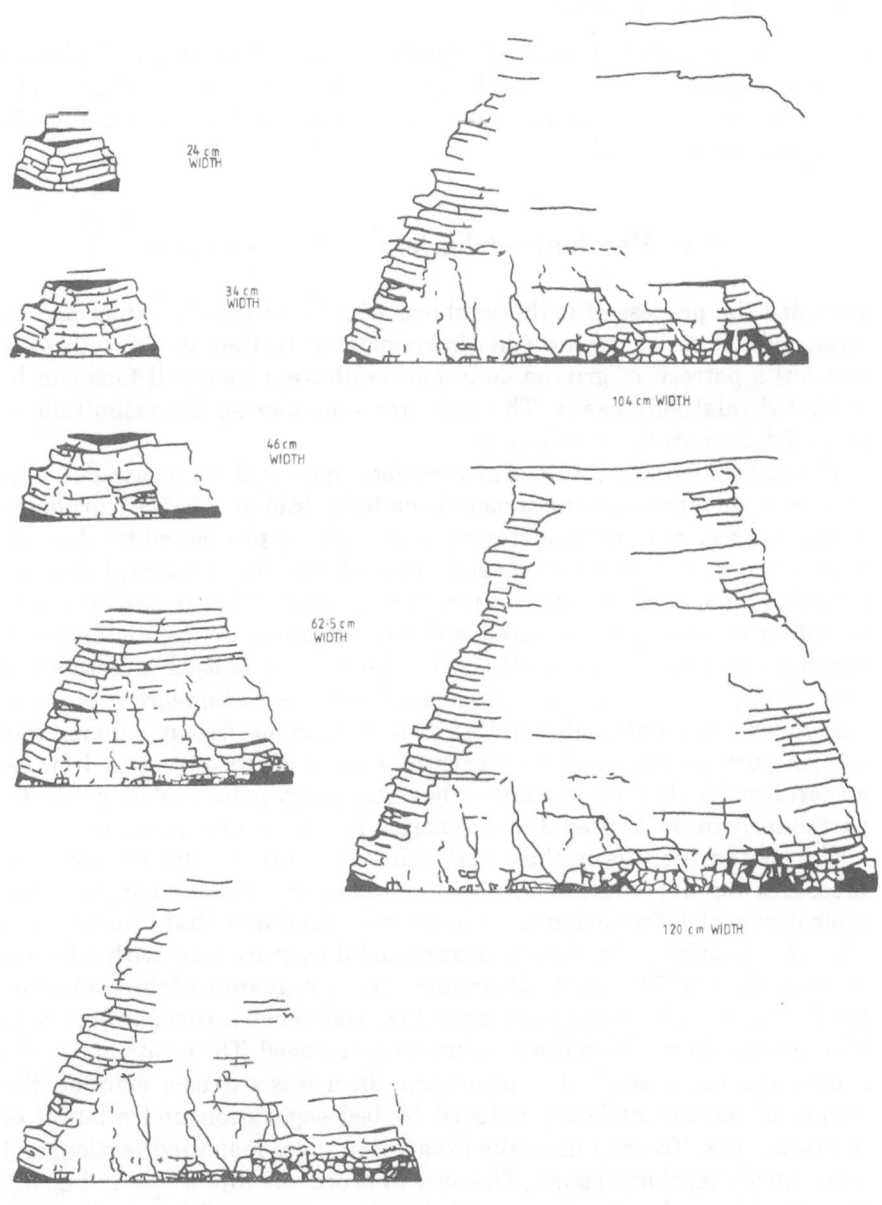

Fig. 5.7 Fracture zones above a 1/100th scale model as in Fig. 5.6 developing with extension of the face (after Harwood, 1980). The fracture zone reaches the surface – or in this case the sea bed – at a depth of 85 m when the scaled extension is 120 m. Note the fractured strata close to the face, the flexured and fractured beds at greater heights.

events (see Chapter 8). It is also interesting to note that the upward extent of the caved zone is rather less than the width of the excavated area. The ratio between collapse height and width is consistently of the order of 0.75 in the case of the model.

5.3   Vertical settlement and subsidence

Provided the process of collapse above a longwall face is unimpeded by strata which resist caving (this is considered further in the following section) a pattern of ground deformation above a longwall face can be predicted relatively easily. The next step – predicting the magnitude of ground deformation – is less easy.

The surface manifestation of ground deformation above a caved tabular underground mine working is usually called subsidence. Since *subsidence* is the only aspect of the underground coal mining process which directly affects the rest of the world, and since claims for structural damage resulting from surface deformation can be substantial, it has been subjected to considerable research and investigation. A useful historical summary is given by Shadbolt (1977). The result has been a plethora of case history data – the most important being the *Subsidence Engineers' Handbook* (National Coal Board, 1975) which summarizes in a unique way British surface subsidence observations over 30 to 40 years. It is so comprehensive that occasionally it has a tendency to be used irrationally, particularly in the United States, where it is not usually applicable.

The reason for this is that, with minor exceptions, the British Coal Measures do not contain strong limestone or sandstone layers. The typical type of deformation is therefore very similar to that illustrated in Fig. 5.7, comprising, in section, a trapezoidal fracture zone with a height equal to about 0.75 times the face width. After excavation of the coal seam, the fractured rocks in this zone move downwards in a *vertical* direction, to fill the space from which the coal has been removed. The strata above the excavation move vertically downwards in a less extreme manner, the degree of settlement being reduced by bed separation and dilation of fractured rock. To the side of the excavation there is limited vertical and associated lateral movement. The overall process is illustrated in Fig. 5.8. This illustrates also the changing shape of the profile. Within the fracture zone the profile shape is usually termed *super-critical* and the subsidence or settlement profile has a flat bottom and a curved edge about the ribside. Above the fracture zone the profile is referred to as *sub-critical*, and the profile peaks over the centre line (in the case of flat deposits) of the excavation. The depth–face width ratio at which the change occurs is quoted in the *Subsidence Engineers' Handbook* as 0.71 (or a face width–

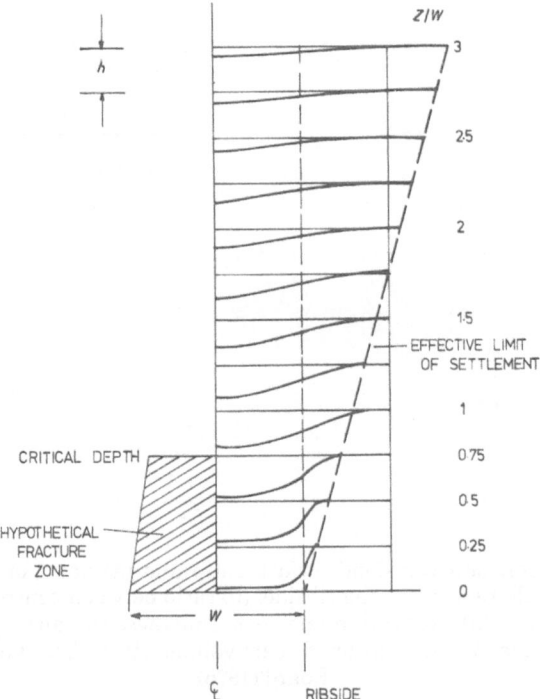

Fig. 5.8 Settlement profiles at different depths (z) above a section normal to the centre line behind a face of width (w) where the settlement is assumed fully developed (based on National Coal Board, 1975).

depth ratio of 1.4) and this is very close to the height of the postulated fracture zone in Figs 5.3–5.7.

It is possible to define the geometry of the subsidence profiles quite accurately for British conditions from the *Subsidence Engineers' Handbook*. In Fig. 5.9, for a horizontal deposit, the position of the point of inflexion relative to the ribside, and the effect of depth on centre-line subsidence and seam thickness are summarized. The most important observations are that the point of inflexion usually lies over, or close to the ribside, and that subsidence reaches an insignificant level at a face width–depth ratio of about 4 to 5.

With this information it should be possible to predict settlement relatively easily if an equation can be fitted to the subsidence curve. There is, fortunately, good precedence for this type of curve fitting (see for instance Martos, 1958; Marr, 1959; and Schmidt, 1969), and the most suitable curves are an inverted Gaussian distribution or a harmonic curve, illustrated in Fig. 5.10.

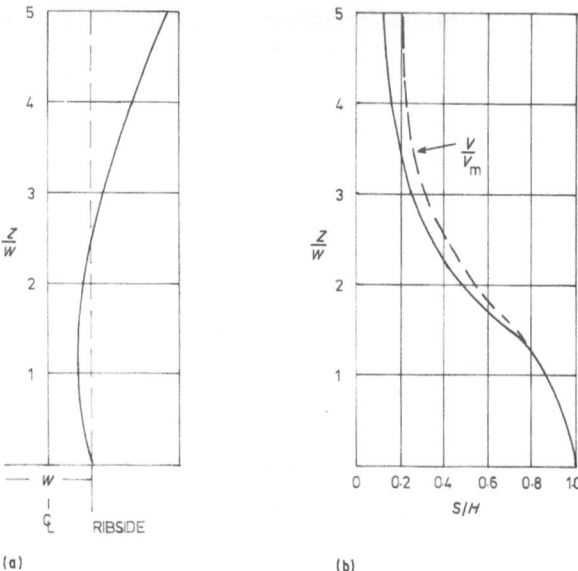

Fig. 5.9 Relation between depth–width ratio and (a) point of inflexion of the subsidence profile relative to the ribside, (b) ratio between centre line settlement (*s*) above a horizontal excavation and seam thickness (*h*), and the ratio between settlement volume (*V*) and extracted seam volume (*V*$_{m}$) – based on National Coal Board (1975).

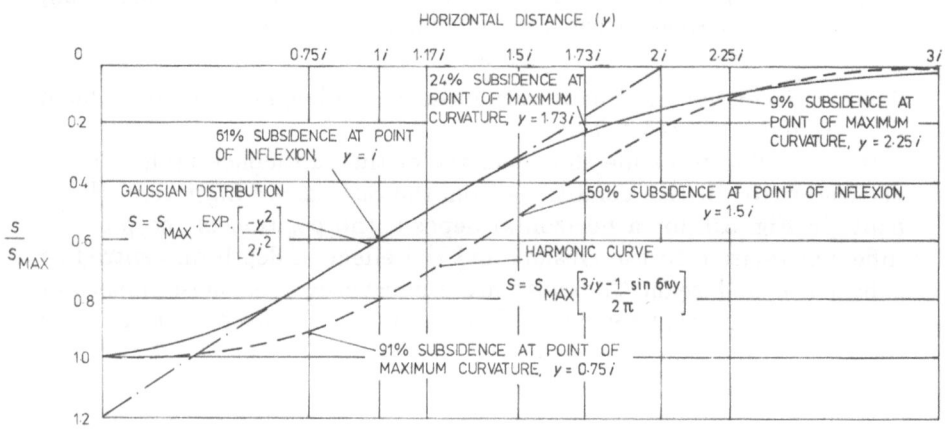

Fig. 5.10 An inverted Gaussian distribution and harmonic curve, characterized as a subsidence semi-profile (after Farmer and Attewell, 1975).

Although a wide range of profile functions can represent mining subsidence (see Brauner, 1973), generally speaking the Gaussian distribution is a good fit for the 'sub-critical' case; the harmonic curve for the 'super-critical' case; the inflexion point being modified according to Fig. 5.9(a). Then for the Gaussian case, the settlement at any point will be given by:

$$S = S_{max} \exp\left(\frac{-y^2}{2i^2}\right) \tag{5.1}$$

where S is the settlement at distance y from the centre line, S_{max} is settlement over the centre line and i is the horizontal distance from the centre line to the point of inflexion.

For British conditions i and S_{max} can be obtained from Fig. 5.9.

Tilt, curvature and strain can be calculated relatively easily by differentiation, whence:

$$\text{Tilt } T = \frac{dS}{dy} = \left(\frac{-y}{i^2}\right)S \tag{5.2}$$

$$\text{Curvature } C = \frac{d^2S}{dy^2} = \left(\frac{1}{i^2}\right)\left(\frac{y^2}{i^2}-1\right)S \tag{5.3}$$

$$\text{Horizontal strain } \varepsilon_h = \left(1-\frac{y^2}{i^2}\right)\frac{S}{2i} \tag{5.4}$$

$$\text{Horizontal displacement } \delta = \int_{-\infty}^{y} \varepsilon_h \, dy = \left(\frac{y}{2i}\right)S \tag{5.5}$$

The extension of this analysis using Litwiniszyn's (1964) theory of 'stochastic media' by Sweet and Bogdanoff (1965), Attewell (1977) and Glossop (1982) has some interesting aspects.

It can be seen from Figs 5.10 and 5.11 and from the equations that maximum strain will be associated with maximum curvature (compression where there is concave curvature, tension where there is convex) and that these points on the curve will be potentially the most dangerous places for surface or underground structures. It can also be seen that underground layouts in one or more seams may be designed to reduce strain at the surface, or underground, to a minimum by locating longwall faces so that overlying tensile and compression zones are superimposed. Conversely incorrect layout design leading to superposition of tensile strain on tensile strain can maximize underground or surface strains.

Since settlement profiles are normally obtained at the surface, there is sometimes a reluctance to extend them to levels below the ground surface. It can be seen from the arguments used to justify the *profile function* approach that the profile shape formed by a type of movement into an

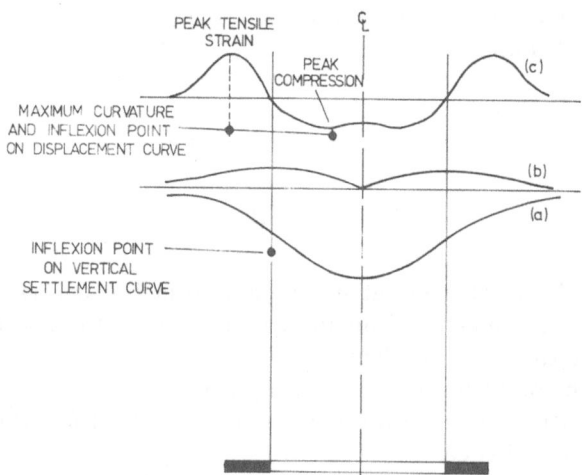

Fig. 5.11 Relative positions of (a) vertical settlement profile, (b) horizontal displacement profile – to the same relative scale as the vertical settlement, (c) strain curve, for a typical sub-critical section above a longwall face.

underground opening will not be affected by depth or confinement. Thus the deformation profile at the surface – subsidence – will only differ from that at depth, in so far as the deformation process differs. This has importance in the consideration of underground interaction effects in Chapter 8, and in groundwater flow problems, in the later part of this chapter.

Although subsidence or settlement may be modelled by many mathematical and numerical methods, the accuracy of resolution has yet to match empirical methods. One further empirical method which is worth mentioning is the *influence function*. This has the notional advantages that it is easily programmed for a computer and considers the whole surface (or underground) plane rather than a section through it. Originally developed in Germany in the 1920s, it is based on the argument the settlement at a point on the surface can be obtained by estimating the influence of all the underground extractions within a cone of influence. This cone has an apical angle equal to an angle of subsidence, and terminates at the lower extraction level.

Marr (1975) has shown that the method can be applied to most mine layouts and proposes that for British conditions an apical angle of 70° should delineate the zone. Projected at seam level (Fig. 5.12) this gives a circular influence area for a flat seam of radius 0.7 times the depth, below a point O. Then any working in this area will cause settlement. Greatest settlement will result from workings vertically below O, say zone *A*, and least settlements from workings closest to the edge of the influence area, say zone *G*. If the influence area is divided into (say) seven zones, each equal to a width of 0.1 times the working depth, then a zone factor (a to g) representing the proportion of settlement at O can be attributed empirically to each zone. Then the total settlement at O will be:

$$Uo = Aa + Bb + Cc + Dd + Ee + Ff + Gg$$

where the higher case represents the proportional area of each zone extracted, and the lower case is the zone influence factor. In addition each of the proportional areas can be raised by a factor *n* to allow for the effect of ribsides, pillars, abutments and other factors. It is a complex (possibly too complex) method and United States examples in Peng and Harthill (1981) should be studied if application is considered.

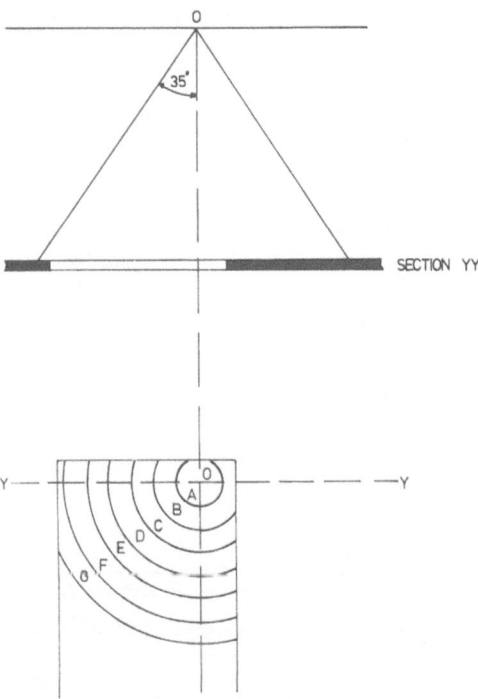

Fig. 5.12 Zone influence area method of estimating settlement at O from an advancing longwall face. In the simplest approach the area of each zone above the face (inside the dense lines) multiplied by an empirically determined zone factor will be summed to give subsidence at O. The process can be repeated for any surface points or underground geometry.

5.4   Concepts of caveability

All conventional approaches to subsidence prediction – with the possible exception of some modifications of the influence zone approach – assume even and full caving of strata above the face. In the British Coal Measures where the sequence contains a preponderance of mudstones and shales, and sandstone beds tend to have limited thickness, this is acceptable. In the United States and in various other coalfields in the world, strong thick sandstone and limestone beds immediately overlying the seam are not uncommon. In such conditions – particularly where the workings are shallow, longwall caving may not be feasible and a series of shortwall faces separated by chain pillars may be a more viable alternative. An example of strain contours resulting from caving over shortwall faces at Hendrix mine in Kentucky, described by Howell, Wright and Dearinger

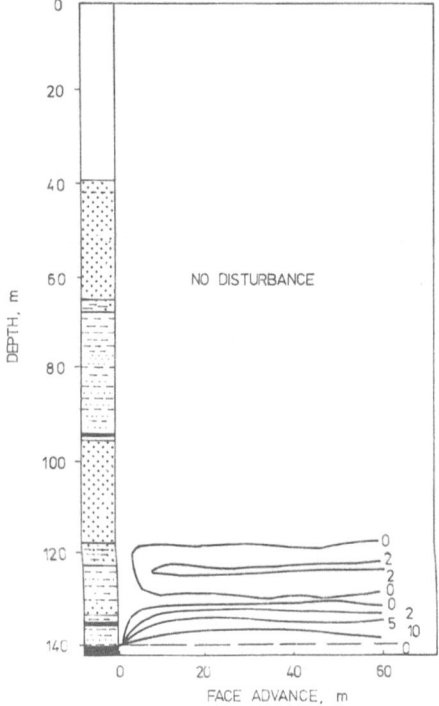

Fig. 5.13   Strain ($\%$) contours parallel to and vertically above the centre line of a 52 m wide shortwall face at Hendrix Mine, Kentucky, United States (after Howell *et al.*, 1977). The contours are computed from relative vertical settlements between fixed anchors in a borehole drilled from the surface.

(1977), is plotted in Fig. 5.13. These were determined from relative anchor settlements in a vertical borehole drilled from the surface above the face centre line. The shortwall faces were 52 m wide at centres of 82 m, the remaining width comprising 9 m pillars and 6 m development entries. It can be seen that caving effectively stops 20 m above the face, immediately below a thick sandstone bed.

While this may not be surprising in the case of a shortwall face, a sandstone bed can have a similar inhibiting effect on caving in the roof of a longwall face. Figure 5.14 illustrates strain contours at two points above a 183 m wide retreating longwall face at Somerset No. 60 mine, Pennsylvania (Barla and Boshkov, 1978). The Coal Measures sequence contained strong continuous sandstones immediately above the seam and similar limestones further in the succession.

Some of the settlement data are difficult to analyse, possibly due to slippage of anchors and rotation of near-seam, blocks. Strains have only been contoured where they are quite clear and it can be seen that both the sandstone and limestone appear to inhibit caving. It is interesting to note that in this case history piezometric levels were also measured in the near surface sandstone aquifer and drops in piezometric level occurred at two points over the face centre line 10 m behind and 3 m in advance of the face.

Since relatively minor perturbations are required in discontinuous rock to increase permeabilities by a significant amount, this need not indicate a fracture zone extending to the surface. It does, however, confirm the general rectangular shape of a zone of increased permeability.

Styler (1980) in a quite creative interpretation of these data concluded that the caving height was between 15 and 23 m above the seam, and since this is a major factor in modern longwall mining it is worth considering further. There are two concepts which are important – *caving height* and *caveability*.

The caving height – the height to which the immediate roof strata cave after removal of face supports – has considerable importance in support design. If the nether, or immediate, roof rocks cave easily and increase in bulk during caving by a significant amount, then the broken rock will partly support the overlying strata, allowing it to subside gently and preventing potentially heavy roof weights on the supports. If the nether roof rocks do not cave easily and slab without bulking, then unsupported cantilevered rocks in the waste will tend to transfer heavy stresses to the front and side abutments, leading to potential overloading and collapse of face and roadway supports. This is considered at length in Chapter 7.

The height at which broken rocks will give support is related to the *bulking factor* B (equal to the ratio of fractured strata volume to intact strata volume) and seam thickness h, so that the height of fragmented strata (h') before temporary stability is reached is given by $h' = h/(B-1)$,

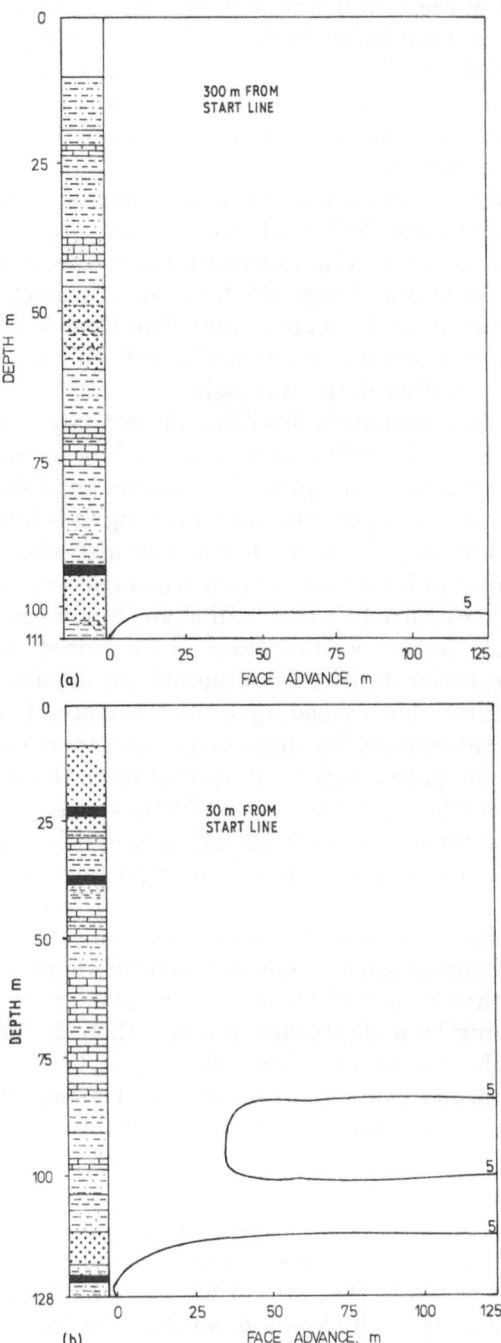

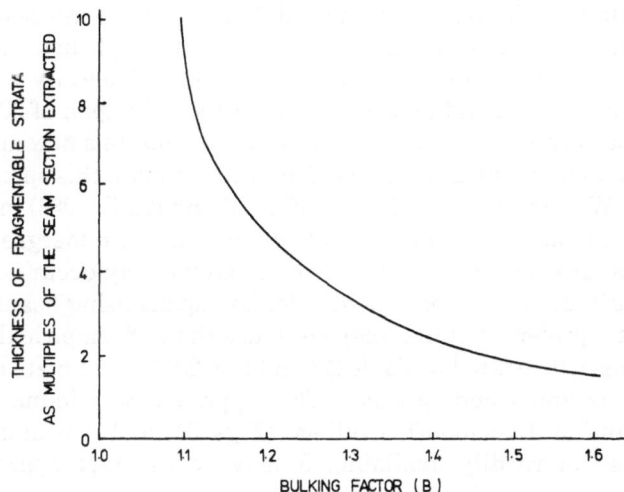

Fig. 5.15   Relation between the ratio of caving height *h'* to seam thickness *h* and the bulking factor *B*.

a relation illustrated in Fig. 5.15. This demonstrates that, for fairly rapid support of the nether roof of the type assumed in Wilson's (1975) calculations for support design, a bulking factor of the order of 1.5 is needed. This is only likely to occur in very friable rocks such as shales or marls (see Jacobi, 1976; Stunya and Meyer, 1960). In layered mudstones and sandstones it can approach much lower levels (see Gorrie and Scott, 1970), and in practice these are the sort of levels which probably occur. Thus a volumetric strain increase of 10% will be equivalent to a bulking factor of 1.1 and for a 2 m seam, a minimum caving height of 20 m or more. It is difficult to argue a lower potential caving height in the cases in Figs 5.13, 5.14. In Figs 5.3–5.5 it is probable that a much higher immediate bulking factor has had the effect of mitigating face loading, and helping to reduce face convergence. In Figs 5.13, 5.14 there would appear to be a gap which has led to some cantilevering, albeit intentional in the case of the shortwall face in Fig. 5.13.

Where cantilevering does occur the concept of *caveability* becomes important. If the strata will not cave at all then shortwall faces become a

Fig. 5.14   Strain (%) contours parallel to and vertically above the centre line of a 183 m wide retreating longwall face in the Pittsburgh seam at Somerset No. 60 Mine, United States (after Barla and Boshkov, 1978). The contours are computed from relative vertical settlements between fixed anchors in two boreholes drilled from the surface.

preferred alternative. Where caving is difficult, correct choice of longwall face length in order to mitigate the effects of weighting, particularly after the start of the face, becomes important. Generally speaking, a cantilever of 30–50 m and an equivalent first weight span of 50–80 m will cause serious problems in modern longwall systems. In a new mine, where this is of critical importance, it is difficult to estimate this span. A recent example at Wistow mine, Yorkshire (Fig. 4.3 and Knill, 1983) in Britain is a classic example of this and also of the problem of underground water inflows. The prediction of spans at which caving may occur is, however, extremely difficult. It is possible to calculate spans using beam and plate theory, but a preferred option may be to use the very empirical NGI rock quality index (illustrated in Table 2.6 and Fig. 2.17) to estimate a combination of spans and stand-up times. This approach was found to predict caving spans at Lynemouth Colliery (Figs 5.2 and 5.3) and although evidence is not readily available, it may be the best approach. It is certainly recommended in the current state of the art.

5.5   Water flow induced by strata deformation

An aspect of strata deformation which is not often considered in detail is the increased flow – often of immense proportions – which can occur through the zone of fractured strata above a longwall face. The increased porosity and permeability of the fractured rock means that if fracture zones of the type illustrated in Fig. 5.7 intersect an aquifer or surface water body, either directly or indirectly, water can penetrate the workings.

The relevant aspects of groundwater flow are considered elsewhere (see Attewell and Farmer, 1976) but the basic dynamics of water flow can be expressed through D'Arcy's law:

$$Q = AKi$$

where Q is the discharge rate ($m^3 s^{-1}$), K is the permeability coefficient ($m s^{-1}$), A is the discharge area (m^2) and i is the hydraulic gradient – a dimensionless ratio of driving head over flow path length.

The effect of strata caving is to increase the discharge area by fracturing the aquifer; to increase the hydraulic gradient by creating vertical and horizontal flow paths and particularly to increase permeability. The permeability coefficient is related in intact rocks to the square of the effective pore diameter and the porosity and ranges from about $10^{-6} m s^{-1}$ for Bunter Sandstone and Magnesian Limestone to $10^{-10} m s^{-1}$ for shales and seatearths. The permeability coefficient of fissured and fractured rocks is related to the *cube* of the dimension of

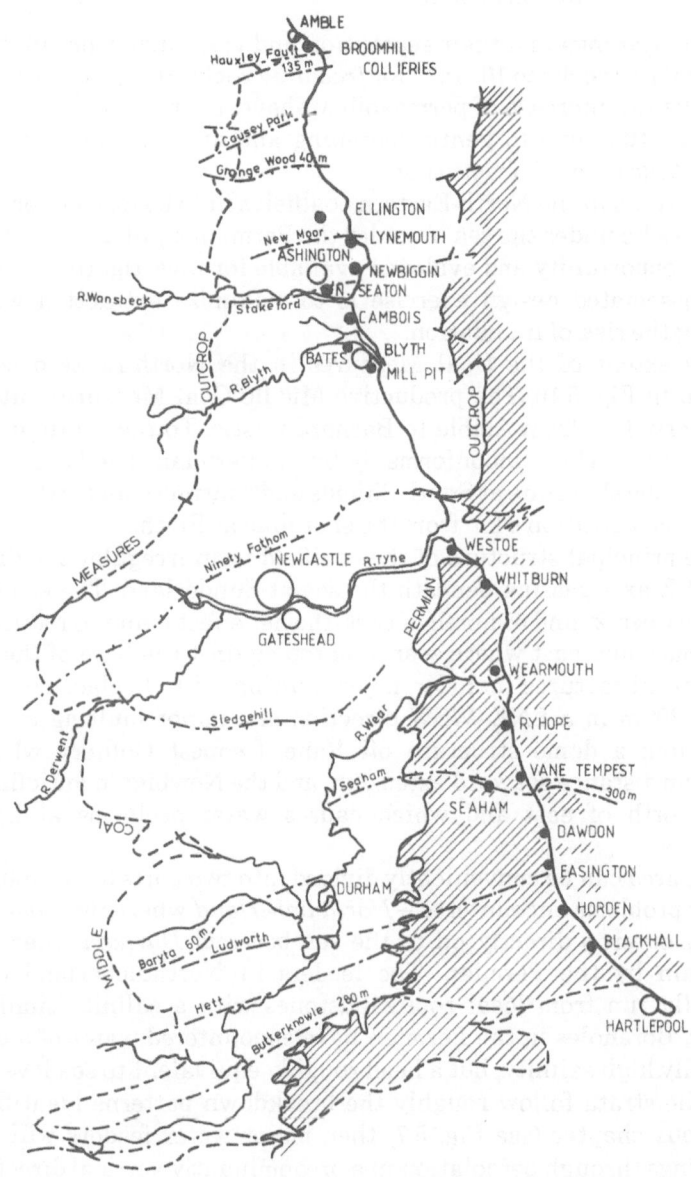

Fig. 5.16 The North-Eastern coalfield of England showing the position of existing and previously worked mines with undersea workings and workings under the (shaded) Permian aquifer.

fissure openings and their separation and can range from $10^{-2}\,\mathrm{m\,s^{-1}}$ for *open jointed* rocks to $10\,\mathrm{m\,s^{-1}}$ for *fractured* rocks. It is possible therefore to visualize an increase of permeability above a mine working of 3–4 orders of magnitude due to gentle loosening and anything up to 10 orders of magnitude through fracturing.

Where, as in the North-Eastern coalfields in Britain, a major part of the reserves lie under the sea or under the Permian aquifer (Fig. 5.16) there is ample opportunity and evidence available for investigating water inflow. The associated design exercise is to maximize extraction while minimizing the risk of inundation.

The extent of the Coal Measures in the North-Eastern coalfield is shown in Fig. 5.16. The productive Middle Coal Measures outcrop on a southerly line from Amble to Barnard Castle. To the south and the east these are overlain unconformably by the Permian. The Permian outcrop crosses the shoreline at South Shields and continues under the North Sea, reaching a position 6 km from the shoreline at Blyth.

The principal structure of the coalfield is an irregular syncline with a NW–SE axis passing beneath the sea at Sunderland. The eastern limits dip between 2° and 5° north of east, the net effect being to reduce cover to the south and east where seams incrop against the base of the Permian. Structural features include: major faulting with displacements ranging up to 300 m in a ENE–WSW direction and minor faulting in a NW–SE direction; a dome structure off Vane Tempest Colliery which limits medium distance seaward extension, and the Newbiggin anticline with an axis north of east and which causes water problems at Lynemouth Colliery.

The area can be conveniently divided into two parts by the nature of the water problems encountered – *Northumberland* where the Coal Measures strata outcrop directly on to the sea bed and *Durham* where they are overlain by the Permian. The feeders in Northumberland characteristically run from local roof sandstones with a salinity similar to sea water. Boreholes in the Permian have encountered water of a characteristically high salinity, but a head roughly equivalent to sea level.

If the strata follow roughly the breakdown patterns identified in the previous chapter (see Fig. 5.7) then major water feeders will enter the workings through percolation in a predominantly vertical direction down breaks above the face abutment as this advances into undisturbed ground. Essentially these breaks will create a series of new discharge areas and flow paths, which will discharge water at a peak rate in the face area. Unless there is significant recharge (as in the case of, say, the sea) the rate of flow will reduce with time and distance into the waste area. Major inflows will be associated with the first break which penetrates the aquifer after the start of the face. It is possible to identify the major factors which will affect water inflow as:

(a) Cover to principal aquifers
(b) Face and excavation geometry
(c) Roof support
(d) Workings in nearby seams
(e) Fracture and deformation of rocks in the aquifer – essentially a combination of (a)–(d)

In order to analyse data in detail, for both Northumberland and Durham, all actual and historical cases of water inflow at all the mines under the aquifer and the sea bed were collected through a current and archival search (Garritty, 1981, 1982).

In each case the inflow rate was noted and the following data were collected:

(a) Cover to Permian or cover to sea bed
(b) Face width/height of Carboniferous cover ratio or face width/sea bed cover ratio
(c) The maximum tensile strain at the base of the Permian or at the sea bed, calculated from the *Subsidence Engineers' Handbook* (National Coal Board, 1975)
(d) The aggregate tensile strain at the base of the Permian or at the sea bed including estimates of strains from adjacent workings in the same seam
(e) The total tensile strain at the base of the Permian or at the sea bed including estimates of strain from all workings in the same and adjacent seams
(f) The ratio between the head of water at the base of the Permian and the height of Carboniferous cover above the face
(g) Strata lithology, and in particular the percentage of sandstones in the succession beneath the base of the Permian or the sea bed
(h) The percentage of sandstone in the vicinity of working below limiting heights – chosen arbitrarily as 20 and 45 m – and also above these heights but below the water source was examined particularly as a contributory factor in water flow.

In the latter two cases emphasis was placed on sandstone since it was considered that sandstone strata represented both potentially strong roof rocks which would be less likely to collapse and water-bearing aquifer rocks which would transmit water.

It can be seen that there are potentially many combinations of strata which would lead to water ingress, ranging from weak rocks above the seam collapsing to intercept an aquifer; to aquifer rocks close to the seam fracturing and causing increased flow through increased permeability.

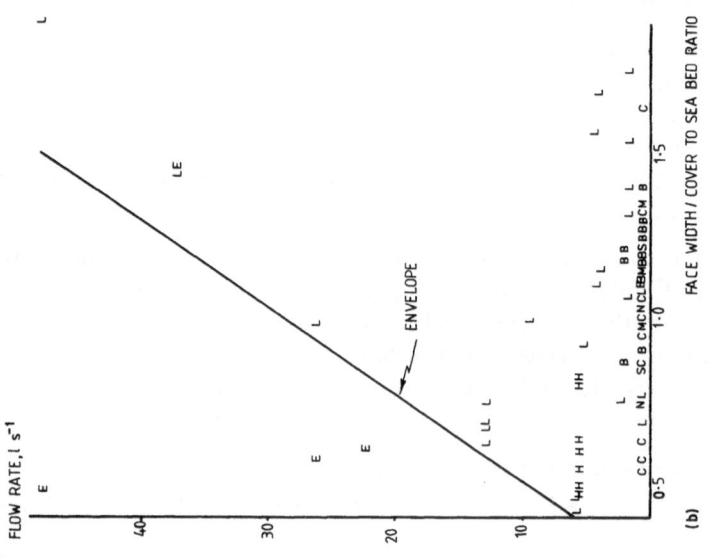

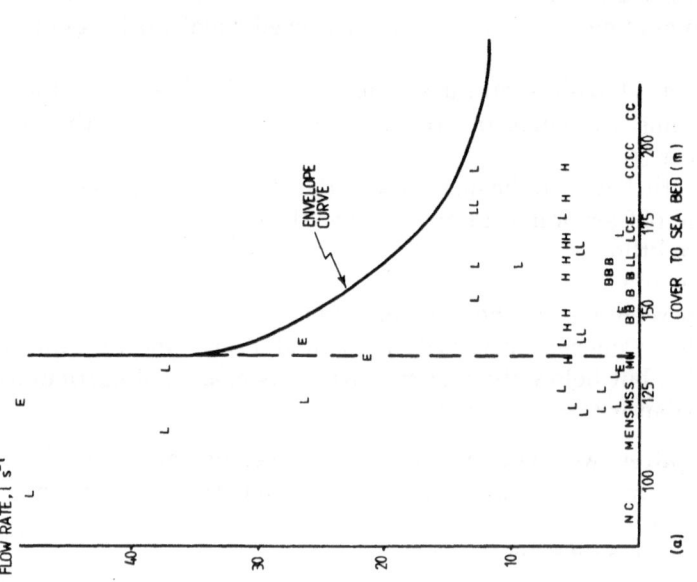

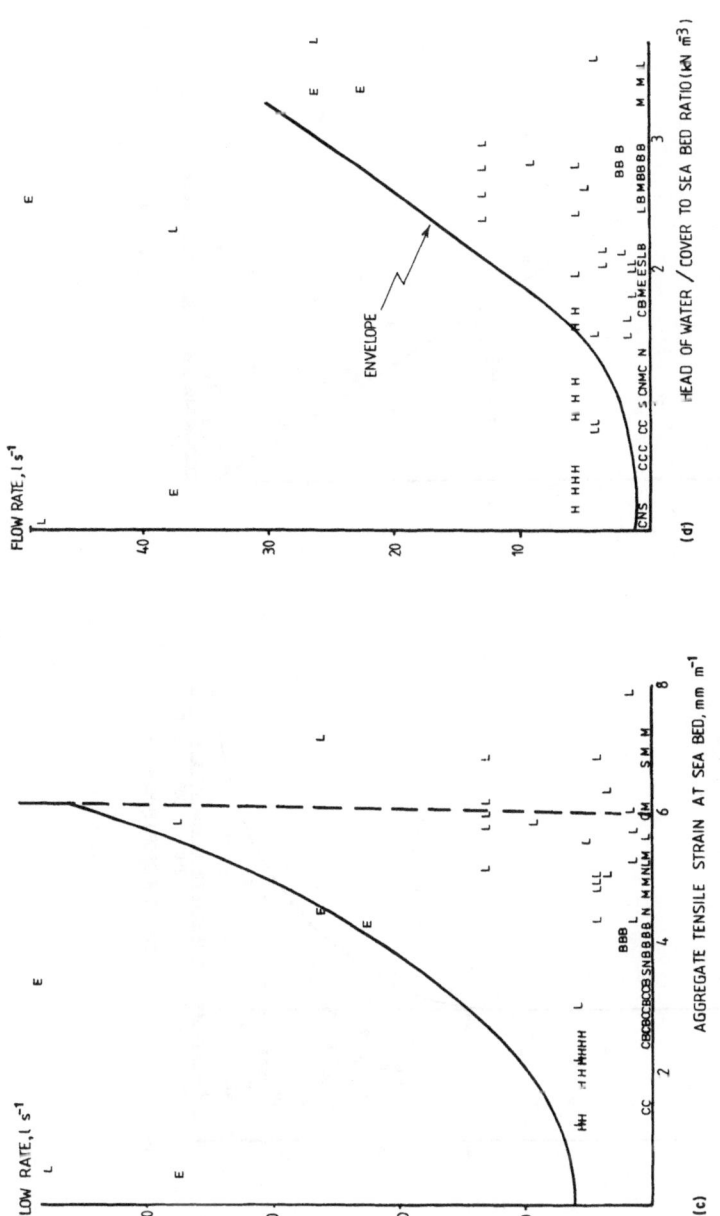

Fig. 5.17 Flow rate into longwall workings in the undersea Northumberland coalfield related to:

(a) Height of cover above the face to the sea bed
(b) Ratio between face width and height of cover to sea bed
(c) Aggregate tensile strain at the sea bed
(d) Ratio between head of water at the sea bed and the height of cover to the sea bed.

Collieries (Fig. 5.16) are identified as: M Mill Pit; B Bates; C Cambois; S North Seaton; N Newbiggin; L Lynemouth; E Ellington; H Broom Hill (after Garrity, 1981).

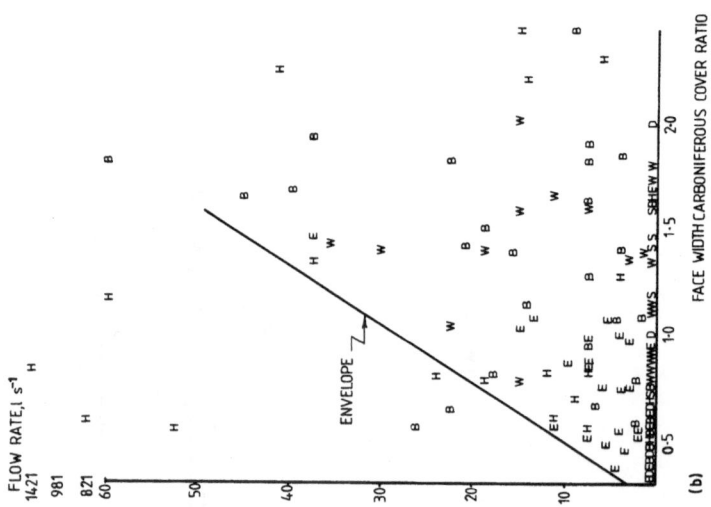

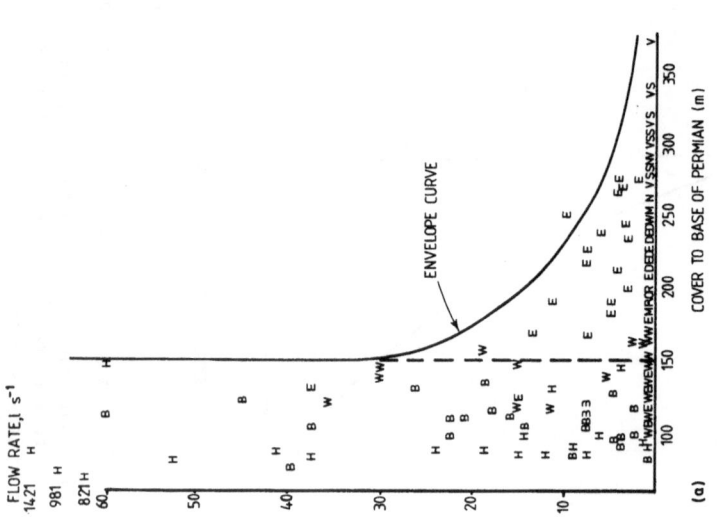

182

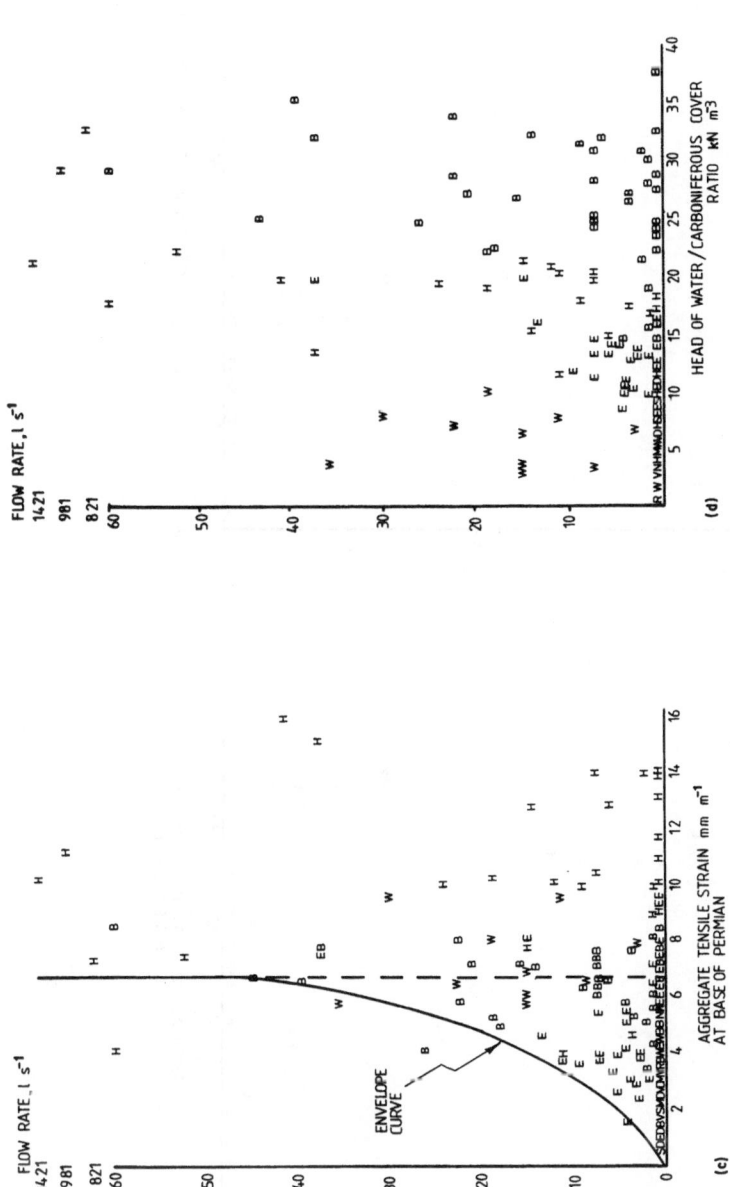

Fig. 5.18   Flow rate into longwall workings under the Permian aquifer in the Durham coalfield related to:

(a) Thickness of Carboniferous strata between the face and the base of the Permian
(b) Ratio between the face width and the thickness of Carboniferous strata above the face
(c) Aggregate tensile strain at the base of the Permian
(d) Ratio between the head of water at the base of the Permian and the thickness of Carboniferous strata above the face

Collieries (Fig. 5.16) are identified as B Blackhall; H Horden; E Easington; D Dawdon; S Seaham; V Vane Tempest; R Ryhope; N Wearmouth; M Whitburn; W Westoe (after Garrity, 1981).

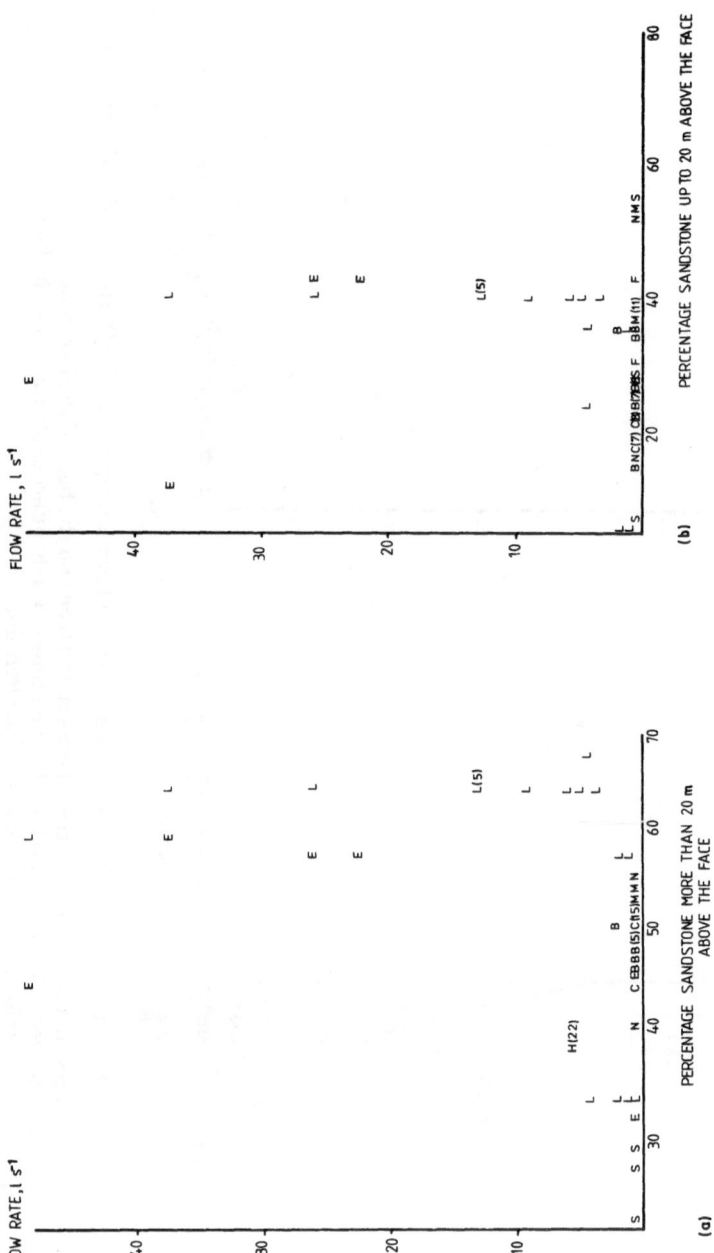

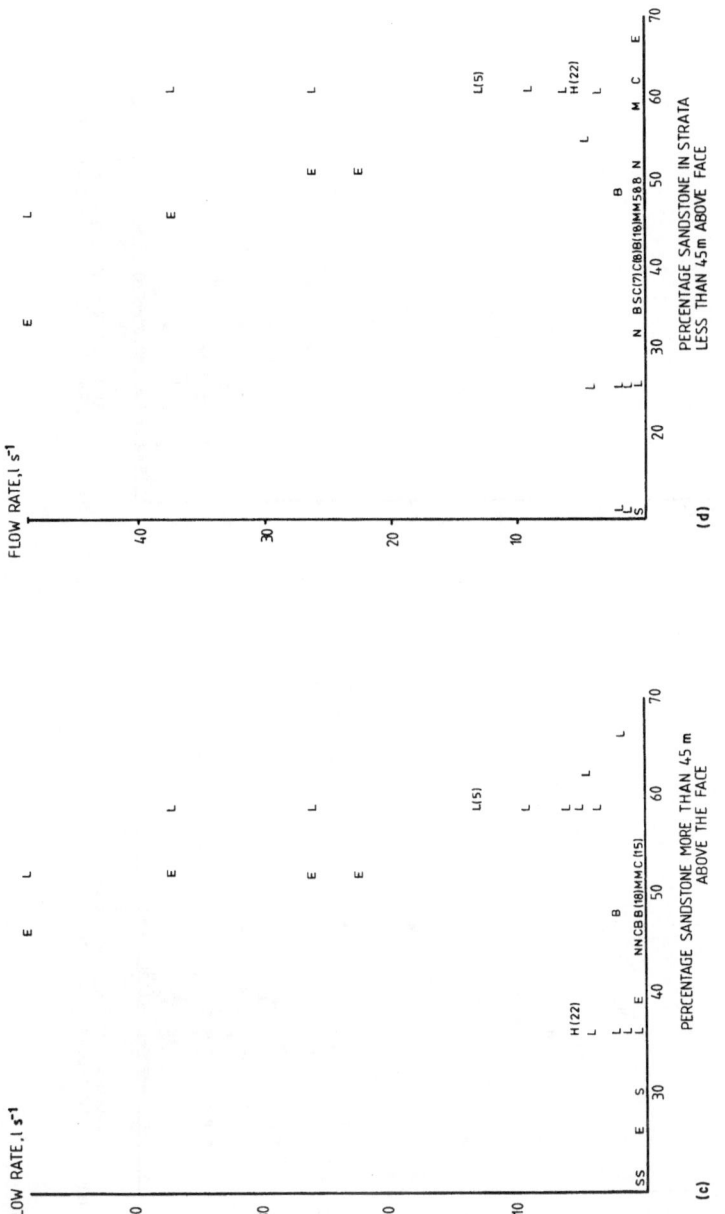

Fig. 5.19 The effect on flow rate of the percentage of sandstone strata in the succession above and below a line 20 m and 45 m above faces in the Northumberland coalfield beneath the sea bed. Collieries are identified in Fig. 5.16 and beneath Fig. 5.17 (after Garrity, 1981).

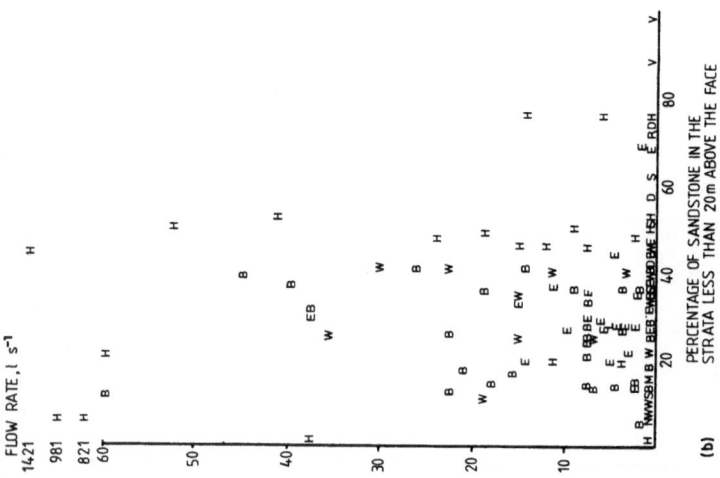

(a) PERCENTAGE OF SANDSTONE IN STRATA MORE THAN 20 m ABOVE THE FACE

FLOW RATE, l s⁻¹

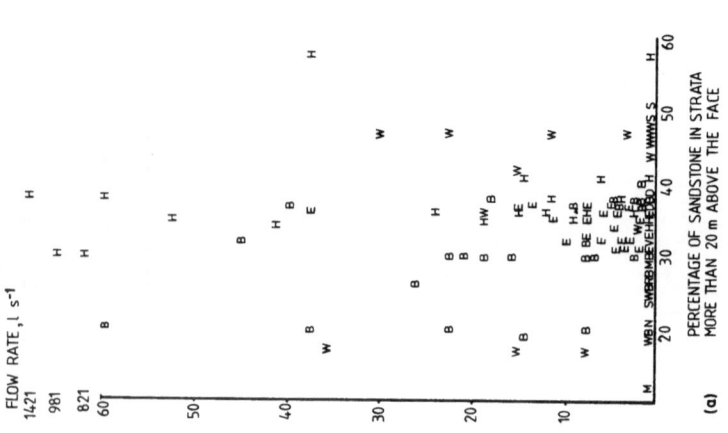

(b) PERCENTAGE OF SANDSTONE IN THE STRATA LESS THAN 20m ABOVE THE FACE

FLOW RATE, l s⁻¹

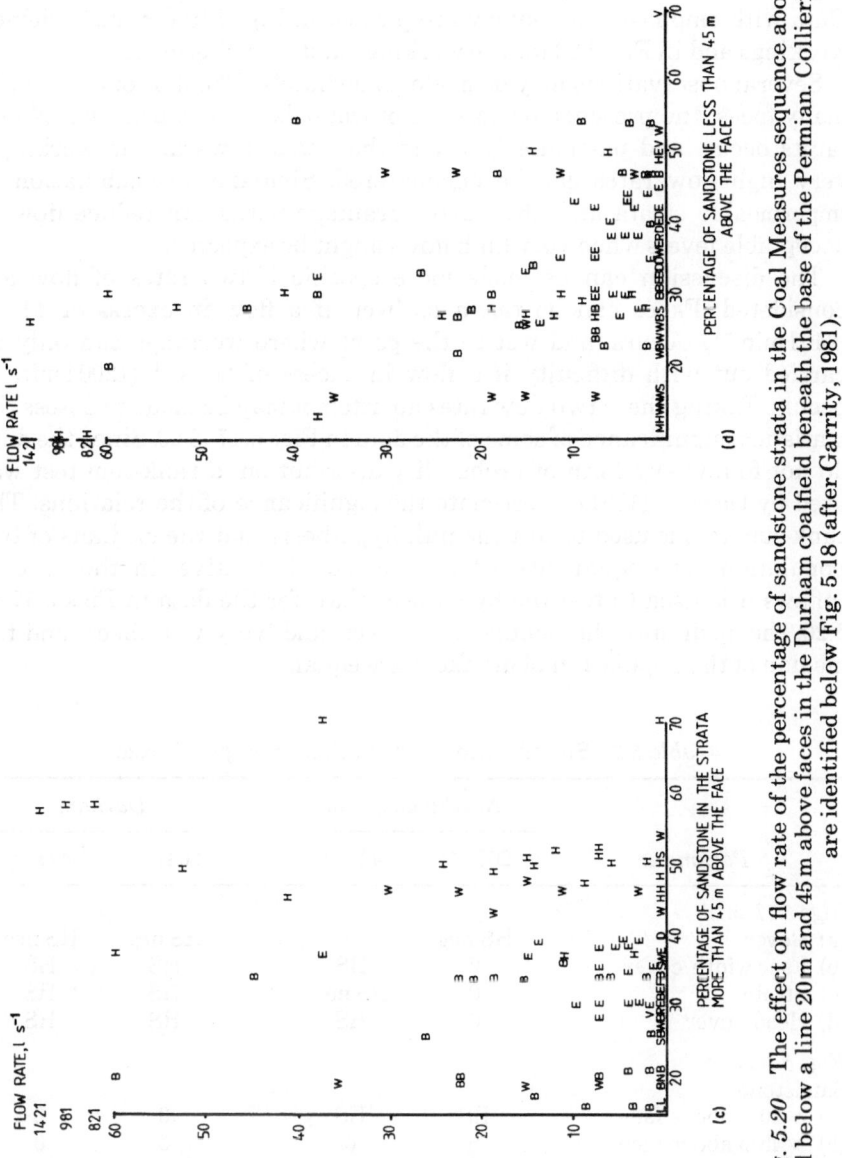

Fig. 5.20 The effect on flow rate of the percentage of sandstone strata in the Coal Measures sequence above and below a line 20 m and 45 m above faces in the Durham coalfield beneath the base of the Permian. Collieries are identified below Fig. 5.18 (after Garrity, 1981).

Data with particular emphasis on geometry are plotted in Fig. 5.17 for the undersea workings in the Northumberland part of the coalfield, and in Fig. 5.18 for the workings under the Permian in the Durham coalfield. Data with emphasis on geology are plotted in Fig. 5.19 for the undersea workings and in Fig. 5.20 for the workings under the Permian.

Several observations may be made immediately. The first one is that in many faces, the presence or absence of water is quite capricious. Where faults occur, and particularly where they hade towards the workings, very high flow rates can be encountered. Similarly a combination of impermeable strata and alternative drainage paths can reduce flow to acceptable levels when very high flows might be expected.

The discussion can be made more specific if two rates of flow are considered. Faces can be taken as 'wet' if a flow in excess of $4 \, \mathrm{l \, s}^{-1}$ $(240 \, \mathrm{l \, min}^{-1})$ occurs, and wet to the point where workings can only be carried out with difficulty if a flow in excess of $20 \, \mathrm{l \, s}^{-1}$ $(1200 \, \mathrm{l \, min}^{-1})$ occurs. Taking these two flow rates an attempt may be made to assess the statistical significance of some of the data in Figs 5.17–5.20. Since the data do not follow any form of probability distribution, a rank-sum test was used by Garrity (1981) to estimate the significance of the relations. The rank-sum test is used to test the null hypothesis that the medians of two populations are equal against a specified alternative. In the case of inflows it is used to test the hypothesis that, for the data in Figs 5.17 to 5.20, the median of the population of 'wet' and 'very wet' faces, and the median of the population of dry faces are equal.

Table 5.2 Significance of correlations in Figs 5.17–5.20

	Northumberland		Durham	
Parameter	$20 \, \mathrm{l \, s}^{-1}$	$4 \, \mathrm{l \, s}^{-1}$	$20 \, \mathrm{l \, s}^{-1}$	$4 \, \mathrm{l \, s}^{-1}$
Figs 5.17, 5.19				
(a) Cover	HS neg	0	HS neg	HS neg
(b) Face width/cover	0	HS	HS	HS
(c) Strain	0	HS neg	HS	HS
(d) Head/cover	0	HS	HS	HS
Figs 5.18, 5.19				
Sandstone %				
(a) > 20 m above face	HS	HS	0	0
(b) > 45 m above face	S	0	0	0
(c) < 20 m above face	0	HS	0	HS neg
(d) < 45 m above face	0	HS	S neg	0

HS – highly significant, S – significant, 0 – not significant, neg – negative relation (after Garritty, 1980, 1981).

In Table 5.2 the significance of correlation for each of the relations in Figs 5.17 to 5.20 is listed. A highly significant correlation is taken as correlating at the 1% level, and a significant correlation at the 5% level. Garritty (1981) lists these in league tables. A similar approach is used in Table 5.2.

In the case of the *Northumberland* coalfield incidence of inflows in excess of $20\,l\,s^{-1}$ appears to be principally a function of the lithology of the overburden (Fig. 5.19), although the cover to the sea bed (Fig. 5.17(a)) would also seem to be of some significance. Parameters related to the magnitude of the disruption induced in the strata between seam and sea bed have comparatively little correlation with incidence of major inflows. Operations appear to be at special risk of encountering inflows in excess of $20\,l\,s^{-1}$ where:

(a) Cover to the sea bed is less than $140\,\mathrm{m}$ (Fig. 5.17(a)). This probably reflects the comparative ease with which overlying Coal Measures aquifers are replenished by the sea, and the increased porosity and permeability of aquifers subject to erosional stress relief.
(b) Where major Coal Measures aquifers exist within $45\,\mathrm{m}$ or so of workings (Fig. 5.19(c)).
(c) Where the competence of the immediate seam roof is comparatively low; that is where the first 20 to $30\,\mathrm{m}$ of overlying strata contains less than about 35% sandstone (Fig. 5.19(a)).

Incidence of inflows of $4\,l\,s^{-1}$ or more seems to be a function of the presence of sandstone aquifers in proximity to workings. Notable influxes of water are associated principally with aquifers within $45\,\mathrm{m}$ or so of the seam being worked, which for many operations in multi-seam workings in the coalfield implies the major aquiferous sandstone sequence which occurs above the horizon of the Ashington seam. This does not mean, however, that longwall operations cannot trap the waters of aquifers at greater height. It is interesting that feeders have been encountered at depths below the sea bed of up to $200\,\mathrm{m}$, which indicates that sandstones of the offshore Northumberland coalfield are water bearing at depth. Total extraction systems will generally liberate feeders from nearby Coal Measures aquifers, a circumstance which should be borne in mind during layout planning and design. Feeders encountered at depth and in the absence of faults and dykes, will, however, usually amount to no more than a nuisance.

In the case of the *Durham* coalfield, parameters related to the magnitude of the deformations induced in the strata between the longwall workings and the Permian have the most significant correlation with incidence of water inflows. However, not all these parameters can be considered useful indicators of probable risk. Of secondary importance appear lithological considerations although there is general correlation between lower proportions of overburden sandstone, especially in the immediate seam roof, and incidence of water inflows.

Consideration of this, and individual case histories, suggests that incidence of notable water inflows is essentially a function of dynamic strata fracturing at the time of major roof breaks which induces open, and effectively continuous discontinuities within the overburden. These are capable of transmitting water into workings from overlying aquiferous formations, notably the Permian, and aquiferous sandstones in proximity to it. The need for adequate strata control is, therefore, apparent. It is possible that a calculated tensile strain of 6 or $7\,\mathrm{mm\,m^{-1}}$ induced at the base of Permian reflects a 'critical' degree of fracturing for 'average' conditions. An over-simplified approach to a study of the incidence of water in mine workings based on simple, or Darcian permeability theory cannot, therefore, be expected to adequately describe the phenomenon.

Workings in Durham appear to run a comparatively high risk of inundation where:

(a) The cover to the base of Permian is less than 150 m (Fig. 5.18(a)).
(b) Aggregate or individual tensile strains calculated at the base of Permian exceed $6\,\mathrm{mm\,m^{-1}}$ (Fig. 5.18(c)).
(c) The overburden comprises less than 35 % of sandstone.
(d) Adequate roof control is not maintained and major strata breaks are allowed for form.

The incidence of water has often been associated with discontinuities in the Durham coalfield. In particular considerable water problems have been found in the vicinity of igneous dykes.

Water in quantity has not generally been associated with faults, especially depositional faults, having throws ranging up to 1 m, although when water is encountered the make tends to be greatest in the vicinity of any fault. A number of inundations have, however, occurred while mining operations have been taking place in the vicinity of faults with throws of more than 1 m. Risks are apparent where workings approach or encounter:

(a) Faulting approximately parallel to the face line which aggravates strata control difficulties
(b) Faults which hade over workings
(c) Faults of increasing throw
(d) Faults whose nature or siting suggests a significance beyond that indicated by consideration of throw, along for example, faults against which smaller faults terminate.
(e) Large faults where strata disruption and dislocation are greatest.

Feeders in Durham have also been encountered by operations in the vicinity of the following geological features:

(a) Swalley banks
(b) Areas of steeply dipping strata, or dip workings in faulted blocks
(c) Areas of rapidly increasing dip
(d) Monoclinal structures
(e) Lenticular sandstones
(f) The incrop to the Permian of major sandstone aquifers
(g) Massive aquiferous roof sandstones

6

ACCESS ROADWAYS AND ROADSIDE PACKS

Access roadways or entries are the tunnels giving access to longwall faces. In *retreating* longwall faces these are driven before mining, and the useful part of the roadway is always in front of the retreating face. Because of this the roadway is generally in a zone of relatively low stress. In the case of *advancing* longwall faces, roadways are formed from the extraction zone at the sides of the longwall excavation, and the useful part of the roadway is generally in a zone of high stress.

Both cases can be illustrated by examining the notional stress distribution in the abutments of an advancing longwall face illustrated from models and measurements in Fig. 6.1. Although the numbers may not be totally reliable, they do give an indication of the positions of peak and lower stresses, confirming the general observations of Figs 2.6, and 5.3 to 5.5. Several observations may be made which can have importance in face and roadway design.

(a) There is a peak stress at the centre of the face abutment reducing to a minimum but not insignificant value at the face end. This applies to both retreating and advancing faces.

(b) In advancing faces there is a peak stress level at the side abutment reached about one-half the face width back from the face line. This is in close proximity to the roadways.

(c) Face and side abutment peaks are 10–20 m (or one-twentieth to one-tenth of face width) away from the face or ribside line. This can sometimes be attributed to crushing of the coal abutments but as demonstrated by the model of Fig. 6.1(a) with rigid abutments, more may properly be attributed to disturbance of the overlying strata in the vicinity of the abutment.

(d) There is a low stress zone with a high degree of longevity over the worked area, resulting from permanently dilated rock. This is sometimes utilized when *gob-scours* (Fig. 6.2) are constructed to replace gate-roadways in weak highly stressed country rocks.

In the United States the picture is complicated by the use of entry pillars – which will be discussed later. Initially though, the classic case of

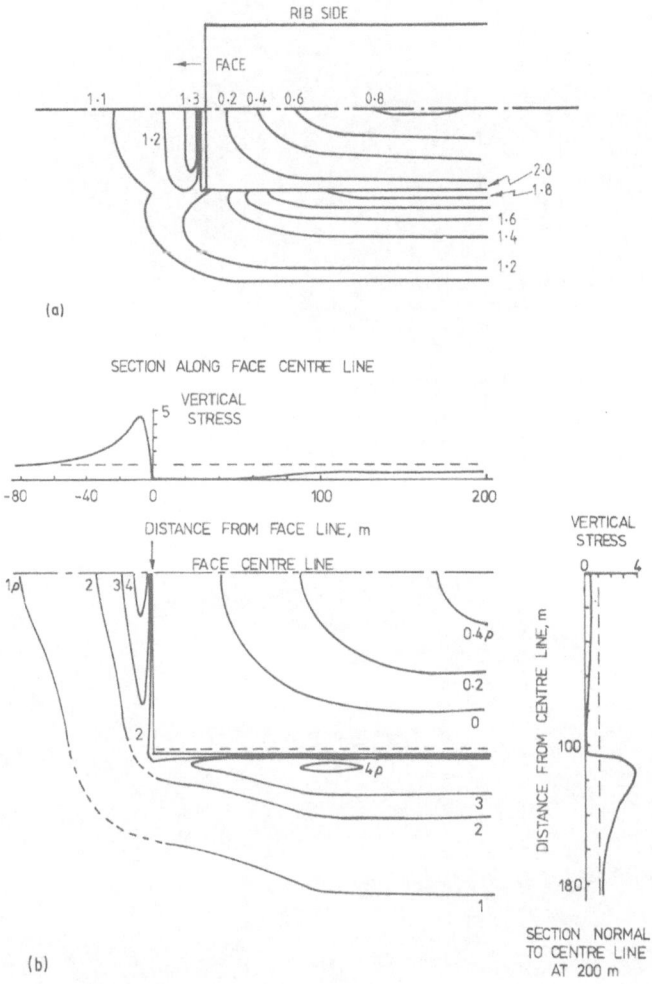

Fig. 6.1 (a) Vertical stress distribution around a sandbox model of a longwall face measured by Harris (1974). The stresses are expressed as a ratio of vertical geostatic stress. The face modelled was based on a 200 m prototype at a depth of 220 m. (b) Vertical stress distribution around a longwall face proposed by Whittaker (1979) and based on field observations in the East Midlands coalfield. Stresses are expressed as a ratio of vertical geostatic stress. The face is 210 m wide at a depth of 470 m.

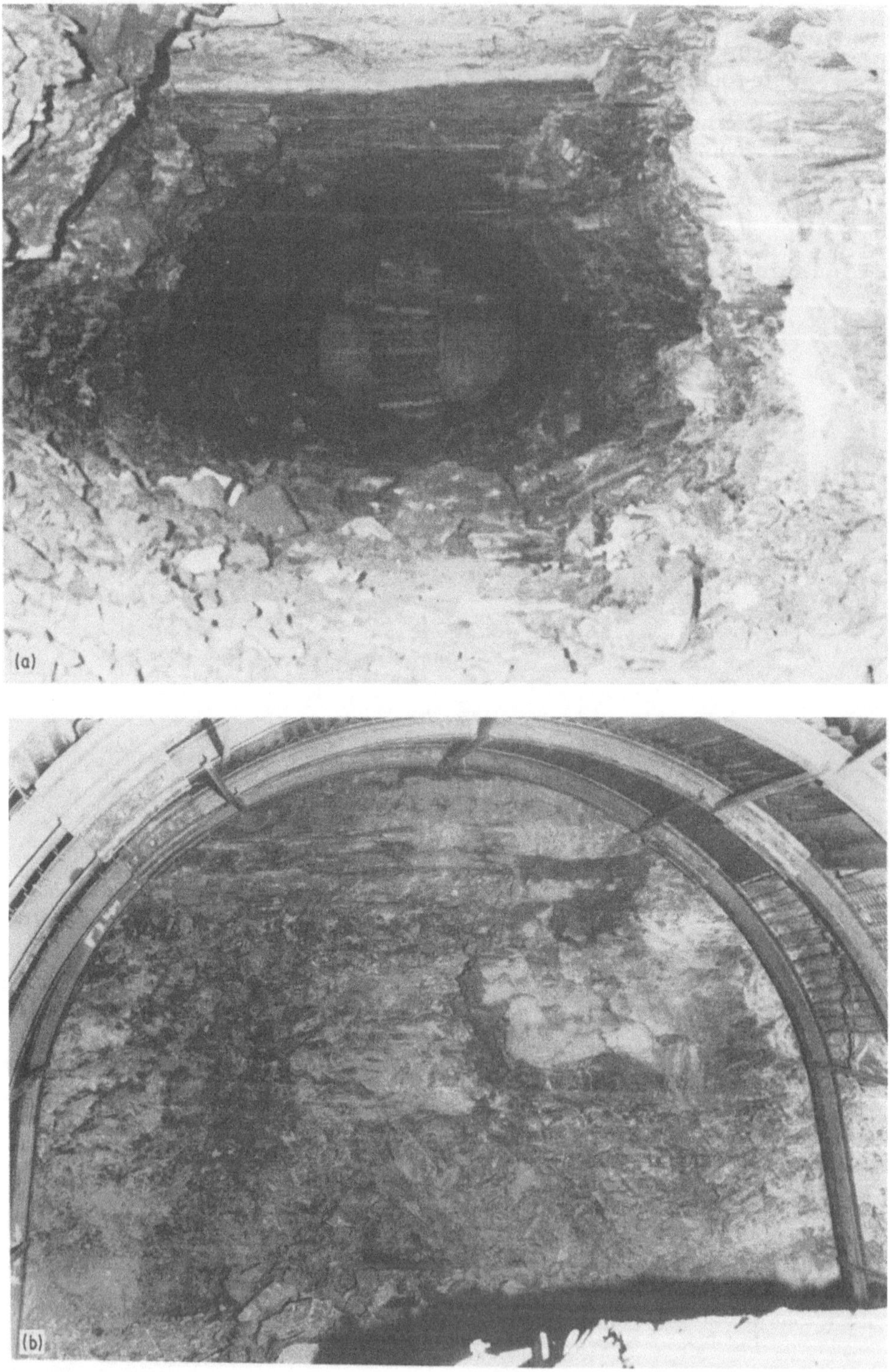

Fig. 6.2 A gob scour constructed in a low stress, caved zone. Note the low level of
physical disturbance of the caved rocks (photograph J. Coggan).

an access roadway to an *advancing* longwall face will be considered. This
has the added complication of an open side which must be packed to
prevent air leakage and to stabilize the roadway, which together with the
high ribside stress makes design of access roadways an interesting
problem – although, with the increasing advent of retreating faces, a
reducing one.

6.1   Access roadway deformation

Although several systematic attempts have been made to isolate the
factors affecting roadway deformation (see, for instance, Johnson, 1973;
Gotze and Kammer, 1976; and Whittaker, 1976), these have not yet been
successfully combined into a satisfactory model which, through an understanding of roadway deformation mechanisms, can be used successfully in
roadway design. It is generally agreed that the main factors determining

the magnitude of roadway deformation are:

(a) The stresses acting on the roadway, resulting from redistribution of geostatic stresses during excavation of coal from the longwall working or from adjacent workings in the same or other seams – as illustrated in Fig. 6.1

(b) The seam height and the compressibility of the face-side pack

(c) The strength of roof and floor rocks, and the effect of water on strength

(d) Proximate geology including the presence of major discontinuities, direction of cleat and strong bands in the roof strata which resist caving

(e) The method of forming the roadway, particularly between drill and blast and a machine cut profile, the degree of roadway support and the treatment of the ribside

(f) The layout of faces and the presence of *ribside pillars*

Of these, none will be important unless the stresses acting on the roadway are sufficiently high to cause convergence and/or floor lift to a significant degree.

During longwall working, although some form of waste or goaf support is feasible, caving is normally adopted. As has been shown in the previous chapter, depending on the lithology of the Coal Measures cyclothem above the extracted seam, the height of collapse of the roof rocks will increase with the area of roof exposed. Although this will be affected by the dilational limits imposed by the seam height, the maximum height (see Fig. 5.7) is usually about 0.8 times the face width. The collapsed rocks will then compact, partly under self weight and partly by the action of a surcharge load exerted by the overlying uncollapsed strata. Wilson and Ashwin (1972) have, however, shown that pillars (see Fig. 2.6) adjacent to the caved area retain redistributed stresses.

Because of the wide variability of geology and face and roadway geometry it is difficult to isolate the exact mechanism of access roadway deformation. In an attempt to examine some of the more obvious variables, a series of investigations including roadway and pack deformation measurements were carried out (see Farmer (1982)) and compared with published and unpublished case histories available from National Coal Board reports (Jones, 1980; Mallory, 1980) and other sources. In the case histories studied a new approach to *pack* instrumentation was used. To measure stresses a Glötzel or Soil Instruments hydraulic pressure cell was used instead of the traditional type of pack load cell. These cells are much slimmer than traditional pack load cells and are filled with mercury in an enclosed system connected to a sampling by-pass valve. This construction means that the cells are less subject to bridging and because of their low

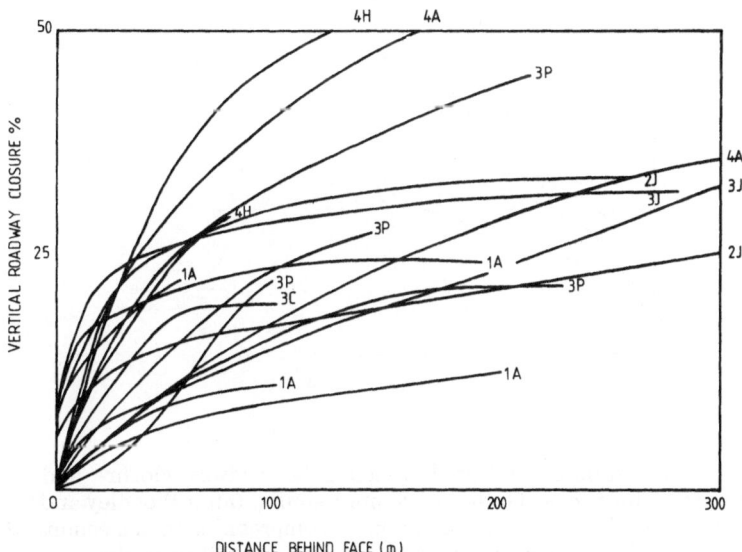

Fig. 6.3 Vertical gate roadway closure behind an advancing longwall face expressed as a percentage of original roadway height against distance behind the face (after Mallory, 1980). The pack types are designated A anhydrite, J jet stowed, C coal, H hand built rubble, P pump pack. The mines are 1 Parkside, Trencherbone seam; 2 Kellingley, Beeston seam; 3 Daw Mill, Warwickshire Thick seam; 4 Parkside, Wigan Four Foot seam.

compressibility are able to sample more accurately pack stresses. A modified strut was used to measure pack closure, and in addition roof lowering and floor heave were measured.

Typical examples of roadway closure are given in Fig. 6.3. These were all obtained by Mallory (1980). The basic geometry was:

Parkside Colliery, Lancashire, Britain. A 180 m longwall face in a 1.6 m seam with a mudstone roof above a fire clay on coal floor at a depth of 690 m and advancing to full dip at 1 in 4.5.

Kellingley Colliery, Yorkshire, Britain. A 300 m advancing longwall face in a 3.1 m (2.8 m extraction) seam with a mudstone roof above a fire clay on siltstone floor at a depth of about 600 m. The dip was not significant.

Daw Mill Colliery, Warwickshire, Britain. A 200 m advancing longwall face taking a 2.2 m extraction in a seam section of about 8 m with 1 m of roof coal at a depth of about 600 m. The dip was not significant.

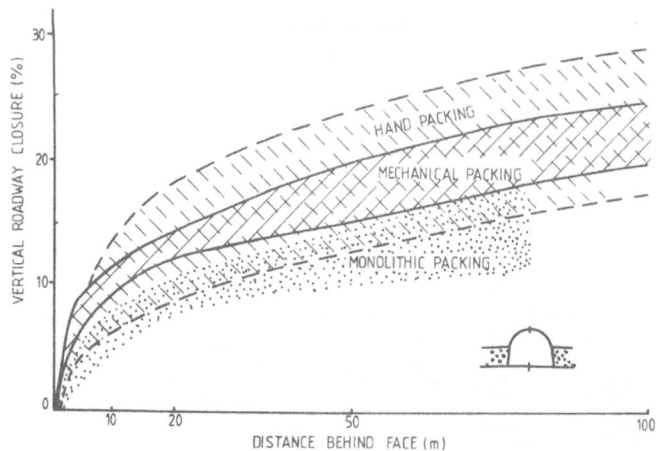

Fig. 6.4 A generalized relation between gate roadway closure and distance behind the face for packs of different compressibility (after Woodley and Osborne, 1980). It is assumed that hand packs are more compressible than mechanical packs which are more compressible than monolithic packs.

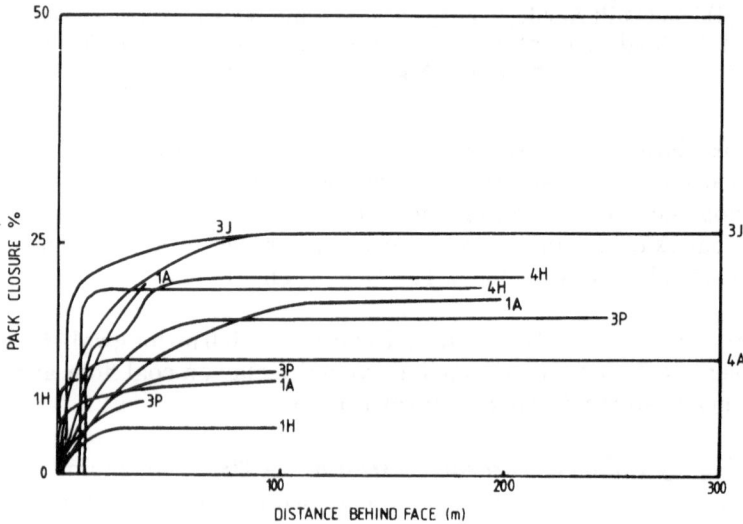

Fig. 6.5 Vertical pack closure expressed as a percentage of original pack height against distance behind an advancing longwall face (after Mallory, 1980) for cases 1, 2, 4 in Fig. 6.3.

Parkside Colliery, Lancashire, Britain.   A 180 m longwall face in a
1.7 m seam with silty mudstone roof and 0.9 m coal floor over 2 m seatearth,
at a depth of about 750 m and advancing to full dip at 1 in 4.5.

Woodley and Osborne (1980) have summarized these and other results
in the form of a general graph (Fig. 6.4) indicating the effect of pack type
on vertical roadway closure. It is unfortunate that none of the other
variables listed above, (a)–(f), are included and comparison with Fig. 6.3

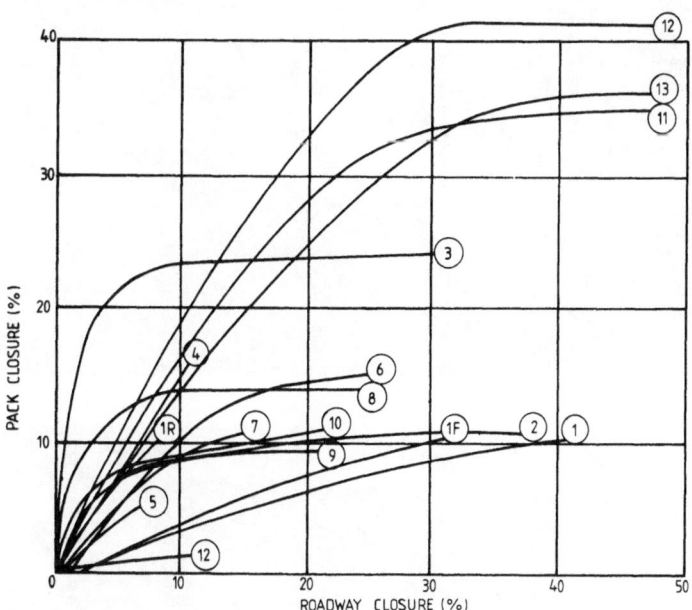

Fig. 6.6   Pack closure against roadway closure, both expressed as a percentage of
original height. Case history details:

 1   Easington mine, Low Main seam (Dudley, 1977): anhydritic pack, 1.6 m
       seam, 550 m depth, 174 m advancing longwall face
 2   Parkside mine, Wigan Four Foot seam (Mallory, 1980): anhydritic pack,
       1.7 m seam, 808 m depth, 180 m advancing longwall face
 3, 4   Abernant mine, Red Vein seam (Jones, 1980): pump pack, 2.1 m seam, 395 m
       depth, 160 m advancing longwall face
 5–7   Merthyr Vale mine, Seven Foot seam (Jones, 1980): high pressure stowing
       with (5, 6) added cement, 140 m advancing face, 2.3 m seam, 700 m depth
 8, 9   Abernant mine, Red Vein seam (Jones, 1980): as 3, 4 but reused pack
       adjacent to 190 m retreating longwall face
 10   Wolstanton mine, Great Row seam (Evans, Hogan and Vallis, 1941): hand
       built pack, 330 m advancing longwall face, 1.8 m seam, 594 m depth
 11–13   Kellingley mine, Beeston seam (Mallory, 1980): 2.6 m seam, 630 m depth,
       300 m longwall advancing face with low pressure stowing.

indicates that without further information the results are seriously misleading. However, they illustrate a point which is *intuitively* correct – namely that a stiffer pack will reduce roadway closure.

However, it can also be shown that pack closure is responsible for only a proportion of roadway closure. Figure 6.5 plots Mallory's (1980) data for closure for cases 1, 3 and 4 in Fig. 6.3 and it is evident that these data although variable include about half of the vertical settlement in Fig. 6.3. What is interesting is that these results show that pack closure occurs relatively rapidly, but that once the pack has compressed, there is continuing roadway closure over a considerable distance of face advance and period of time.

An indication of the relation between pack convergence and vertical roadway convergence can be obtained from the relations in Fig. 6.6, based on a slightly wider series of case histories. These show that initially, pack compression is directly related to roadway convergence, but that roadway convergence continues after pack convergence ceases. There is evidence (see for instance case 1 in Fig. 6.6) that while initial convergence is a direct result of roof lowering following pack convergence, later convergence is caused mainly by floor penetration. In the case of strong roofs and floors or roadways at shallow depths this pack convergence is likely to represent the limit of roadway convergence. In weak rocks, in the vicinity of major discontinuities or shear zones, or at depth, considerable additional deformation may occur.

It is interesting to note that roof lowering appears in many cases to mobilize shear breaks (Fig. 6.7) above the gate roadway. These have a tendency to hade both towards the ribside and the face at the seam level although there is evidence (see Fig. 5.7) that at a higher level, the break line will hade predominantly towards the face. It is possible on the basis of these observations to propose some directions of movement around an idealized gate roadway containing a pack on one or both sides (Fig. 6.8). The type of deformation around roadways such as those illustrated in Fig. 6.7 with a solid ribside and compressible face side pack can be represented in Fig. 6.8(b) using the Rankine plastic state concept as active and passive zones. The former, adjacent to the peak abutment pressure zone, will expand in reaction to the abutment stress in the direction of the roadway. The latter including the pack will compress. Separating these zones will

---

Fig. 6.7   Back ripping lips in gate roads, showing evidence of shear breaks in the roof. In (a) the solid ribside is on the right-hand side of the figure and a double shear break can be seen in the centre roof hading to both face and ribside. In (b) the solid ribside is also on the right-hand side and a similar pattern can be seen although the break hading over the face is dominant.

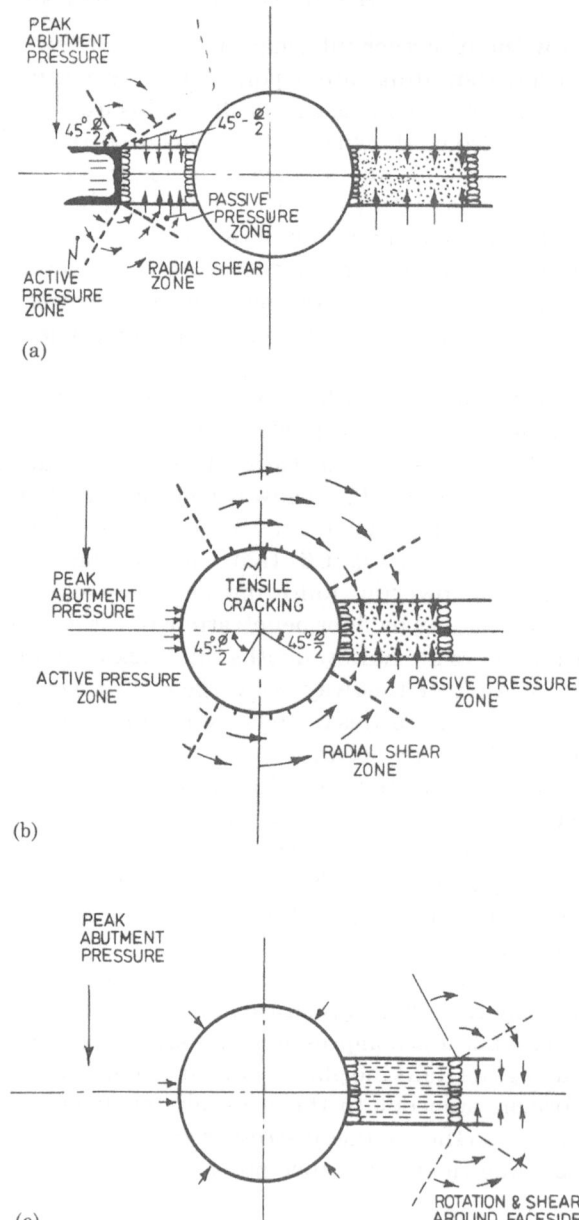

Fig. 6.8 Idealized representation of strata movement around a gate roadway with (a) compressible ribside and face-side packs; (b) solid ribside and compressible face-side pack; and (c) solid ribside and face-side pack of low compressibility (after Farmer and Robertson, 1975).

be a zone of rotational shear creating tensile cracks – principally in the roof of the roadway. The basic idea was proposed for stope deformation by Wiggill (1963) but is more applicable to this case. The limits of the zones will be marked by an active shear break at $45 + (\phi/2)°$ to the horizontal axis and hading over the ribside and a passive shear break at $45 - (\phi/2)°$ to the horizontal and hading over the face. While very much an approximation, the concept of an expanding active wedge in a horizontal plane resulting from a remote abutment pressure resisted by a compressive passive wedge at the pack or waste interface, does help to explain the observed directions of movement in Fig. 6.7. It is used in Fig. 6.8 to develop an alternative support hypothesis. The basic argument assumes controlled ground movement in a fractured zone around the main excavation, rather than cataclysmic shear failure at the edges of the excavation, an argument which is supported by observation.

The purpose of the face-side pack in conventional practice is two-fold. Initially the stone pack walls are designed to give sufficient resistance to the roof to support the dangerous face-end area. As the face advances, however, the pack walls crush allowing the roof to converge, and the weight of the strata is taken gradually by the yielding pack infill material. Eventually after final consolidation – by up to 50% of seam height – the centre core of the pack is sufficiently strong to sustain the tangential stress concentration around the roadway.

This approach has been extended through local experience, based on a knowledge of roof and floor conditions, into a satisfactory method of strata control in which flexible roof supports are allowed to lower as the strata converge. Eventually, the ground around the roadway is able to support the redistributed stresses, forming a stable but much reduced opening. This process is assisted in conventional practice by a ribside pack (Fig. 6.8(a)) of similar compressibility to the faceside pack which removes the effects of the abutment pressure away from the roadway side.

Difficulties occur with this traditional approach when increased rates of advance and face-end mechanization lead to elimination of the ribside pack and reduction in length of the face-side pack. The major effect of a solid ribside is of course, to move the side abutment pressure peak closer to the roadway side, the exact position depending primarily on the strength of coal and floor. With a strong seam and strong strata, the side abutment will be close to the gateside and shear breaks may result, as illustrated in Fig. 6.7. A second and complementary effect may occur as reduced convergence of the face-side pack, leading initially to uneven and insufficient compaction. In fact Spruth (1966) quotes cases in Germany where face-side packs associated with solid ribsides were so ineffective that they were eliminated entirely without any noticeable effect on roadway supports. The effect of the solid ribside may be exacerbated in the case of a shortened face-side pack, for it is evident that under optimum conditions the fully compacted centre core of a pack will only comprise a relatively minor part of the whole length, and it is probable that a pack length several times the seam height would be required to give any useful support to the roadway.

Thus the ultimate effect of a solid ribside and a shortened and/or compressible face-side pack, will be to create zones of radial movement around the roadway (Fig. 6.8(b)) rather than around a more remote ribside, where ribside packs are used, causing tensile failure in the roof and floor. This may be controlled initially by roof supports, but these are no substitute for overall structural stability. For even if the ribside is capable of resisting the abutment stresses, the weakening through tensile fracture of the roadway surrounds may lead to excessive convergence partly through floor lift under restored equilibrium or cover stress conditions, and total collapse if subject to restressing by adjacent longwall workings.

The problem of increased roadway damage resulting from the non-applicability of conventional roadway design philosophies to rapid advance mechanized faces has been recognized quite widely, and two types of solution may be suggested. Carr *et al.* (1967) and Hobbs (1969) have suggested retaining as far as possible conventional packing methods, inducing a degree of stress relief at the ribside by holing or slotting. A more positive approach, however, has been the installation of face-side packs of relatively low compressibility designed to give immediate support to the roadway, and reduc:ng to a minimum roadway deformation and subsequent tensile fracture of roof and floor.

The reason for the effectiveness of the method appears obvious if proposed movement directions in Figs 6.8(b) and 6.8(c) are compared. Irrespective of the exact positions of the peak abutment pressure, in the case of the compressible pack the direction of deformation will cause damage to the rock in the roadway annulus, inducing fractures and reducing its competence as a structural unit. The results will be spalling and collapse of the roof and floor-lift.

In the case of the pack of low compressibility, the direction of deformation will be determined by the cutting-off line of the pack side, and deformation of the actual roadway will be limited to minor radial compression.

6.2 Compression of pack materials

Figure 6.9, based on the observations in Figs 6.3 to 6.6, illustrates the roadway convergence in a simpler way. This suggests that an estimate of

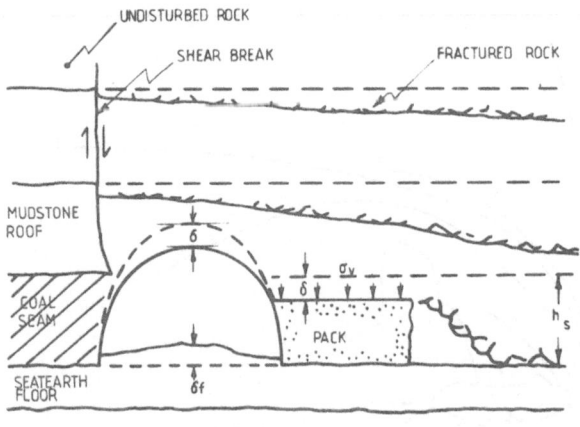

SEQUENCE OF ROADWAY CLOSURE

1. PACK COMPRESSES δ̄ UNDER STRESS σ_V

2. ROOF LOWERS AS PACK COMPRESSES

3. AS σ_V INCREASES PACK PENETRATES FLOOR

Fig. 6.9 Idealized representation of the main variables affecting roadway closure.

roadway closure could be obtained provided that an estimate of pack stress, pack compressibility and roof and floor strength could be obtained.

Hobbs (1968) in a series of model tests on mine roadways with a ribside and face-side of equal compressibility has empirically equated percentage roadway convergence (δ) to applied pressure and rock strength:

$$\delta = 20\sigma_\mathrm{v}/(\sigma_\mathrm{CR}, \sigma_\mathrm{CF})^{\frac{1}{2}} \qquad (6.1)$$

where σ_v is the vertical geostatic stress, σ_CR, σ_CF are respectively the uniaxial compressive strength of the roof and floor rocks.

This is probably too over-simplified and ignores the importance of pack compressibility.

Large numbers of pack materials (see Blades, 1975) have been introduced in the past ten years in order to give added support to and in some cases to reduce the permeability (where spontaneous combustion of waste coal can occur through air leakage across the goaf) of the face-side of access roadways to advancing longwall faces. Materials have usually been selected arbitrarily for use with various packing systems developed to eliminate manual packing. To date, research has been concentrated on the method of packing, and to a lesser extent, on the interaction of the packing material and the method. Less emphasis has been placed on the

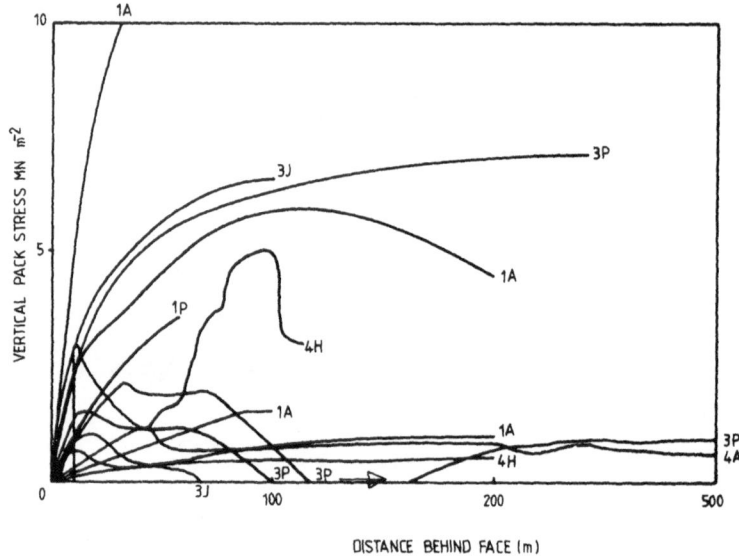

Fig. 6.10 Vertical stress at the centre of a pack against distance behind face (after Mallory, 1980). Pack types and locations are listed under Fig. 6.3.

mechanical reaction of packing materials to the redistribution of forces following mining and the mechanics of side support and its effectiveness in improving roadway conditions.

Data on pack stress build up collected by Mallory (1980) during the case histories of Fig. 6.3 is plotted in Fig. 6.10 and a generalized summary of this and other data by Woodley and Osborne (1980) is reproduced in Fig. 6.11. As in the case of Fig. 6.4 it is difficult to justify the simplified summary profiles, and the stress magnitudes are misleading. However, the characteristic of the stiffer pack materials – exhibiting the strain-softening behaviour discussed in Chapter 1 – is typical of the behaviour of strong materials, or materials with low confinement at strength failure.

In practice a face or ribside pack is subjected to a degree of confinement and its characteristics can only be correctly defined by taking this into account. A pack is confined longitudinally by adjacent packs, quite rapidly on the goaf side by the caving roof and on the roadway side by the arches and by lagging between the arches. If it is assumed that the pack material is completely confined, then it should closely follow the fundamental compression characteristics of other drained granular materials. On a plot of void ratio (e) against the logarithm of the vertical pressure (log σ_v) this should (depending on the initial void ratio) follow a line parallel to the stress axis to between 10 and 100 kN m^{-2}, a constant slope

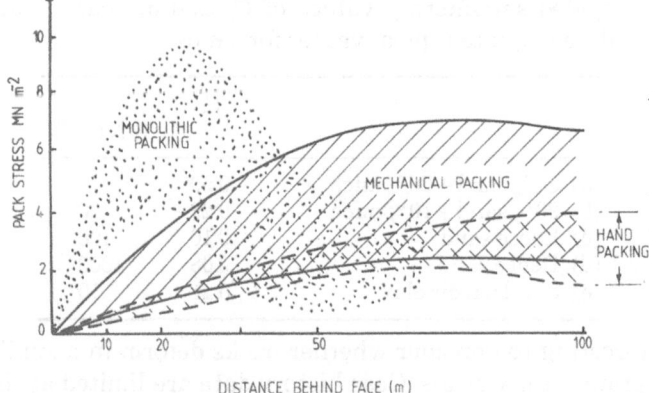

Fig. 6.11   Notional mechanical behaviour of pack materials with distance behind longwall face (after Woodley and Osborne, 1980). Note that the monolithic pack is assumed to behave in a brittle manner and the hand pack in a ductile or compressible manner.

straight line curve to about 10 000 to 100 000 kN m^{-2} and a line parallel to the stress axis beyond this. Illustrations of sigmoidal compression curves of this type for particulate granular materials are given in Terzaghi and Peck (1967). The first part of the curve at constant void ratio represents a pressure too small to compress the material. During the second part grains are crushed and rearranged, and during the third part the material is too dense for further significant deformation. The constant slope part of the curve defines a fundamental reaction to stress of a granular material in confinement and can be described by the form:

$$e = e_0 - C_c \log_{10} \frac{\Delta\sigma_v}{\sigma_{v0}} \tag{6.2}$$

where σ_{v0} is the pressure at the start of the constant part of the curve, $\Delta\sigma_v$ is a pressure increment within the part of the curve during which e_0 the initial void ratio is increased to e and C_c is a dimensionless ratio known as the *compression index*.

It can be shown that the compression (δ) of a layer of material of initial thickness h_s can be given by:

$$\frac{\delta}{h_s} = \frac{C_c}{1+e_0} \log_{10} \frac{\sigma_v}{\sigma_{v0}} \tag{6.3}$$

Expressed as a percentage δ/h_s will be equivalent to the pack closure for a given σ_v provided satisfactory values of C_c and σ_{v0} can be determined. Terzaghi and Peck quote typical values for sands as:

	C_c	σ_{v0} $(\mathrm{kN\,m^{-2}})$
(a) 80% sand + 20% mica	0.50	10
(b) 90% sand + 10% mica	0.29	10
(c) Loose sand	0.10	100
(d) Dense sand	0.08	1000
(e) Soft Detroit clay	0.21	10

It is interesting to consider whether packs deform in a similar way to confined granular materials. Case history data are limited at the moment and the cases quoted in Fig. 6.6 which are plotted in Fig. 6.12 as pack closure percentage void ratio against the logarithm of measured pack stress, designated σ_v, are the best available. Even so some of both the stress data and deformation data must be queried, and generally speaking, data refer to the first 100 days face advance or less. In several cases, the pack, or, more probably, the floor, has failed.

Where the data are plotted as void ratio, e, against $\log \sigma_v$, initial void ratio data were not available and e_0 of 0.6 was assumed for rubble packs, 0.5 for monolithic (except in the case of 6) and 0.4 for anhydrite packs.

Two curves in Fig. 6.12(b), cases 8 and 10, follow what Terzaghi and Peck (1967) describe as a typical compression curve, but all curves have a significant straight line portion, and many exhibit parts of the typical compression curve. The conclusion must be that if pressures below $100\,\mathrm{kN\,m^{-2}}$ could have been measured and above $10\,000\,\mathrm{kN\,m^{-2}}$ had been monitored, then all pack materials would have exhibited typical deformation characteristics for confined granular materials.

If this is correct, then the ideal pack material will be one which has a low C_c value at a relatively high pressure threshold. The results are not sufficiently reliable or complete for detailed analysis, but from cases studied, suitable values for substitution in the compression equation can be summarized approximately as:

	e_0	C_c	σ_{v0} $(\mathrm{kN\,m^{-2}})$
Mechanized pack	0.7	0.20	10
Hand pack	0.6	0.15	50
Stowed pack (low pressure)	0.6	0.15	50
Stowed pack (high pressure)	0.6	0.2 (0.10)	1500
Pump pack (Bentonite)	0.5	0.4 (0.05)	3000
Monolithic (Cement or anhydrite)	0.4	0.4 (0.05)	2000

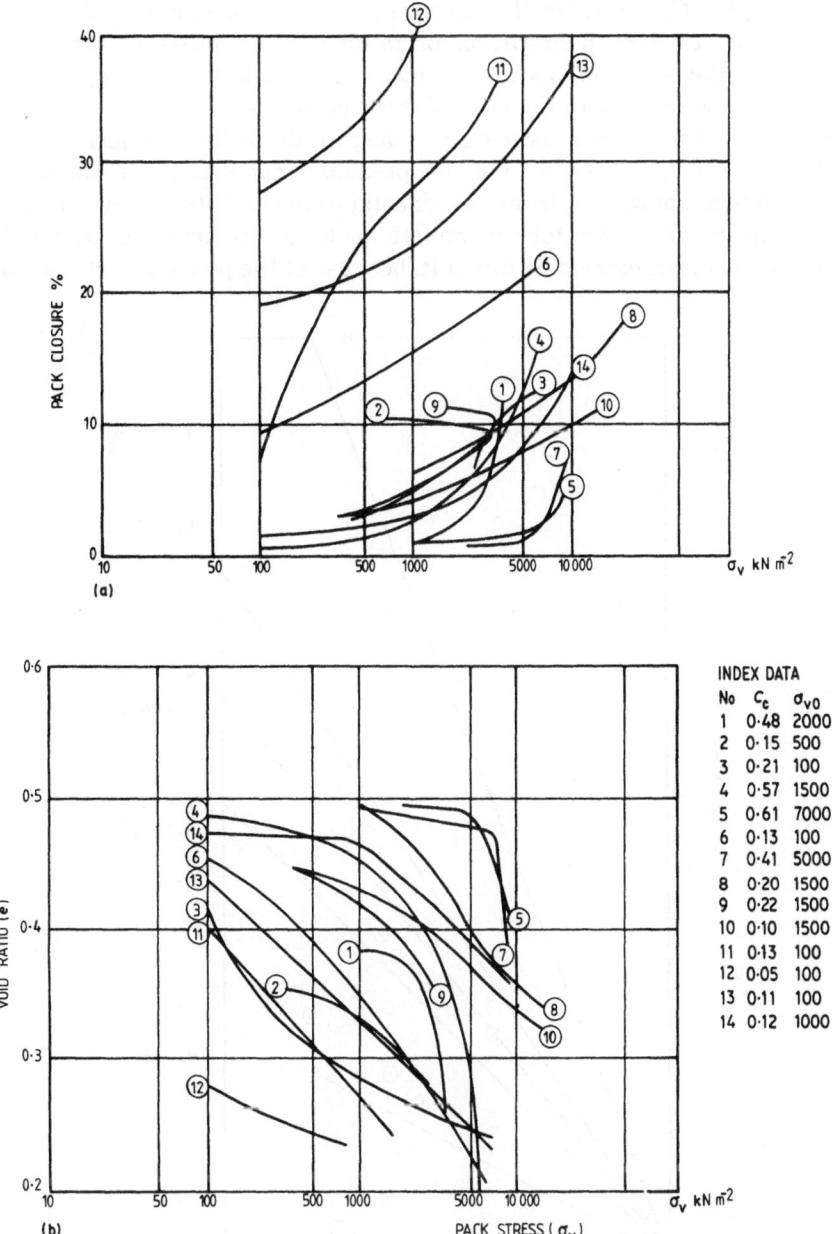

Fig. 6.12 (a) Pack closure (δ/h_s) against the logarithm of measured vertical pack stress σ_v for the cases in Fig. 6.6. No. 14 is a laboratory compression test on mine rubble. (b) Void ratio (e) against the logarithm of σ_v.

The higher C_c values for the stronger packs are unrealistic and indicate a tendency to floor penetration or failure at the onset of rapid convergence. Realistic values are indicated in brackets. The approach is probably less reliable in the case of these packs than in rubble or stowed packs although it does indicate the magnitude of floor penetration. In rubble packs, provided σ_v can be predicted accurately, it should be possible to estimate δ/h_s from the consolidation equation although allowance may be necessary for incomplete packing. Prediction of σ_v for the pack is, however, extremely difficult, because of the proximity of the edge

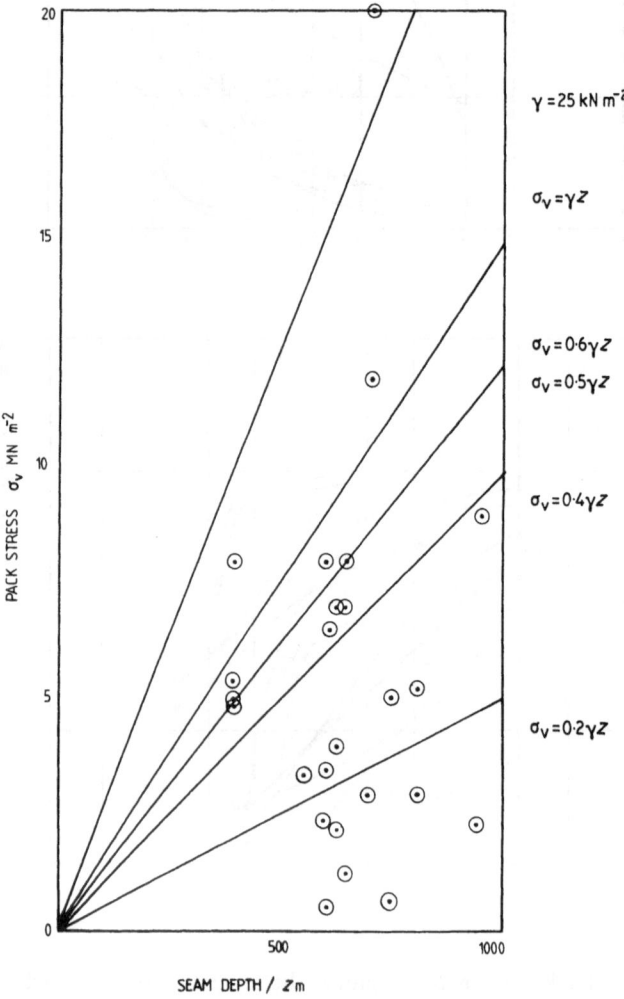

Fig. 6.13   Measured pack stress against depth for the case histories in Fig. 6.6.

of the excavation and other factors. σ_v was plotted for the available data against seam height, rock strength, face width and seam depth (Fig. 6.13) without finding any significant relations. In Fig. 6.13 it can be seen that in all except one exceptional case (Merthyr Vale) σ_v is less than the vertical geostatic stress, σ_z. However, in all except eight cases (mainly failed anhydrite packs, and questionable measuring equipment) σ_v is greater than the $0.2\sigma_z$ predicted by Harris's model (Fig. 6.1). A reasonable limit would appear to be $0.5\sigma_z$ and a rule of thumb, that the maximum pack stress should not exceed by a large amount, half the geostatic stress in most longwall mining, appears to be an acceptable criterion for pack design.

6.3   Effect of proximate geology

The effect of roof and floor rocks and structures such as cleat is generally agreed to have a strong influence on roadway stability. Quantification is, however, difficult, and any comments are best confined to observation. The most serious factor in roadway stability is *floor heave* and this is the most easily approached – usually on the basis of simple empirical methods developed for foundation engineering.

Floor rock in the Coal Measures cyclothem usually contains clay minerals. As has been seen in Chapter 1, if the floor clay when wet behaves in a ductile manner, prediction of the magnitude of closure is difficult. For instance where thick layers of weaker seatearths are confined and subjected to high deviatoric stresses deformation of a strain-hardening type (as in Fig. 1.4) will lead to flow of the deforming rock which will exert considerable pressures on any unconfined layers in the immediate floor area. This type of deformation is common in arch-supported access roadways, and has been illustrated in Figs 1.16, 3.10 and 3.17. Figure 6.14, based on measurements from Easington Colliery (Dudley, 1977), shows comparative arch and floor deformation. The continuing deformation of the floor and the virtual stability of the arch in the stronger roof rocks is a feature of this type of strata sequence (see Fig. 1.16). The simplest approach to design is to take some of the rules of thumb developed by civil engineers for estimating foundation settlements. The first of these is that, for penetration to occur, the pack stress must exceed the bearing capacity of the floor material. *Undrained* conditions may be used, since consolidation, even of wet floor clays, is unlikely to lead to sufficient penetration to cause large scale floor heave.

Then from Terzaghi's modified bearing capacity equation penetration will occur when:

$$\sigma_v = q_{ult} = N_c C_u \tag{6.4}$$

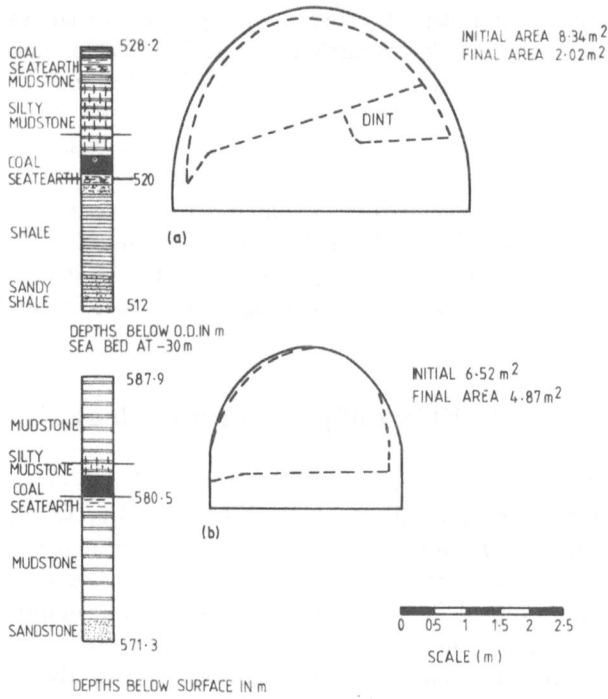

COAL
SEATEARTH
MUDSTONE
528·2

SILTY
MUDSTONE

COAL
SEATEARTH —520

SHALE   (a)

SANDY
SHALE   512

DEPTHS BELOW O.D.IN m
SEA BED AT −30 m

INITIAL AREA 8·34 m^2
FINAL AREA 2·02 m^2

DINT

587·9

MUDSTONE

SILTY
MUDSTONE

COAL
SEATEARTH —580·5

(b)

MUDSTONE

SANDSTONE
571·3

INITIAL 6·52 m^2
FINAL AREA 4·87 m^2

0 05 1 1·5 2 2·5
SCALE (m)

DEPTHS BELOW SURFACE IN m

Fig. 6.14 Roof and floor deformation profiles in gate roadways at (a) Dawdon
Colliery and (b) Wearmouth Colliery.

where σ_v represents the pack stress, N_c is a bearing capacity factor having
a magnitude between 6 and 9, C_u is the undrained shear strength of the
floor clay and q_{ult} is the bearing capacity of the floor clay.

Since σ_v (Fig. 6.13) is likely to be about half the vertical geostatic stress
σ_v, then a factor of safety against floor heave can be defined as:

$$F = \frac{20C_u}{\sigma_v} \tag{6.5}$$

C_u values for clays quoted in various sources follow the range:

	kN m^{-2}
Very soft clay	25
Soft clay	25–100
Medium clay	50–100
Stiff clay	100–200
Very stiff clay	200–400
Clay shales	500–1000
Strong seatearths	500–1000

Since it is possible to relate strength to moisture content, the implication for coal mining is that the weaker clays are saturated. Thus heave in a seatearth with a low natural moisture content ($C_u = 1000\,\mathrm{kN\,m^{-2}}$) may not occur until depths exceed 800 m, while in a seatearth with a high natural moisture content ($C_u = 200\,\mathrm{kN\,m^{-2}}$) heave will occur at depths less than 200 m. In the case of active or swelling clays, further information on the plasticity index would be required.

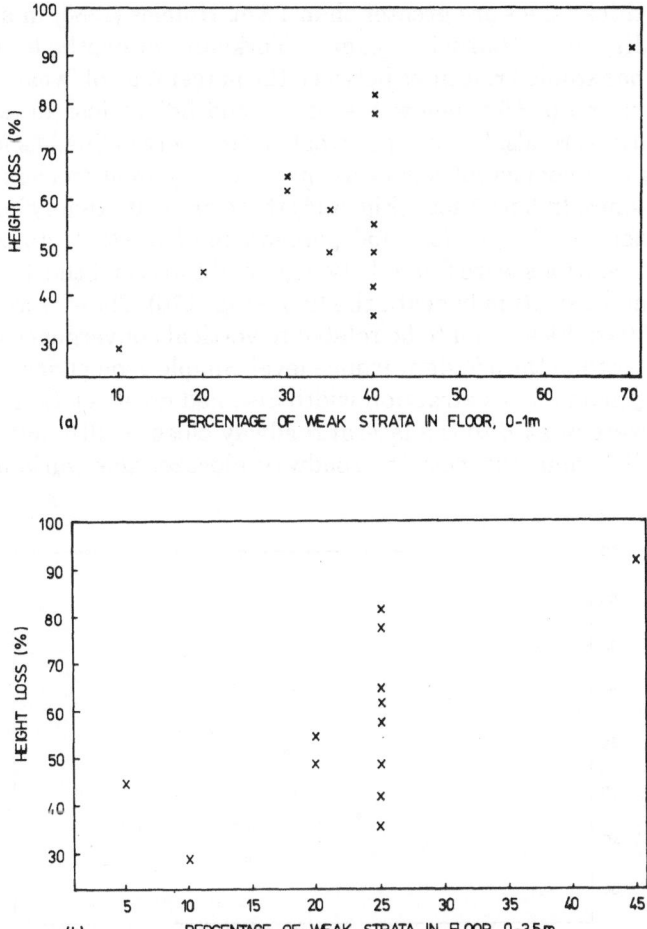

Fig. 6.15 Height loss expressed as a percentage of original gate roadway height against percentage of weak strata in the floor: (a) between 0 and 1 m, (b) between 0 and 3.5 m, of the roadway. All observations by Holmes (1983) in gate roadways to advancing longwall faces in the Barnsley seam, Yorkshire, at depths of 700–800 m.

The depth of penetration, once initial failure has occurred, will depend on the strength of underlying layers, and will be difficult to predict without a knowledge of local conditions and/or observations and experimentation. In most cases it is a reasonable assumption that where penetration occurs it will be to the full potential depth of the floor clay layer, with the limitation that in thicker layers, mobilization of residual strength of the fractured surface layers will inhibit penetration. Speck (1981) in a study of seatearths in Illinois has shown that floor heave is greatest where floor clays have high natural water contents and where floor clay thicknesses are greater than 1.8 m. Holmes (1983) in a study of floor stability in the Barnsley seam in Yorkshire at depths between 700 and 800 m has studied relations between the percentage of 'weak' strata in zones 0–1 m and 0–3.5 m below the seams and height loss in advancing longwall access roads. In this case weak strata were defined as mudstone seatearths and competent strata as massive silty mudstones, siltstones and sandstones. In both cases (Fig. 6.15) there are statistically significant relations between height loss and percentage of weak strata. Similarly significant relations were found between height loss and the depth to the first competent stratum beneath the floor (Fig. 6.16). These were the only geological variables found to be related to vertical convergence or height loss at better than the 5 % significance level. Simple geometrical variables such as depth and face extraction width also had no effect. Some of these variables were related to the lateral roadway closure, although this was much smaller than the vertical roadway closure. One variable which

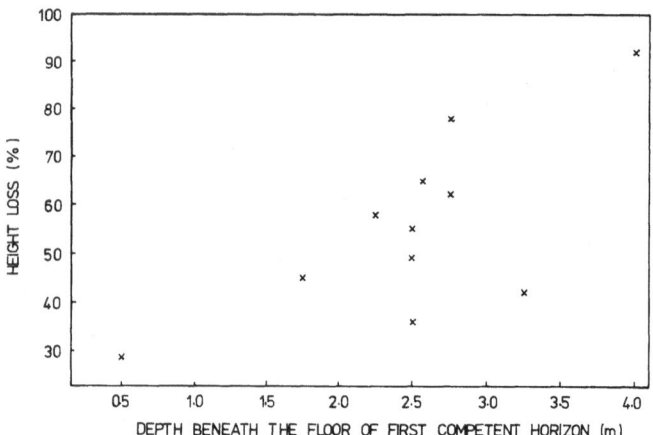

Fig. 6.16 Vertical closure expressed as percentage height loss against depth below the floor of the first competent horizon in gate roadways to advancing longwall faces in the Barnsley seam, Yorkshire coalfield (after Holmes, 1983).

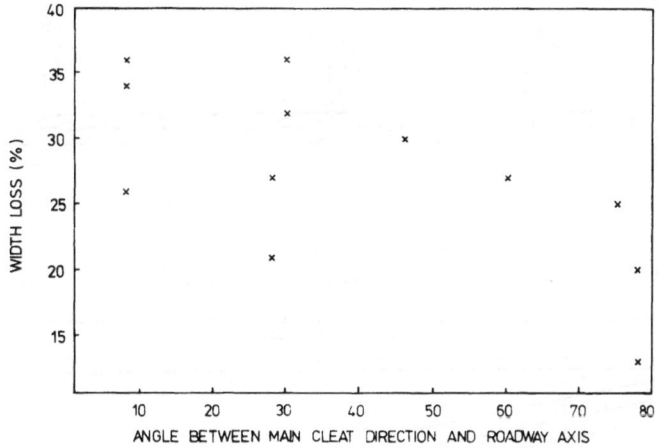

Fig. 6.17 Lateral closure expressed as a percentage of width loss against the angle between the major cleat direction and the roadway axis in roadways to advancing longwall faces in the Barnsley seam, Yorkshire coalfield (after Holmes, 1983).

showed a significant relation to the width loss or lateral closure was cleat direction, where there was an inverse relation (Fig. 6.17) between closure and the angle of cleat to the roadway. This will be further considered in the much more important case of the relation between coal face and stability and cleat direction in the next chapter.

6.4 Prediction of roadway deformation

Although it is possible to explain the mechanics of deformation of gate roadways, the complexity of the factors involved in access roadway stability makes prediction of the magnitude of deformation particularly difficult. Ultimately there will be a reliance on experience of mining operations at a particular horizon. Two approaches which have already been mentioned, while not necessarily within the scope of the present work, can be used to illustrate such an empirical approach.

Gotze and Kammer (1976) have demonstrated a method of estimating the vertical convergence of access roadways based on observations on arched roadways in the Ruhr coalfield. They claim that the method can be used to estimate convergence of roadways on single unit longwall faces within very high probability limits. They isolated four main convergence factors: seam *thickness*, the nature of the *floor* rocks, the *depth* of the seam (natural factors) and the nature of *packing*.

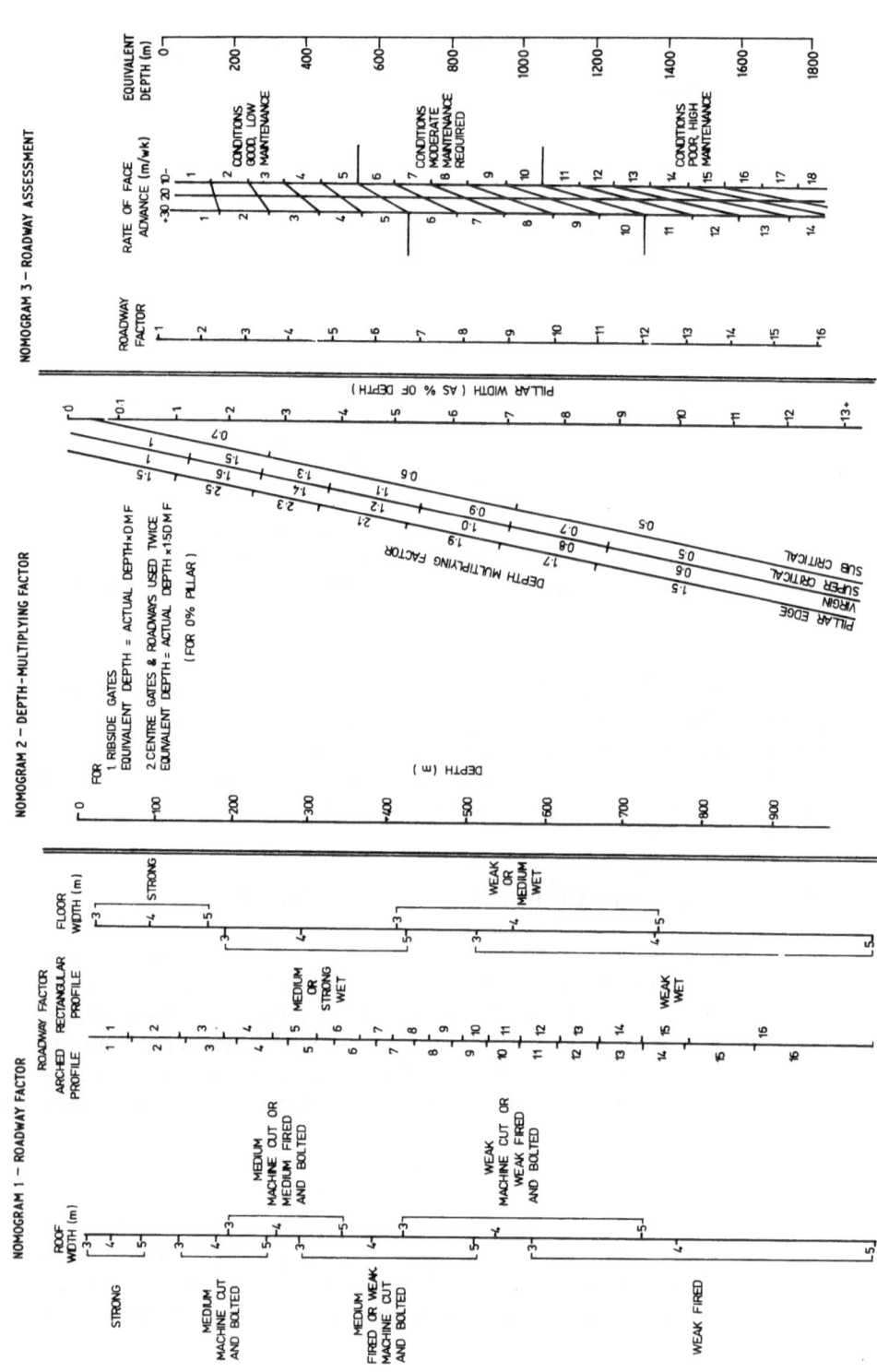

Then through statistical evaluation of data from mines in the Ruhr they proposed the following equation for convergence:

$$\text{Convergence \%} = -78 + 0.066z + 4.3h_s \, (\text{PI}) + 7.7 \quad [10(\text{FI})]^{-\frac{1}{2}}$$

where z, h_s are the depth and seam thickness in metres, FI is the floor index ranging from 1 for hard floor rocks to 6 for weak floor rocks and PI is the pack index ranging from 1 for stiff (anhydritic type) packs, 2 for rubble or wood chock packs and 3 for no packs.

Observations covered faces at depths between 500 and 1200 m deep. Convergence ranged from 10 % in thin shallow seams with hard floors and stiff packs to 70% in thick deep seams with soft floors and wooden chock packs. Examination of Fig. 6.15 – where the seam thickness is about 2 m – indicates that the approach would lead to similar magnitudes of predicted closure in Britain.

Interesting observations from the work were that there was a 10 % improvement in convergence where the roadway was driven at the face rather than in advance of the face and a 20 % improvement if ripping was delayed for 7 m behind the face. The type of arch support has no effect on convergence although some types of arch support were better able to accommodate convergence than others. Thus rigid arches could accommodate up to 15 % convergence; lightweight arches of 26–29 kg m^{-1} section up to 30 % convergence and yielding arches greater than 30 % convergence (see Table 3.2).

A similar approach for British conditions has been proposed by Johnson (1973) using a series of nomograms. These are reproduced in Fig. 6.18. The first nomogram allows computation of a *roadway factor* from *roof* and *floor rock strength*, method of *construction*, presence of floor *water* and the shape of the roadway.

The second nomogram allows computation of a *depth multiplying factor* from the seam *depth*, the width of *pillar* adjacent to the access roadway and the presence of other workings and the location of the surface above the face. As can be seen, the worst case is where a pillar edge occurs immediately above or below the roadway or working. The reasons for this are obvious and will be expanded in the final chapter. The best case is where an opening overlies the face and this is represented by the sub-critical (see Fig. 5.8) case.

Fig. 6.18 Nomograms to allow assessment of access roadway conditions based on observations in British mines (after Johnson, 1973). *Weak rocks* (unconfined compression strength 0–30 MN m^{-2}) include mudstones, seatearths and clays. *Medium rocks* (30–70 MN m^{-2}) include mudstones and siltstones. *Strong rocks* (> 70 MN m^{-2}) include sandstones and other well cemented rocks.

The third nomogram allows assessment of roadway conditions from the roadway factor and the equivalent depth – obtained from the product of the actual depth and the depth multiplying factor. Since the nomograms are based on observations of roadway deformation, they indicate the important factors affecting deformation in Britain. One which is worth comment is the width of the pillar between longwall faces.

In single unit advancing and, to a lesser extent, retreating longwall faces, it is conventional practice to leave a barrier pillar between the faces to protect the access roadways. The width of barrier pillar for full protection is difficult to determine (as can be seen from the uneven stress distribution in Fig. 2.6). Wilson (1981) has proposed a detailed analytical approach which is worth further consideration. Drawing on the experience of observations such as those summarized in Figs 2.6 and 6.1 he postulates (Fig. 6.19) a triangular or trapezoidal stress distribution over the goaf area (Fig. 6.1) which can be characterized in terms of a stress deficiency A_w which is ultimately transferred to the ribside abutment. Here it is characterized in terms of two stress distributions A_b in the yield zone and A_s (Fig. 6.20) over the rest of the abutment, so that the stress

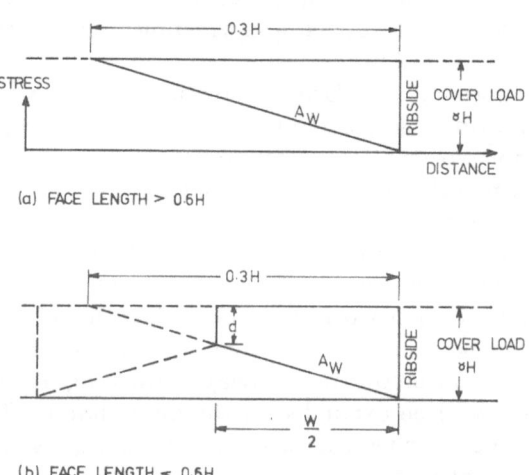

(a) FACE LENGTH > 0.6H

(b) FACE LENGTH < 0.6H

Fig. 6.19 Hypothetical stress distribution (after Wilson, 1981) over a section normal to a longwall face where a stable stress regime has been reached. This can be characterized as a stress deficiency A_w transferred to the ribside abutment where:

$$A_w = 0.15\gamma z^2 \ (w > 0.6H)$$
$$= 0.5w\gamma(z - w/1.2) \ (w < 0.6H)$$

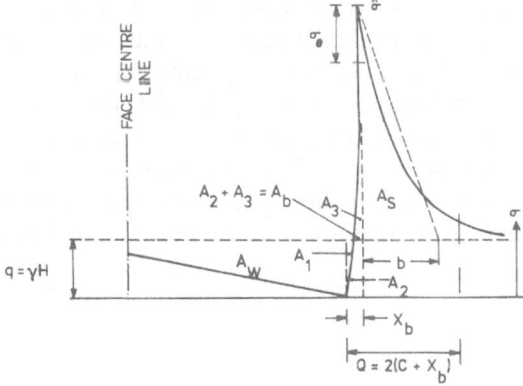

Fig. 6.20 Hypothetical stress distribution (after Wilson, 1981) over a ribside abutment based on distribution of stress deficiency A_w, where A_b (the stress in the yield zone) $= (m/F)k(\sigma_z - p_i)$, A_s (the stress decay over the abutment) $= c(\hat{\sigma} - \sigma_z)$, m = height of extraction, $k = (1 + \sin \phi)/(1 - \sin \phi)$, $\sigma_z = \gamma z$ = vertical geostatic or cover stress, p_i = support pressure including that from broken rock, $\hat{\sigma} = k\sigma_z + f\sigma_{cf}$ is the peak abutment stress, $c = (A_w + \sigma_z x_b - A_b)/(\sigma - \sigma_z)$ is a constant called the exponential decay factor, $F = (k-1)/\sqrt{k} + ((k+1)/\sqrt{k})^2 \tan^{-1} \sqrt{k}$ is a constant, $x_b = (m/F) \ln (\sigma_z/p_i)$ is the peak abutment stress, σ_{cf} is the compressive strength of the strata, and $f = 1$ (for massive unjointed rock) to 6–7 for fault and shear zones is a rock mass factor.

distribution at any distance x into the abutment can be represented by:

$$\sigma = \sigma_z + (\hat{\sigma} - \sigma_z) \exp \frac{(x_b - x)}{c} \tag{6.7}$$

Wilson suggests that at a distance $b = 2c + 2x_b$ into the ribside (Fig. 6.20), the stress will have decayed to a sufficient level to have a minimum effect on a roadway to an adjacent face. This distance is therefore selected by Wilson as a stable pillar width for access roadway protection.

This distance can be compared with the rule-of-thumb rules for Britain – that the width of pillar should be equal to one-tenth of the depth plus 15 yards (13.7 m); or for North America, fixing the allowable extraction percentage at $1 - z/2100$ where z is in metres. Both of these rules give similar pillar widths to a depth of about 600 m. Below this they tend to overestimate.

There is no reason why this expression should not be used for North American layouts incorporating chain pillars, provided an allowance is made for the extracted area of the entries in the pillar. It is, however, interesting to speculate on the effect of the position of the entries in relation to the stress distribution in Fig. 6.20. Peng and Su (1983) show that interaction of abutment stresses over chain pillar layouts can lead to interesting entry convergence patterns. It must however be emphasized that pillars – as will be illustrated in Chapter 8 – are an undesirable and wasteful aspect of longwall mining. Whilst they may be justified in advancing longwall mining they cannot be justified in retreating longwall mining. There are so many other advantages for retreating longwall mining that it is difficult to find a justification for advancing longwall mining apart from incompetent forward planning and mine design.

7

COAL FACE STRATA DEFORMATION AND SUPPORT

In modern coal mining, the longwall coal face is the focus of the mining operation. It is the place where the success or failure of the whole process of investment and design in a mine is determined. Although there are many factors ranging from machinery and services to manpower which come together to determine the productivity of the mining operation, the most important is the stability of the sensitive rock structure which comprises the coal face. Needless to say it is the one which often receives the least attention.

The potential problems associated with this structure are extremely complex. In simple terms it is an excavation about 200 m long by 3 m wide and 2 m high. The front of the excavation is advanced at a rate of about 30 m per 5 day week. The supports are advanced, and the rear roof of the excavation is expected to collapse, at the same rate. The rocks around the face structure are usually weak with, in the case of coal, a very low unconfined compression strength (Fig. 1.5). They are initially subjected, depending on depth, to vertical horizontal geostatic stresses of the order of 12.5 MN m^{-2} (at 500 m) to 25 MN m^{-2} (at 1000 m) which will increase by two to ten times in a vertical mode and decrease by a similar order amount in a horizontal mode in the face and ribside abutments (Figs 2.6, 6.1 and 6.20) as the coal is extracted. There is no corresponding rise in the waste stresses where dilation leads to a permanent stress deficiency (Figs 5.3 to 5.5). A section through the centre line of a coal face (Fig. 7.1) parallel to the direction of advance gives an indication of the problems likely to be encountered. These are:

(a) The peak abutment stress (Fig. 6.1) will induce sub-vertical fractures parallel to the face line as strata dilate laterally into the zone of reduced confinement created by the collapsing waste.

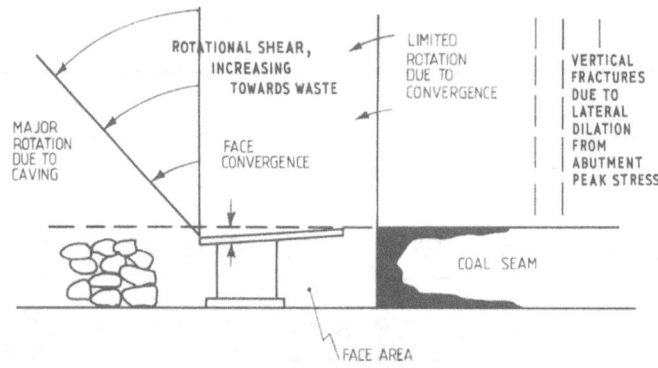

Fig. 7.1 Simple representation of the main strata movements in the vicinity of the coal face.

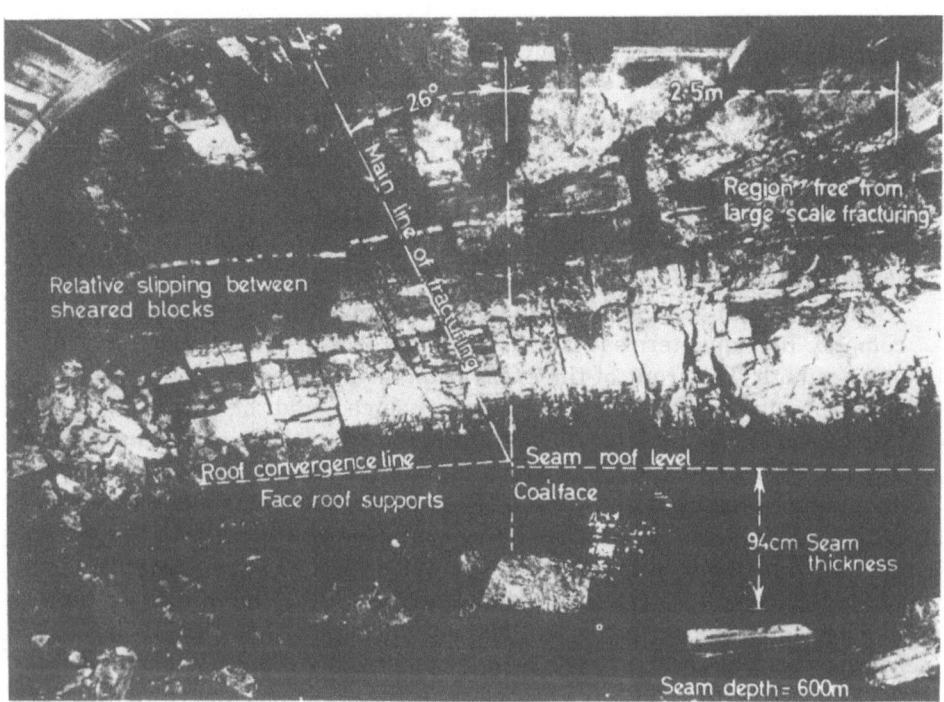

Fig. 7.2 Whittaker's (1975) interpretation of deformation zones in an exposed section through a coal face. There is evidence of fracturing and cambering ahead of the face, but the main zone of rotation occurs at the face line. This indicates that in an advancing face the degree of rotation and subsequent fracturing will depend on the degree of convergence permitted by the face supports.

Fig. 7.3 An illustration of caving behind a longwall face from the model in Fig. 5.6.

(b) Strata blocks delineated by these fractures will camber and rotate about the face line towards the waste area. This process will increase with increasing convergence of the face supports.

Both of these aspects are illustrated in Figs 7.2 and 7.3, and it can be seen immediately that they are the basis of face support design.

7.1   Roof and face stability

The abutment pressures ahead of a longwall face can induce fractures. The term *induced cleavage* was used by Faulkner and Phillips (1935) to describe this phenomenon and it is a useful term contrasting with the natural cleats, joints and fissures which occur in the Coal Measures and which can also affect stability. The mechanics of induced cleavage are those of classical rock mechanics. Predominantly vertical compression stresses are induced by stress redistribution close to and parallel to the ribside and face abutments. At the same time lateral restraint is reduced by removal of coal and collapse of the caving waste. This allows, as

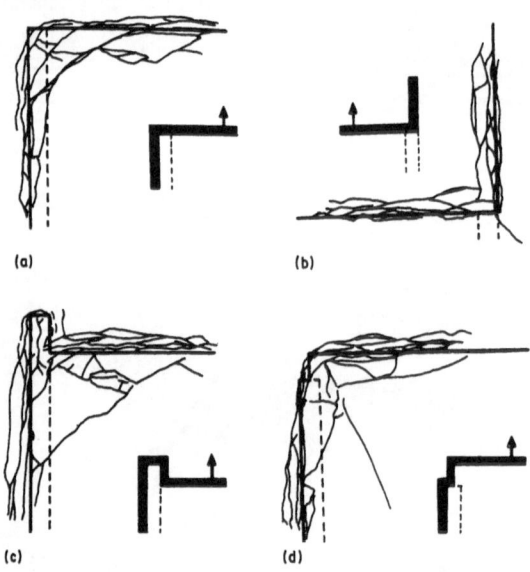

Fig. 7.4 Directions of fracturing induced at the face end by longwall mining as postulated by Sziwlski and Whittaker (1975) from model studies: (a) gate-roadway formed at face line with solid ribside; (b) as (a) but with no rib-pillar; (c) advanced heading; (d) half heading.

Griffith's theory postulates and observation confirms, the formation of sub-vertical cracks parallel to the major compressive stress direction and normal to the minor stress direction. Thus cracks will be formed parallel to the face direction and parallel to the ribside direction. This will lead to changes in the angle of induced cleavage at face ends, where a combination of low stress and indeterminate fracture directions can cause difficult support problems which are considered later. Some possible face end induced fracture distributions suggested by Sziwlski and Whittaker are illustrated in Fig. 7.4. A more detailed series of observations obtained on faces in the Barnsley seam in the Yorkshire coalfield by Holmes (1983) is illustrated in Fig. 7.5. Here the dip and dip direction of induced fractures is plotted on a lower hemisphere equal area stereographic projection, the planes being represented by their poles. The orientation of the fractures was measured on several faces at different collieries at depths between 600 and 800 m. The orientation of the fractures is plotted relative to the direction of face advance.

Three groups of fractures can be identified:

(a) A group hading at between 70° and 90° over the goaf and having a strike within 45° of the face line. These are typical of the induced cleavage of Faulkner and Phillips (1935). Their spacing is related to lithology, being 250–500 mm in sandstone and less than 50 mm in weak sandstones. Also included in this group are a small number of fractures striking over the face. These occurred during the initial stages of mining and may be related to the increased incidence of roof falls during the early stages of face advance.

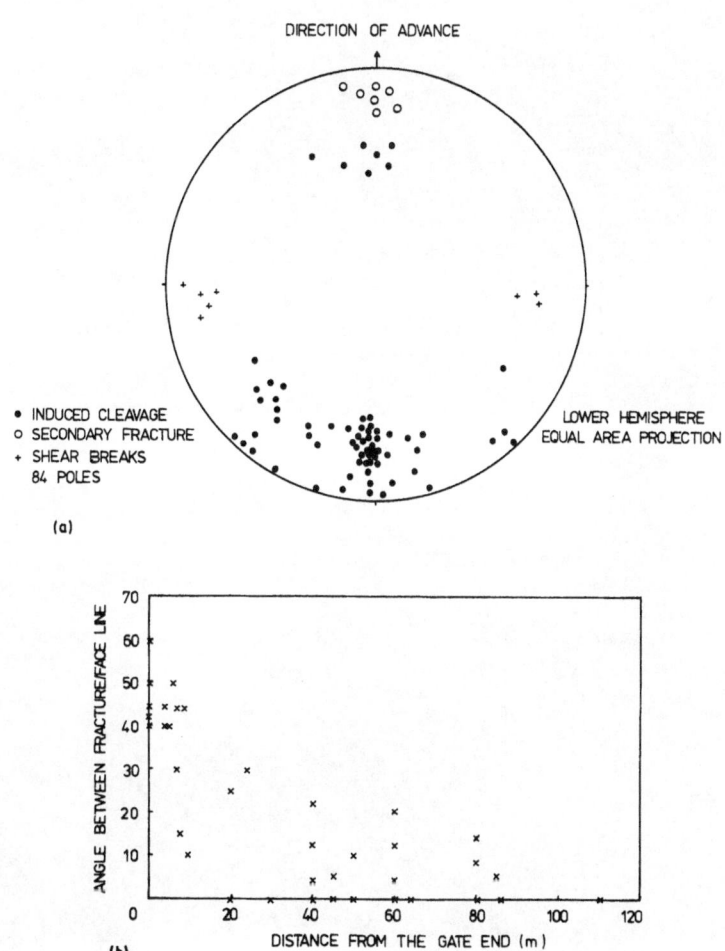

(a)

(b)

Fig. 7.5 (a) The dip and dip direction of mining induced fractures on longwall faces in the Barnsley seam of the Yorkshire coalfield plotted relative to the direction of face advance on an equal area stereographic projection; (b) the angle between the strike of the same mining induced fractures and the face line against distance from the gate end of the face (after Holmes, 1983).

(b) A group hading at about 70° over the face. These are secondary fractures resulting from rotation of blocks (see Fig. 7.2) in the roof towards the goaf, and the shearing of the face side lower corner of these blocks. If such fractures are allowed to develop in the unsupported roof ahead of the supports and intersect with induced cleavage, a triangular section of rock may be detached from the roof. Clarke (1963) describes this type of roof collapse as 'rammel conditions'.

(c) A group similar to the first but parallel to the gate roadway. These only occurred close to the ripping lip. They are arguably induced cleavage parallel to the ribside abutment and are often responsible for falls of ripping lip.

In Fig. 7.5(b) the angle between the face line and the strike of the induced cleavage planes is plotted against the position of the fractures. Within 10 m of the face end the majority of fractures strike at 45° to the face line. This angle decreases to less than 20° within 40 m of the face end. The data presented in this way confirm the observations of Fig. 7.4 that towards the face ends the planes of induced cleavage deviate away from the face line and towards the ribside.

The effect of induced cleavage can range from minor roof flaking to major roof flaking – as illustrated in Fig. 7.6. In order to determine the effect of various mining and geological parameters on coal face stability, it is important to be able to classify the overall magnitude of face deformation. Two indices have been suggested – *roof flaking* and *face spalling*.

Roof flaking is defined using Gupta's (1982) definition as the percentage of total face length occupied by cavities 300 mm deep between the front legs of the supports and the face. This is similar to Jacobi's (1966) *roof sensitivity index*. Both terms can be used to define the length of broken face roof in percentage terms.

Face spalling was measured immediately before the first shear of the first shift of the week. This is because spalling tends to be time dependent. It is expressed as the average magnitude of horizontal face spalling over the face length in terms of distance from the original face line.

Holmes (1983) in his study of the Yorkshire coalfield tested numerous variables against roof sensitivity index and face spalling. He found a high degree of correlation between both roof sensitivity index and average face spalling distance and the angle between the *main cleat* and the face line (Fig. 7.7); between roof sensitivity index and the position of the first

---

Fig. 7.6 Induced cleavage in the roof of the Barnsley seam at mines in the Yorkshire area. The falls are in mudstone roof and the close proximity of the induced cleavage planes can be clearly seen (photograph P. Holmes).

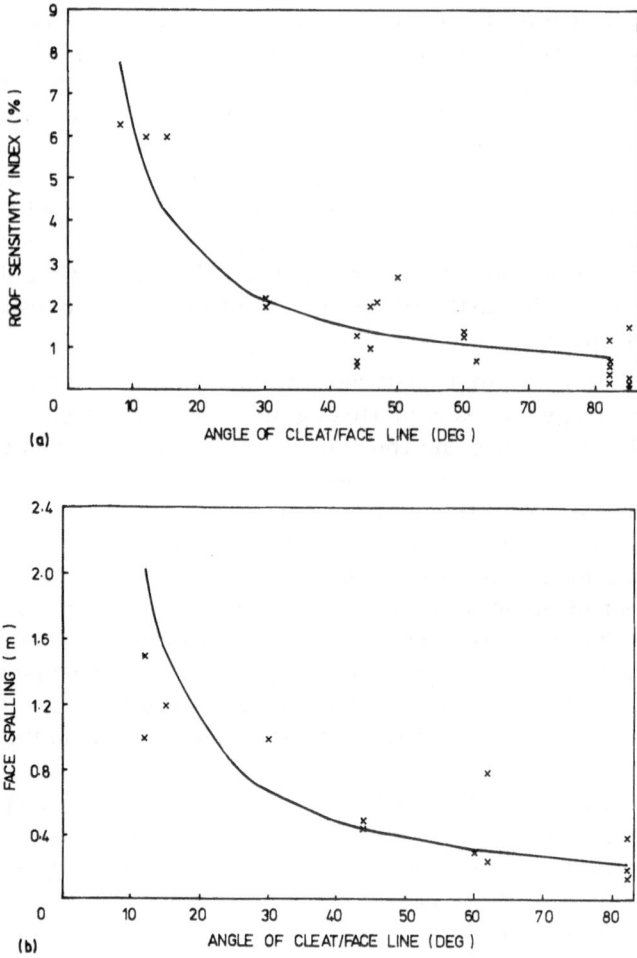

Fig. 7.7 (a) Roof sensitivity index (% of total face length containing cavities greater than 300 mm high in the front 1.5 m) and (b) face spalling (distance in metres ahead of the face before the first cut of the week) related to the angle between the main cleat strike direction and the face line direction. Each datum point represents a face in the Barnsley seam in the Yorkshire coalfield (Holmes, 1983).

competent horizon above the face roof (Fig. 7.8) and between face spalling and the percentage of mudstone seatearth in the face section (Fig. 7.9). The relation between roof sensitivity (flaking) index and setting pressures is considered later in Fig. 7.22. Correlations outside the 5% level and not deemed significant were tested between both variables and the thickness

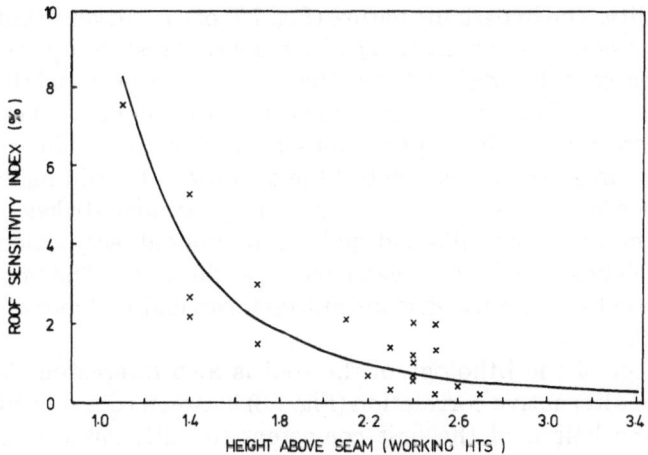

Fig. 7.8 Relation between roof sensitivity index and the ratio between height above the roof of the first competent horizon (Holmes, 1983) and the seam height (usually about 2 m).

of roof coal, the percentage of competent strata in roof zones up to 20 m, the support density and the depth of the seam.

The most important feature is the expected significant correlation, and alternative correlation between *main cleat direction* and both face and

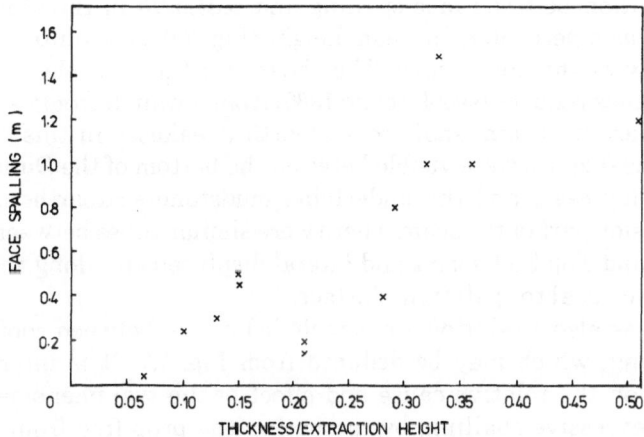

Fig. 7.9 Relation between face spalling and the ratio of mudstone seatearth to coal in the extracted seam section (Holmes, 1983).

roof stability. The hyperbolic nature (Fig. 7.7) of the curves in both cases is such that risk of both excessive spalling and roof instability are considerably increased if the angle between the strike of the cleat and the face line is less than 30°. It is interesting to consider the similarity between this observation and similar observations of planar sliding in rock slopes (Hoek and Bray, 1976) where one of the conditions for sliding is a similar coincidence between slope and preferred discontinuity strikes.

In coal mining instability and spalling are undoubtedly exacerbated by the coincidence of induced breaks and cleat direction. The imposition of induced breaks on natural discontinuities forms major planes of weakness in the coal.

The effect of the lithology of the roof is also interesting. There is a strong and alternative correlation (Fig. 7.8) between roof stability and the ratio of the height of the first competent (usually sandstone) horizon above the roof and the seam height. This indicates that the greatest incidence of roof falls is associated with faces where the thickness of weaker rocks (usually mudstones) is less than 1.5 times the seam height or about 3 m. The reason for this will be discussed in greater detail where roof deformations are considered in Figs 7.23 to 7.25. The simplest explanation is the presence of bedding plane *shear zones* (see Chapter 1) between the base of the stronger sandstones and the top of the weaker mudstones. Separation will undoubtedly tend to occur at this horizon and blocks delineated by natural or imposed discontinuities will be difficult to control if there is insufficient cover above the supports. The data in Fig. 7.8 are a strong indication of the importance of bedding plane shear zones above longwall faces.

The relation between face spalling and *seatearth* in the seam section expressed as a percentage of seam height (Fig. 7.9) says quite a lot about the quality of the coal mined. The shear resistance of cleat or discontinuity planes reduces (see Jeremic 1980) from durain (strongest) through vitrain, clarain, jusain, shale to seatearth (weakest). In this case quite distinct shear zones were visible between the bottom of the Dunsil part of the Barnsley seam and the underlying mudstone seatearths above the Low Barnsley part of the seam. There were similar zones between the Low Barnsley and Top Soft seams and lateral displacement along these shear zones caused coal to spall from the face.

There was also a relation – not included here – between roof stability and spalling, which may be deduced from Fig. 7.7. It is interesting to speculate on the relative cause and effect of the two phenomena. Most probably excessive spalling, by increasing the prop free front distance, may increase the likelihood of convergence and roof instability. On the other hand, the extension of roof fractures and particularly the delays caused by timbering may increase spalling.

7.2   Face roof supports

The emphasis on fractures, flaking and spalling on a longwall face highlights the support requirements to control a dynamic deformation process over a relatively short period. Ideally support is required:

(a) To exert some passive resistance to the roof during *setting* in order to control deformation, flaking and spalling in the face area and

(b) To mobilize some active resistance by building up *yield* resistance to further deformation during the mining cycle.

Unfortunately the emphasis in support design has, until recently, been on yield resistance rather than setting forces, and since deformation is necessary in order to mobilize yield resistance this has allowed some of the movements illustrated in Fig. 7.1, with resultant face instability problems. It is nevertheless important to examine the design methods developed by Wilson (1975) for two main reasons. The first is that they represent one of the first attempts to treat a mine structure as a conventional engineering structure; the second is that while they set out to provide estimates of yield resistance, they can equally well be used to determine setting force. The ideal support is one where setting force and yield resistance are equal.

At the present time there are four main types of hydraulic support in use:

Frame supports, which were developed from the traditional layout of hydraulic props under a single roof bar. They are designed with walking ability, independent of the face conveyor, to which they may or may not be attached. The main disadvantage of such a simple structure is that the frames may tend to be unstable laterally on gradients and on uneven or soft floors, and that fragmented rock from the goaf and cavities may flush into the face area.

Chock supports (Fig. 7.10(a)) comprise four to six hydraulic props connected by a common roof canopy and floor base, which may be rigid or articulated. They are attached to the face conveyor through a horizontal double-acting hydraulic cylinder and are available for working heights between 0.6 and 3.7 m. They have large diameter double-acting legs and in some cases the legs are double telescopic (Shepherd, 1977). Chocks are categorized by the number of legs and the total yield load (in tonnes); for example 4/300, 5/150, 6/240. The specification for chocks varies with the conditions in which they are to be used and they may be grouped into thin, medium and thick seam chocks. The most common types of support suitable for the *thin seam* range of 0.6 to 1.4 m are 5/150 and 6/180. Higher capacity supports have been manufactured for very strong roof conditions.

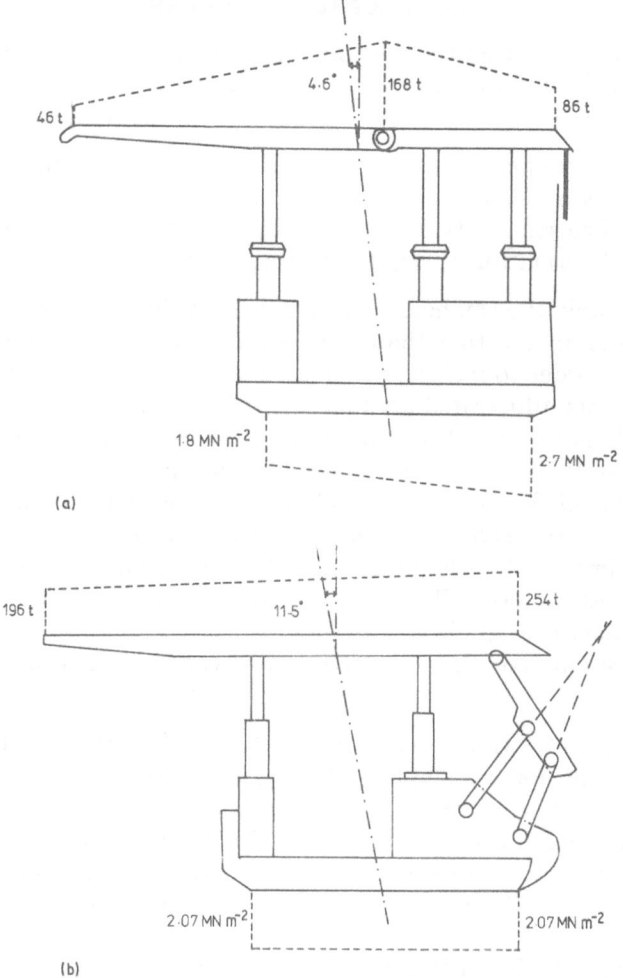

Fig. 7.10 (a) Section through Gullick Dobson, 6/300 tonne, rigid base fully side-shielded chock support with hypothetical roof load and base pressure distribution (after Wilson, 1983); (b) Section through Gullick Dobson, 4/450 tonne, lemniscate chock shield support with hypothetical roof load and base pressure distribution (after Wilson, 1983).

Medium seam supports, suitable for the 1.4 to 2.2 m seam range, cover most of the coal faces in Britain. The most common support capacity in this range is 6/240 but supports up to 720 tonnes capacity are available. They are normally fitted with rubber shear mountings at the leg bases to compensate for any lateral roof-to-floor movement. *Thick seam* supports, suitable for the 2.5 to 3.5 m seam range, are widely used (Morgan, 1982) in Russia, Poland, Czechoslovakia and parts of West Germany. They are normally provided with extensible cantilevers designed to yield at 10 tonnes or more and in a few versions the cantilevers are linked with a load-bearing face 'sprag' in order to stabilize the face coal.

Shield supports comprise a floor beam, caving shield, roof beam, hydraulic legs and additional hydraulic components for control and advancement. Wilson (1980) suggests four types of shield depending on their mode of operation, location, leg disposition and yield. These are illustrated, together with hypothetical canopy force distributions, in Fig. 7.11.

Two-leg simple caliper shields (Fig. 7.11(a)) have a number of limitations, which include variation of load resistance with height, low yield

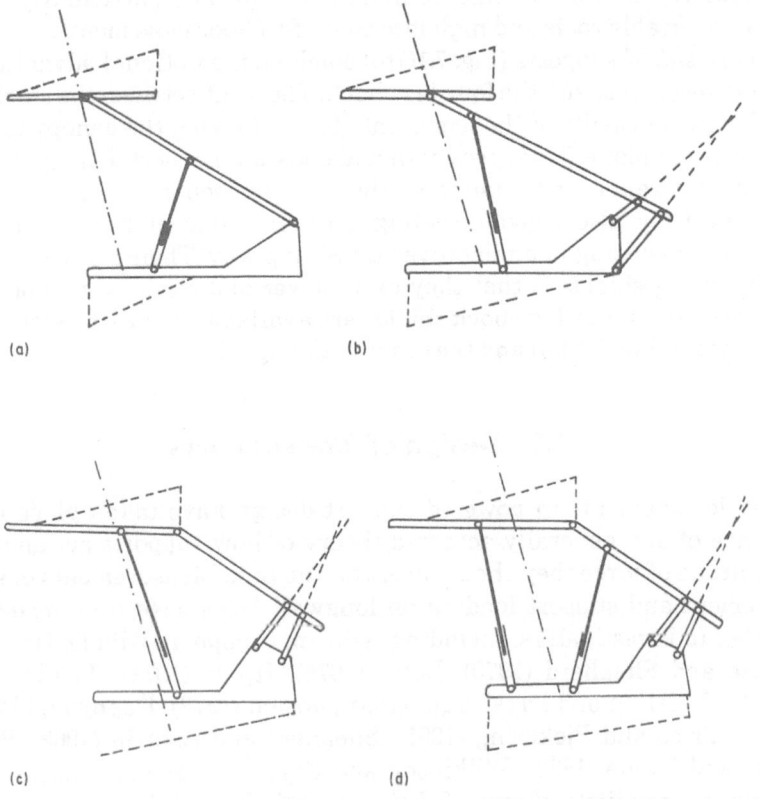

Fig. 7.11 Typical designs of shield support (after Wilson, 1980): (a) two-leg simple caliper; (b) two-leg lemniscate caliper; (c) two-leg lemniscate caliper (legs from base to roof canopy); (d) four-leg lemniscate caliper.

load, variation in canopy tip to face distance, very high base tip loading, limited travelling way, instability of roof canopy and potential ingress of roof material. Similar problems are also associated with two-leg *lemniscate caliper* shields (Fig. 7.11(b)) although the lemniscate linkage keeps the canopy equidistant from the coal face and increases support density for a given leg loading. These limitations can be overcome by attaching the legs directly from the base to the roof canopy (Fig. 7.11(c)) to increase yield loads for a given leg load. This also provides increased roof canopy stability, improved floor loading characteristics and a better travelling way. Further improvements can be obtained by adding two additional legs from the base to the shield (Fig. 7.11(d)). This improves tip loads, increases the travelling way behind the front legs, stabilizes the canopy and provides more even floor loading. Shield supports with capacities up to 600 US tons (5.6 MN) capacity are available. They are particularly useful in difficult mining conditions, particularly where there are friable roofs and high lateral roof-to-floor movements.

Chock shield supports (Fig. 7.10(b)) combine the notional advantages of the two-leg shield and the four-leg chock. The joint between the shield and the base is normally of the lemniscate type, allowing the canopy to move in a vertical plane. This gives high and constant support density over the full height range and eliminates the need for separate leg restoration devices. They also provide sealing between adjacent rear shields and canopies, resulting in an improved travelling way. Their main advantage over two-leg shields is that they exert lower and more even floor loads. Two designs of four-leg chock shields are available, one with vertical legs as shown in Fig. 7.10(b) and the other with angled legs.

7.3   Design of face supports

Most developments in powered support design have taken place in the absence of any generally accepted theory of how supports act and what magnitude of force they should exert on the roof. Measurements of strata movement and support loading on longwall faces have been made by a number of investigators, including Ashwin, Campbell, Kibble, Haskayne Moore and Shepherd (1970), Bates (1978), Hinde (1978), Jacobi (1966), Jacobi, Everling and Irresberger (1964), Josien (1972), Kenny and Wilson (1963), Price and Pickering (1981), Shepherd and Ashwin (1968), Smart, Isaac and Hinde (1980), Wilkinson and Evans (1963) and Wilson (1964). However, very little, if any, of their research findings have been directly incorporated into the design of powered supports.

Early trials with powered supports were based on the assumption that good roof control required a certain amount of convergence. Accordingly, most of the supports were designed for a *rated load density* of between 0.08 and $0.2\,\mathrm{MN\,m^{-2}}$ for thin and thick seam workings respectively. With the change in emphasis from manpower intensive to machine intensive production, it became necessary to concentrate mining operations on fewer faces with higher rates of face advance in order to achieve high production and productivity, maximum machine utilization, and lower production costs. Consequently, more robust supports were designed to meet these requirements. A typical example is the 4/800 US ton (726 tonne) rigid base Gullick Dobson powered support which has been successfully used at Appin Colliery, Australia (Brabbins, 1978). Its rated yield load density is $1.53\,\mathrm{MN\,m^{-2}}$ and the setting load density $0.47\,\mathrm{MN\,m^{-2}}$. If it is assumed that yield load and setting are the most important parameters in support design, then these loads can be defined for a given support system as:

(a) The weight of rock needing support at any time or
(b) The resistance offered by supports which have been found to operate satisfactorily

Wilson (1975), using the latter approach, concluded from observations that roof to floor convergence was minimal above a mean load density of $0.1\,\mathrm{MN\,m^{-2}}$. *Mean load density* (MLD) is defined as the thrust offered by the support to the area of roof supported. However, to incorporate time dependent phenomena and the effects of changes in area due to shearing, Wilson (1964) suggested a method for calculating the overall MLD for the complete cycle as:

$$(\mathrm{MLD})_{\mathrm{cycle}} = \Sigma_n^1 \frac{P_n}{A_n} \tag{7.1}$$

where P is the mean resistance over the cycle time t of each of n supports on the face and A is the area of roof acting on each support.

When MLD fell below $0.1\,\mathrm{MN\,m^{-2}}$ Wilson (1964) observed an increase in convergence. MLD values less than $0.05\,\mathrm{MN\,m^{-2}}$ resulted in unstable roof conditions for seam heights between 0.7 n and 1.7 m. Similar conclusions were drawn by Shepherd (1964) and Shepherd and Ashwin (1968). However, later work by Ashwin *et al.* (1970) indicated that MLD values should be increased in proportion to the height of extraction from $0.14\,\mathrm{MN\,m^{-2}}$ in seams less than 1 m high to $0.34\,\mathrm{MN\,m^{-2}}$ in seams greater than $2.4\,\mathrm{MN\,m^{-2}}$. A statistical survey showed that the majority of MLD values were in the $0.1{-}0.2\,\mathrm{MN\,m^{-2}}$ range, although the nominal yield load density was about $0.35\,\mathrm{MN\,m^{-2}}$. Ashwin *et al.* (1970) suggested that it would be interesting to investigate the effect of variation of load density during the working cycle on face conditions. They pointed out that the only way to be sure that a certain MLD is achieved is to make the *setting load density* (SLD) equal to the mean (yield) load density (MLD). They

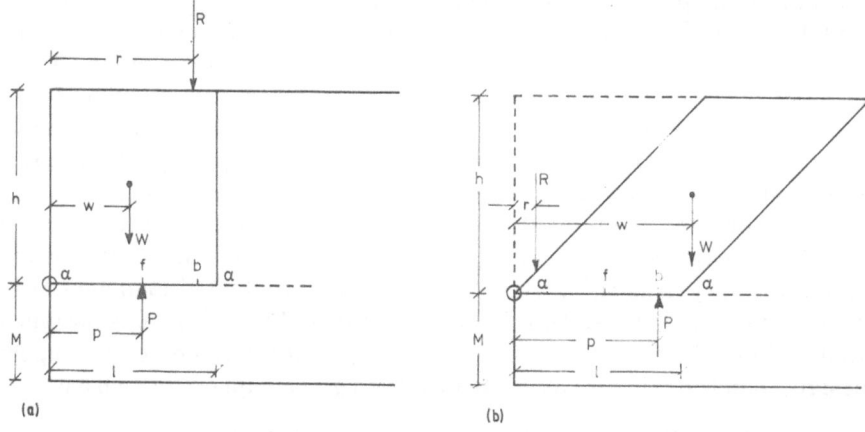

Fig. 7.12 Dimensions of a roof block above a face support having a width including the prop-free front distance, in a seam of height M (after Wilson, 1975): (a) weight forward of support thrust; (b) weight to rear of support thrust. Based on Kenny's (1969) concept of caving height, $h = M/(b-1)$ where b is the bulking factor. If $b = 1.5$, $h = 2M$.

defined setting load density as the support load at the time of setting divided by the roof area supported when the supports are standing back from the conveyor.

Wilson (1975) pointed out that although the results obtained from investigations enabled a minimum load density to be postulated, they gave no indication of the optimum distribution of load between the front and back legs, nor how load densities should be varied for extreme conditions. He suggested that the minimum support thrusts required could be estimated by judging the amount of rock which might fall if the supports were removed. Wilson (1975, 1980) defined this rock body in two dimensions as a *roof block* having dimensions determined by the width of the supported face (Fig. 7.12) held in static equilibrium by a combination of forces:

The *body weight W*, taken as a single vertical force acting at the centre of the block and at a distance w from the face line.

The *combined support resistance P*, acting at a distance p from the face line. p depends on the distribution of load between the support legs, the geometry of the support, and the prop-free front distance.

The *reaction R* between the freed block and the remaining mass of the roof, acting at a distance r from the face line. The presence of force R is necessary for stability unless P and W act along the same line. The reaction R lies on the same side of W as P, although the position of both will depend on the direction in which the roof block will tend to rotate. This direction in turn will depend on the interaction between support, nether and upper roof.

For equilibrium of the roof blocks, two basic conditions must be met; the algebraic sum of the forces on the block must equal zero:

$$P = W + R$$

and moments about any point, say O, must equal zero:

$$Pp = Ww + Rr$$

The resultant thrust P of all the legs of the support unit can be conveniently divided into two parallel forces, F and B, where F is the sum of thrusts on the front leg or legs and B is the sum of the thrusts on the back legs. These can be taken as acting at distances of f and b from the face; for any particular installation these values will be known. Then for F and B to be statically equal to P:

$$F + B = P$$

and

$$Ff + Bb = Pp$$

Substituting for P and Pp gives:

$$F + B = W + R$$

$$Ff + Bb = Rr + Ww$$

$F + B$ will be a minimum when R is a minimum. If W lies between f and b, R can be made zero by putting:

$$F = W \frac{(b - w)}{(b - f)}$$

and

$$B = W \frac{(w - f)}{(b - f)}$$

If w is less than f, R will be a minimum when:

$$F = W \frac{(r - w)}{(r - f)} \quad \text{and} \quad B = 0$$

where r is as large as possible as in Fig. 7.12(a).

Table 7.1 Minimum values for front (F) (MN) and rear (B) (MN) support forces and recommended minimum mean load densities (MLD) (MN m^{-2}) for level seams*

Roof rock	Seam thickness (m)	l = 3.0 m (before cut)			l = 3.6 m (after cut)			l = 4.2 m (wide web)		
		F (MN)	B (MN)	MLD (MN m^{-2})	F (MN)	B (MN)	MLD (MN m^{-2})	F (MN)	B (MN)	MLD (MN m^{-2})
Very strong (α = 30°)	1	—	0.21	0.07	—	0.22	0.06	—	0.22	0.05
	2	—	0.67	0.22	—	0.66	0.18	—	0.66	0.16
	3	—	1.39	0.46	—	1.35	0.38	—	1.34	0.32
Strong (α = 45°)	1	—	0.15	0.05	—	0.13	0.05	—	0.10	0.05
	2	—	0.45	0.15	—	0.46	0.13	—	0.47	0.11
	3	—	0.90	0.30	—	0.89	0.25	—	0.90	0.22
Medium (α = 60°)	1	0.06	0.09	0.05	0.12	0.06	0.05	0.19	0.02	0.05
	2	—	0.32	0.11	0.05	0.31	0.10	0.17	0.25	0.10
	3	—	0.61	0.20	—	0.63	0.17	—	0.66	0.16
Weak (α = 75°)	1	0.10	0.05	0.05	0.17	0.01	0.05	0.25	—	0.06
	2	0.13	0.17	0.10	0.25	0.11	0.10	0.41	0.01	0.10
	3	0.08	0.37	0.15	0.24	0.30	0.15	0.46	0.17	0.15
Very weak (α = 90°)	1	0.14	0.01	0.05	0.21	—	0.06	0.31	—	0.07
	2	0.27	0.03	0.10	0.43	—	0.12	0.61	—	0.14
	3	0.41	0.04	0.15	0.64	—	0.18	0.18	0.92	0.22

* Assume h = 2M, γ = 25 kN m^{-3}, r = 2w −0.5 m (w < f), r = 0.5 m (w > b), s = 1 m, f = l −1.6 m, b = l −0.5 m (after Wilson, 1975).

If w is greater than b, the minimum condition will be achieved by putting:

$$F = 0; \; B = W\frac{(w-r)}{(b-r)}$$

where r is as small as in Fig. 7.12(b).

If the angle of break is assumed equal to the caving angle α, W and w can be found from the relationship:

$$W = \gamma slh = 2\gamma slM \quad \text{if} \quad h = 2M$$

$$w = \frac{l}{2} + \frac{h}{2}\cot\alpha$$

where γ is the unit weight of the roof material and s is the spacing of the support units along the face.

Thus from simple assumptions, it is possible to calculate front and back support forces and hence the minimum mean load density. An example of this, calculated by Wilson (1975), is given in Table 7.1. This uses as a preliminary assumption the relation discussed earlier between caving height, seam thickness and bulking factor ($h = M/(B-1)$) and a notional relation between rock strength and caving angle, both of which may be open to question. The results do, however, illustrate that greater seam thickness, angle of caving and width of extraction require greater support capacity, and the magnitudes of MLD appear sensible. One important observation which can be guessed from Fig. 7.12 is that at low caving angles the front leg force is low or zero and at high caving angles the rear leg force is low or zero. Wilson (1975) suggests that where this is the case minimum MLD values be doubled.

In the case of steep seams (Carver, Cowan and Binns, 1977) where the face line is close to full dip, the component of gravity down the dip tends to create instability. Wilson (1980, 1983) has examined the conditions for rotational instability and shown that for a seam inclined at 30°, the setting force per support must be nearly doubled to prevent movement.

7.4   Face support setting pressures

In Britain it has been, until recently, the practice (Graham, 1978) to have line pressures for hydraulic supports of between 10 and 16 MPa. At these pressures, the setting load on a typical 40 tonne power set leg is between 9 and 14 tonnes, giving a setting load of between 25 and 30% of yield or up to 45% with double telescopic legs. European practice (Graham, 1978; Jackson, 1968; Everling, 1977) has been to aim for setting pressures of

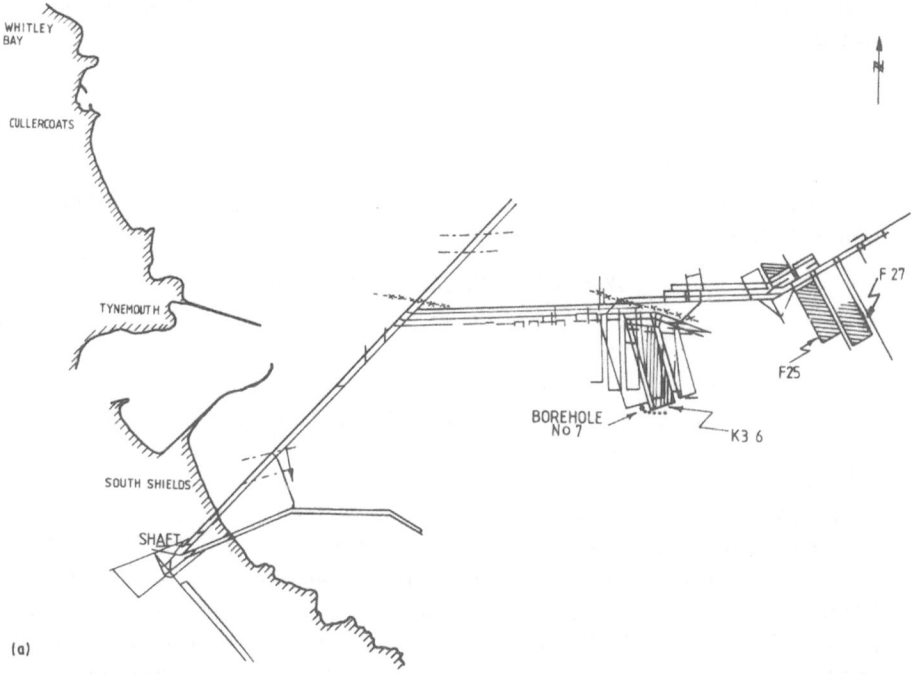

(a)

Fig. 7.13   (a) Plan of faces F25, F27 and K36 at Westoe Colliery, Tyne and Wear, Britain: (b) section through borehole No. 7; (c) detailed section through F25 face. The location of the colliery can be obtained from Fig. 5.16.

around 75 to 80% of yield pressure – usually expressed as a *yield-setting pressure ratio* of 1.25–1.33. This requires (Hess, 1972) a line pressure of 30–40 MPa to give setting loads of 60–100 tonnes per leg. In Australia and the USA (Linden, 1977) line pressures of 20–25 MPa give setting pressures at around 50% of yield (a yield-setting pressure ratio of 2).

It is difficult to justify the British practice. One of the reasons why it may be acceptable is that British mine roofs tend to be weak and to cave relatively easily. It is, however, now generally accepted (see Bates, Butler, Smith and Waring (1975), Waring (1977), Price and Pickering (1981)) that higher setting loads do create better face conditions and there has recently been an increase in the use of high pressure lines and guaranteed positive set values and sequence control operations.

The effectiveness of increased setting pressures can be illustrated through a series of underground observations carried out by Gupta (1982) and described by Gupta and Farmer (1982, 1983). The objective of the observation programme was to examine the effect of setting loads on strata and support behaviour on a longwall face. The faces investigated

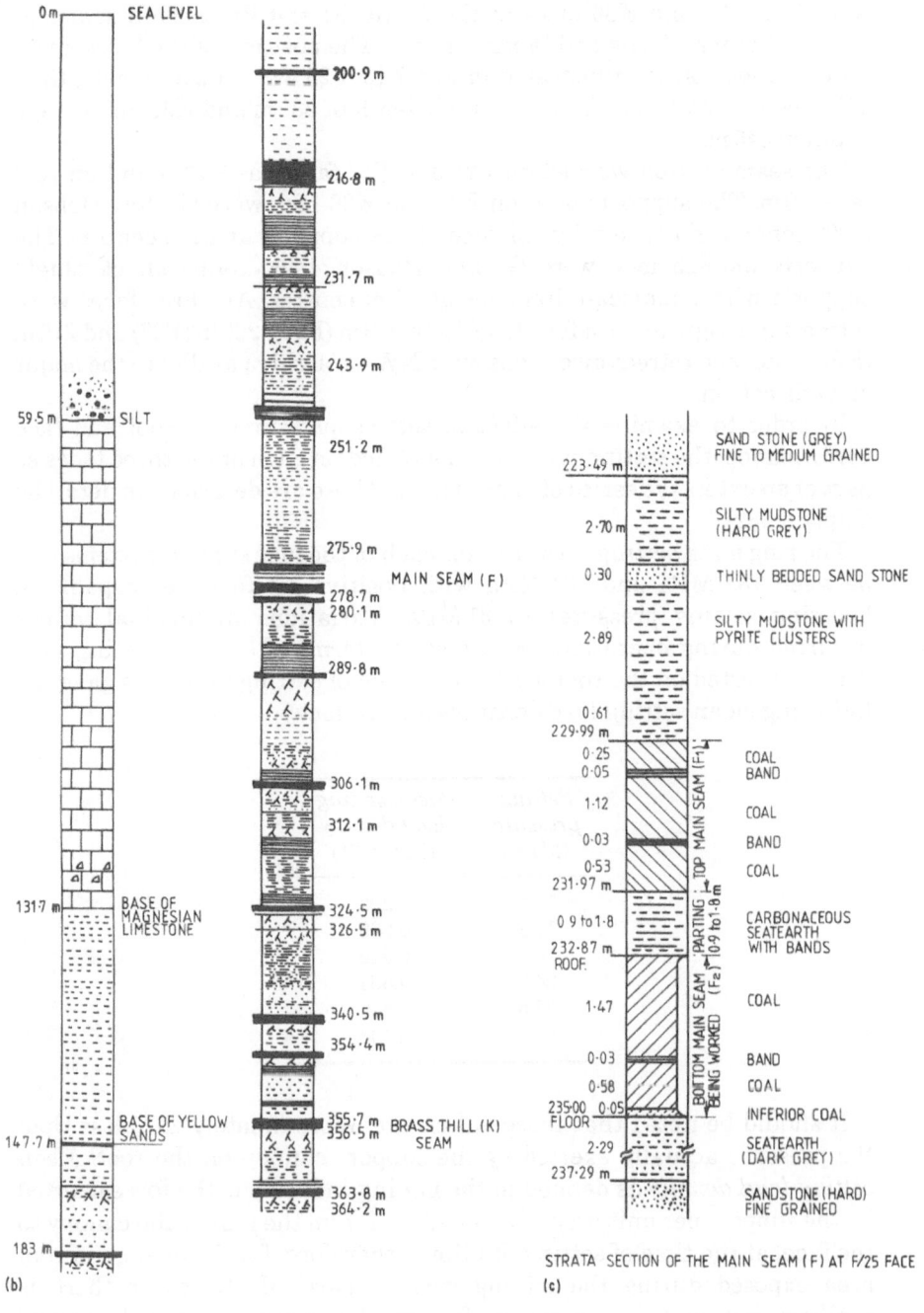

0 m — SEA LEVEL

59.5 m — SILT

131.7 m — BASE OF MAGNESIAN LIMESTONE

147.7 m — BASE OF YELLOW SANDS

183 m

(b)

200.9 m
216.8 m
231.7 m
243.9 m
251.2 m
275.9 m — MAIN SEAM (F)
278.7 m
280.1 m
289.8 m
306.1 m
312.1 m
324.5 m
326.5 m
340.5 m
354.4 m
355.7 m — BRASS THILL (K) SEAM
356.5 m
363.8 m
364.2 m

STRATA SECTION OF THE MAIN SEAM (F) AT F/25 FACE

(c)

Depth	Thickness (m)	Description
223.49 m		SAND STONE (GREY) FINE TO MEDIUM GRAINED
	2.70	SILTY MUDSTONE (HARD GREY)
	0.30	THINLY BEDDED SAND STONE
	2.89	SILTY MUDSTONE WITH PYRITE CLUSTERS
229.99 m	0.61	
	0.25	COAL
	0.05	BAND
	1.12	COAL
	0.03	BAND
231.97 m	0.53	COAL
232.87 m	0.9 to 1.8	CARBONACEOUS SEATEARTH WITH BANDS
ROOF		
	1.47	COAL
	0.03	BAND
	0.58	COAL
235.00	0.05	INFERIOR COAL
FLOOR		
	2.29	SEATEARTH (DARK GREY)
237.29 m		SANDSTONE (HARD) FINE GRAINED

TOP MAIN SEAM (F₁) 0.9 to 1.8 m
PARTING
BOTTOM MAIN SEAM (F₂) 0.9 to 1.8 m BEING WORKED

were F25, F27 and K36 faces in the Main (F) and Brass Thill seams at Westoe Colliery, Tyne and Wear, Britain. The position of the faces and a borehole section are illustrated in Fig. 7.13. F25 was an average depth of 277 m below sea level; F27 an average depth of 232 m and K36 an average depth of 345 m.

The seam section worked on F25 and F27 face was 2.13 m and on K36 face 2.0 m. The supports used on F27 and K36 face were Gullick Dobson 6/300 tonne rigid base fully shielded chock supports at 1.2 m centres. The supports on F25 face were Gullick Dobson 4/450 tonne chock shield supports with lemniscate linkages at 1.5 m centres. All three faces were retreating longwall with face lengths of 219 m (F25), 220 m (F27) and 215 m (K36). The face retreat directions were NW – roughly parallel to the major cleat direction.

In order to examine the effect of setting pressures on roof and face performance, the setting pressure was varied on each of the three faces as part of an extensive series of experiments. These are described in detail by Gupta (1982).

The ring main setting pressures for each of the face support systems was between 13.8 MPa and 16.8 MPa with positive set facilities capable of boosting setting pressures to 31 MPa. Variations in nominal values occurred during operations. Expressed in terms of support setting load density exerted on the roof a selected range of setting pressures gave the following mean setting load densities for F25 face:

Setting pressure (MPa)	Mean setting load density $(MN\,m^{-2})$
13.8	0.208
17.2	0.260
20.7	0.312
24.1	0.364
27.5	0.416
31.0	0.468

It should be noted that these values are approximately 20 % less than the pressure actually exerted by the support canopy on the roof. *Mean setting load density* as defined in the mining industry is the force exerted by the support per unit area of exposed roof from the rear of the canopy to the face, at the time of setting. It allows, therefore, for the change in roof area exposed during the mining cycle – particularly where there is spalling, but may be an unsatisfactory definition of applied support pressure. In the case of F27 and K36 faces mean setting load density was

10 % higher than F25 because the distance between the front of the canopy and the face was shorter.

In designing the experiment, the *setting pressure*, and the related *mean setting load density*, was selected as the independent variable since this is the feature of the face supports which immediately affects the lower roof strata. In the full programme the following observations were obtained at each of the setting pressures listed above:

(a) All leg pressures at each support at the time of setting and immediately before moveover
(b) Leg closures at 10 % of supports
(c) Roof to floor convergence at the same location as leg closures
(d) Support inclinations at these locations
(e) Roof and floor debris compaction at these locations
(f) Lateral movement of roof relative to floor at these locations
(g) Roof flaking and cavity formation throughout the face
(h) Cavities greater than 100 mm high and 30 mm wide were monitored
(i) Face spalling throughout the face. This was considered significant

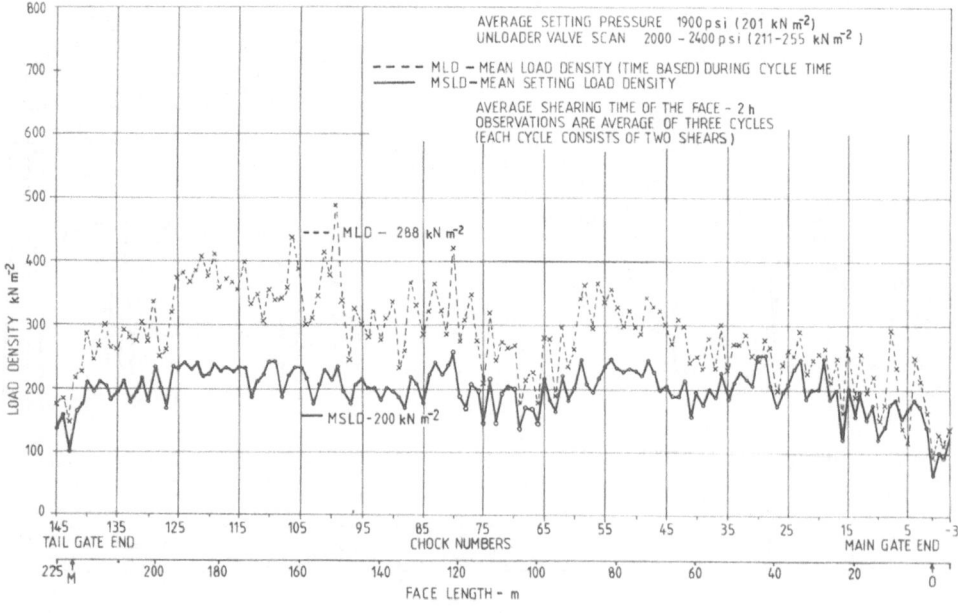

Fig. 7.14 Increase in mean load density (MLD) from mean setting load density (MSLD) on F25 face at Westoe Colliery, during a cycle between two shears over a full face. Nominal setting pressure for each chock was 13.8 MPa (after Gupta, 1982).

and monitored when the distance from the tip of the support canopy to the face exceeded 500 mm.

(j) Lateral movement of the coal face resulting from dilation behind the face.

Each of these observations was obtained over four cutting cycles and two weekend breaks at each setting pressure. The cutting cycles represented an average duration of 2 h 30 min during which the face and supports were advanced 0.45 m on F25 and K36 face and 0.65 m on F27 face. The weekend break was of average duration 60 h from setting to first moveover.

In addition to the main observation programme, at one location at the centre of F25 and K36 face and three locations on F37 face, roof and floor deformations were observed using an anchor wire or magnetic ring extensometer system installed in a borehole vertically above the face supports. The detailed instrumentation has been described by Gupta and Farmer (1981).

In Figs 7.14 to 7.17 the detailed performance of chocks on F25 face at the

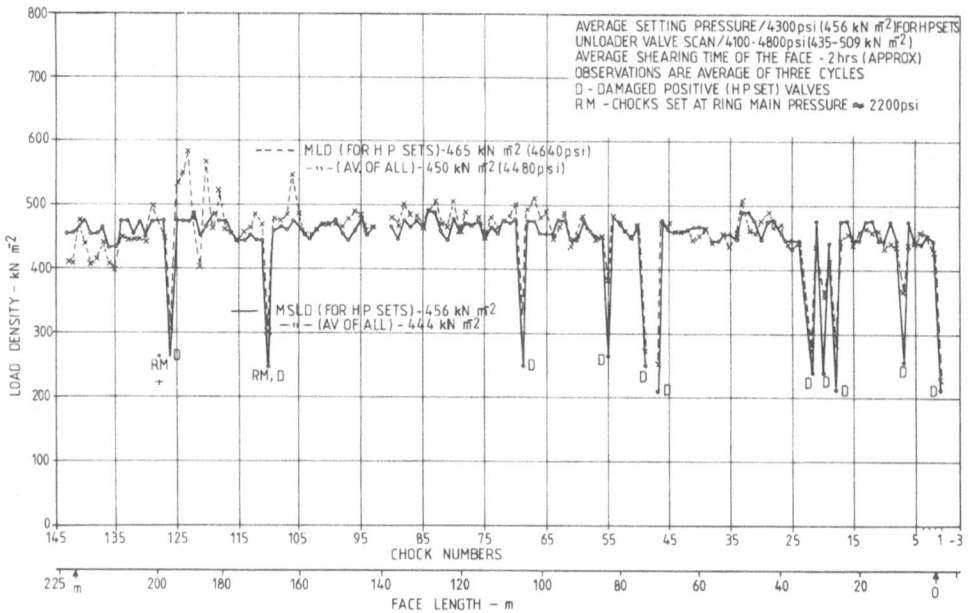

Fig. 7.15 Increase in mean load density (MLD) from mean setting load density (MSLD) on F25 face at Westoe Colliery during a cycle between two shears over a full face. Nominal setting pressure for each chock was 31 MPa (after Gupta, 1982).

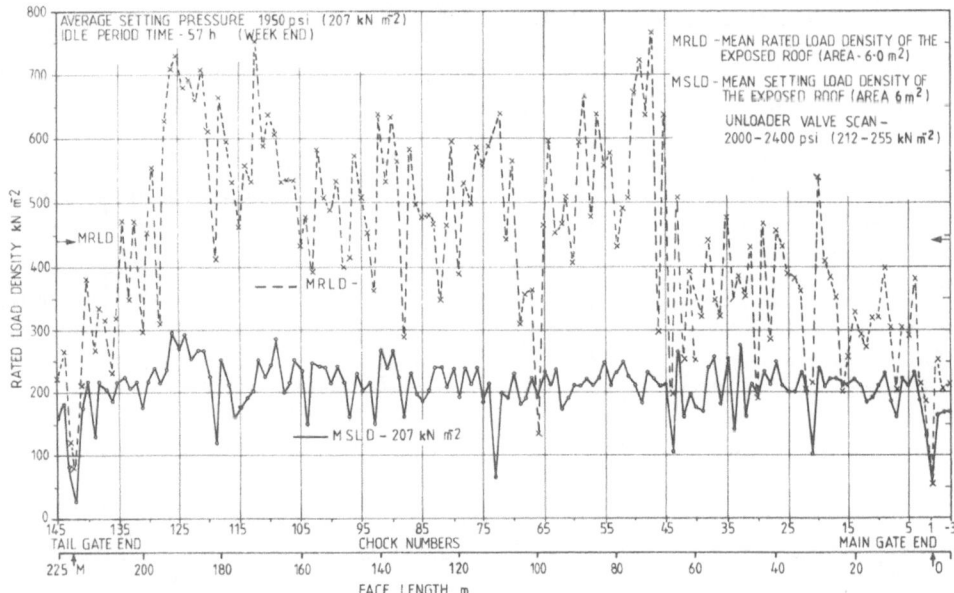

Fig. 7.16 Increase in mean rated load density (MRLD) from mean setting load density (MSLD) on F25 face at Westoe Colliery during a weekend stand period of 57 hours over a full face. Nominal setting pressure for each chock was 13.8 MPa (after Gupta, 1982).

highest and lowest setting pressures selected are considered in detail. The lowest nominal setting pressure was 13.8 MPa, giving an average setting pressure of 13.1 MPa and mean setting load density of 200 $kN\,m^{-2}$. The highest nominal setting pressure was 31 MPa, giving an average setting pressure of 29.6 MPa and mean setting load density of 456 $kN\,m^{-2}$.

Figures 7.14 and 7.15 show the increase in pressures for each chock along the face length over a *single* shearing cycle of approximately two hours duration. Results are based on average measurements over three cycles. It can be seen that chocks set at the lower average setting pressures (Fig. 7.14) of 13.1 MPa showed an increase in mean load density (MLD) over the cycle to 288 $kN\,m^{-2}$. Chocks set at the higher average setting pressure (Fig. 7.15) of 29.6 MPa showed an insignificant increase in MLD to 465 $kN\,m^{-2}$. In the case of many chocks the mean load density did not increase above the mean setting load density.

Figures 7.16 and 7.17 show the increase in pressure for each chock along the face over a *weekend* period of 57 to 60 h. In the case of the lower average setting pressure (Fig. 7.16) of 13.1 MPa, the mean load density more than doubled from 207 $kN\,m^{-2}$ to 442 $kN\,m^{-2}$ over the weekend. In

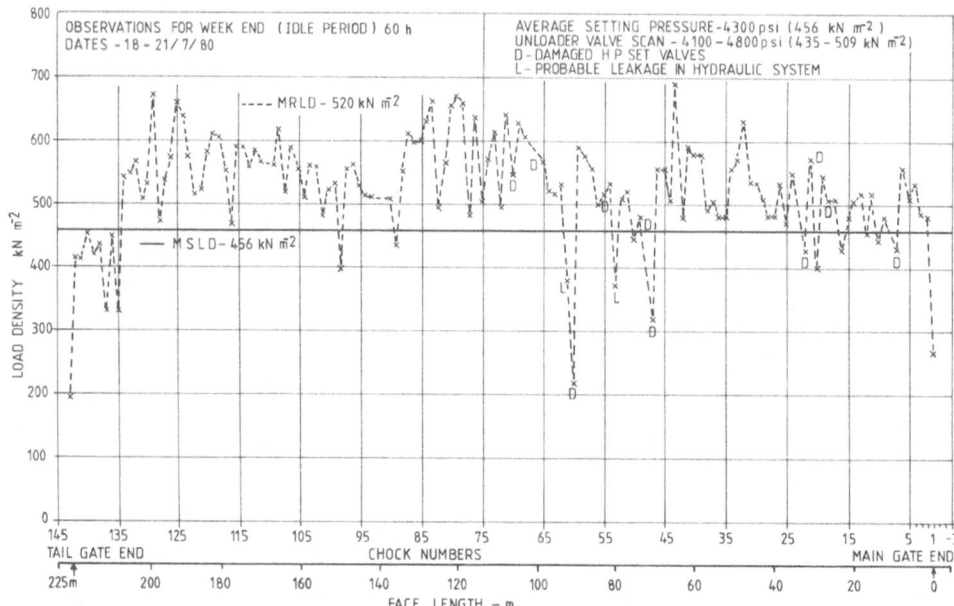

Fig. 7.17 Increase in mean rated load density (MRLD) from mean setting load density (MSLD) on F25 face at Westoe Colliery during a weekend stand period of 60 hours over a full face. Nominal setting pressure for each chock was 31 MPa (after Gupta, 1982).

the case of the higher setting pressure (Fig. 7.17) of 29.6 MPa, mean load density increased from $456\,kN\,m^{-2}$ (individual chock densities are not plotted) to $520\,kN\,m^{-2}$.

Figures 7.14 to 7.17 represent the extreme cases. Results for the range of setting load densities (MSLD) and resultant mean load densities (MLD) are summarized in Fig. 7.18, which plots the difference between MLD and MSLD over a cycle and MRLD and MSLD over a weekend for the full range of pressures. It should be pointed out that MSLD and MRLD are calculated on the basis of the exposed roof area which is continuous over the face. Definitions of these quantities are given under Fig. 7.18.

Figure 7.18 shows an inverse relationship between load increase and setting pressure, which is evidently affected by time. The implication is that load and hence convergence increases with time and with reduced load density.

Overall, the following points may be made from the observations, which are confirmed or expanded by similar observations on F27 and F36 faces:

(a) Because of multiple demands on the ring main hydraulic power supply and distribution system during support advance the average observed setting pressures were 70–80 % of the nominal value and fluctuations in setting pressures were high. Positive set valves enabled face supports to achieve a higher average observed setting pressure of about 90 % of the observed value and to reduce fluctuation of setting pressures.

(b) A support set at a low setting pressure tended to develop a low final pressure when compared with adjacent supports. This is because the

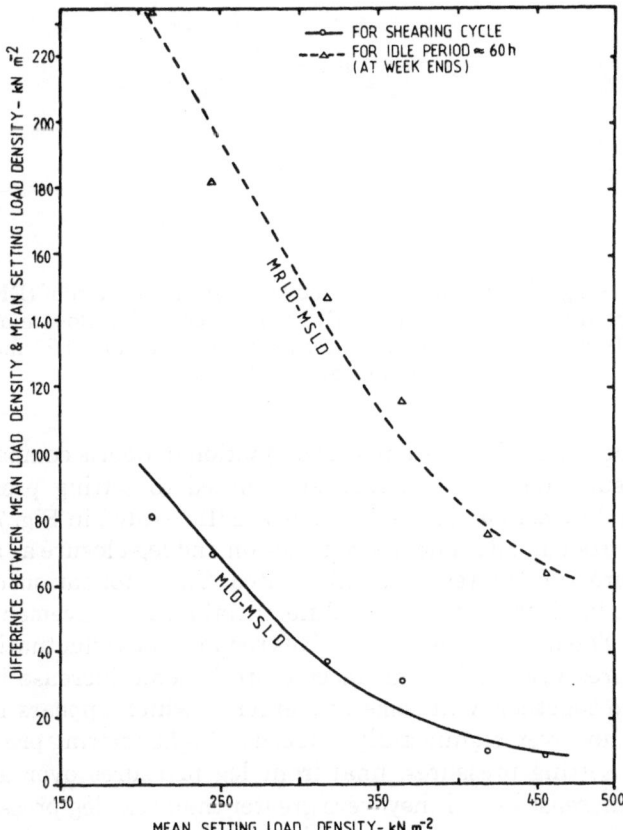

Fig. 7.18 Relation between the increase in mean (MLD) or mean rated (MRLD) load density over a cycle or weekend and the mean setting load density MSLD, for F25 face at Westoe Colliery (after Gupta, 1982). Note that MSLD is calculated from the total setting force in the front and rear legs divided by the exposed roof area with the chocks drawn forward and assumed to be 6 m² per chock. MRLD is used for idle periods and is the total ultimate force in the front and rear legs divided by 6 m². MLD is the final force at the end of a cutting cycle in the front and rear legs divided in this case by 6.8 m² to allow for the increased roof area after shearing and before drawing forward.

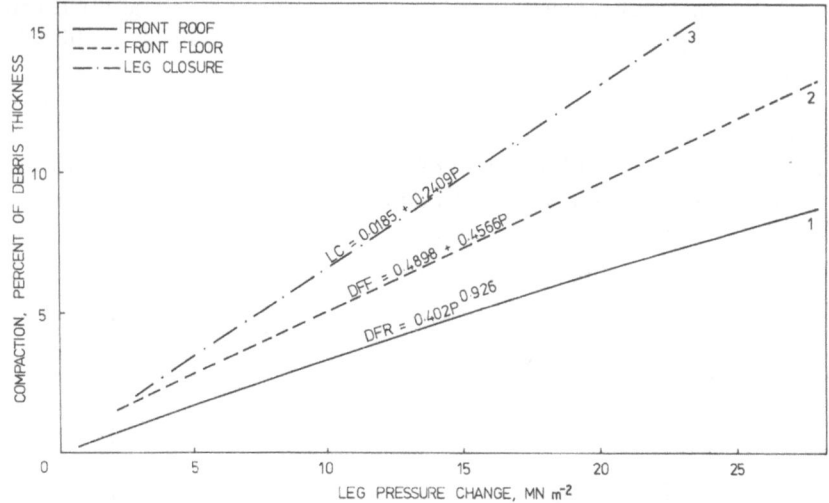

Fig. 7.19 Average debris compaction above the front of the roof (DFR) canopy and below the front of the floor base (DFF) and average leg closure of a Gullick Dobson 6/300 chock related to the increase in leg pressure. F27 face, Westoe Colliery (after Gupta and Farmer, 1982).

resistance offered by a support is a function of debris compaction and leg closure and these in turn are related to setting pressure and support dimensions. These relations are illustrated in Fig. 7.19 where data for roof and floor debris compaction and leg closure as a function of leg pressure increase are illustrated. The implication of fluctuations in final pressure is that differential strata movements may be caused. The use of positive set valves reduces these fluctuations.

(c) At all pressures and on all faces there is some increase in support pressure together with some convergence which appears inevitable. This is, however, significantly reduced at higher setting pressures.

(d) At low setting pressures, final front leg pressures over a weekend period increased until they were greater than rear leg pressures. This was associated with increased flaking and cavity formation, spalling and face extrusion. At higher setting pressures the rear leg pressures remained higher than the front and flaking was reduced.

(e) The frequency of leg circuit pressures reaching a pressure equivalent to 85% of rated yield pressure increased with increasing setting pressure. The frequency of leakage also increased and this depended strongly on the quality of support maintenance.

(f) Face end zone pressures increased by a smaller amount than face pressures at lower setting pressures. This effect was reduced at higher

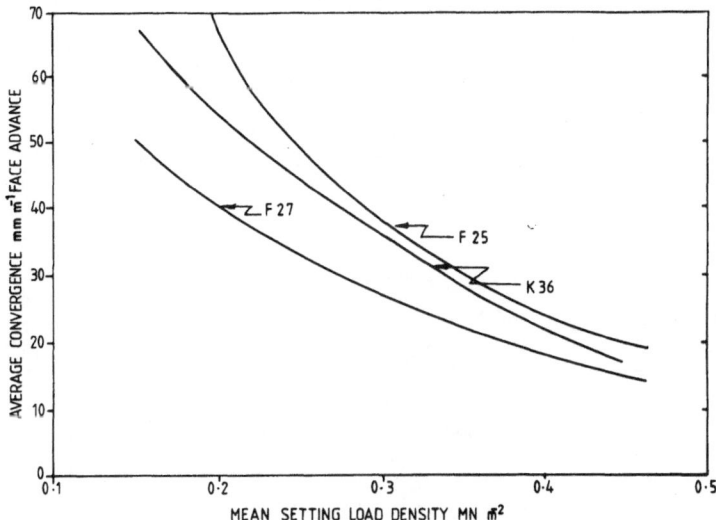

Fig. 7.20 The effect of mean setting load density on roof to floor convergence, normalized in terms of face advance, over a single cutting cycle (after Gupta and Farmer, 1982).

setting pressures. The result of lower face and zone pressure increases was to reduce convergence in the face end zone with consequent improvement in roof conditions. The face end zone was about 40 m from the main gate and 20 m from the tail gate.

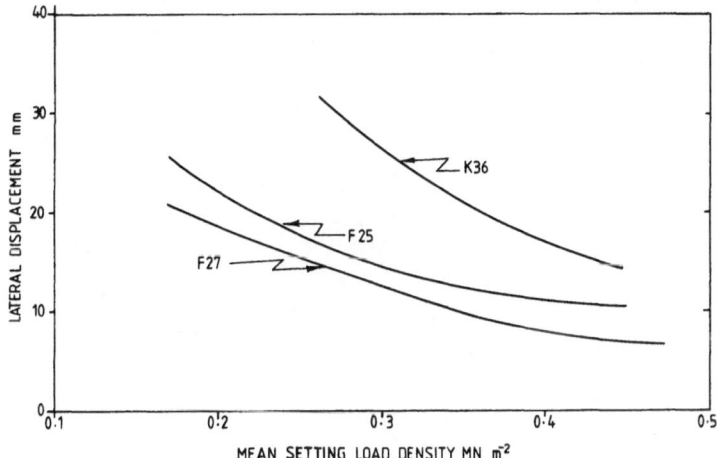

Fig. 7.21 The effect of mean setting load density on lateral movement over a 60 hour period (after Gupta and Farmer, 1983).

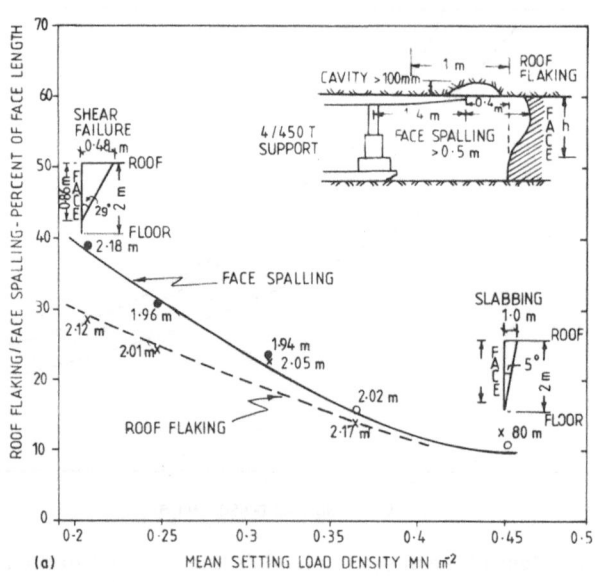

(a)

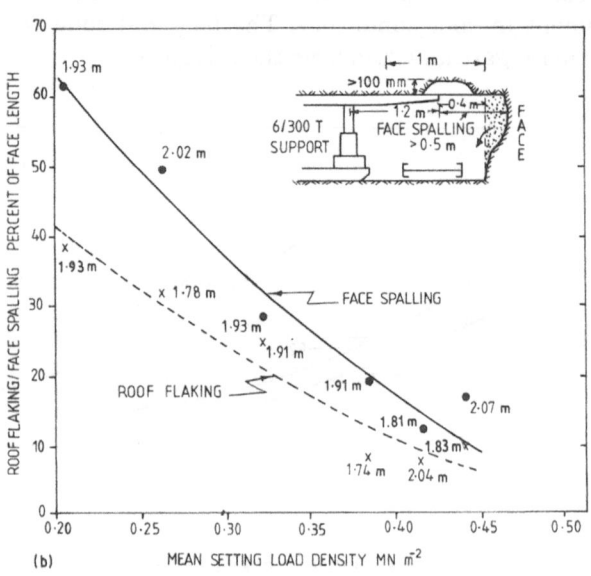

(b)

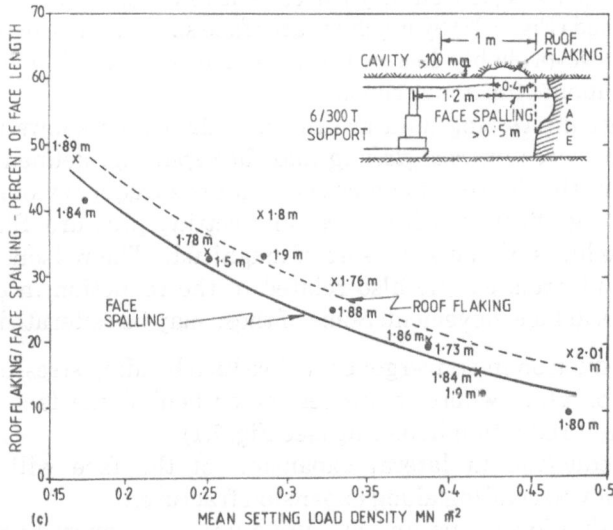

Fig. 7.22   The effect of mean setting load density on roof flaking and face spalling at (a) F25, (b) K36, (c) F27 faces. Note that in this case *flaking* describes roof cavities greater than 100 mm deep and 300 mm long. Spalling is defined as by a distance greater than 0.5 m between the advanced support tip and the face. Distances on the curves are the prop-free front distances (after Gupta and Farmer, 1983).

7.5   Setting pressures and roof stability

From the large amount of data collected by Gupta (1982) during the experimental programme at Westoe Colliery, the most important is that concerning the effect of setting pressures on roof stability. The effect of setting pressure on roof to floor convergence is illustrated in Fig. 7.20 for the three faces. This is the average convergence which takes place over a single extraction cycle, normalized in terms of the distance advanced. It is interesting to note the reduced convergence of F27 face where a wider cut and a resultant increased advance was obtained. If the results were adjusted for this, the convergence per cut would be almost identical for the three faces. There is a similar relation between setting load density and lateral movement of the coal face, which is illustrated in Fig. 7.21. This can probably be explained in terms of the reduced vertical compression transmitted to the coal closest to the face at higher setting loads.

It is interesting to note that whereas lateral displacement – measured in this case over a weekend period – is similar in the case of F25 and F27, the two Main seam faces, it is much higher in the case of K36, in the Brass

Thill seam. This is particularly the case at low setting pressures and it is accompanied (Fig. 7.22) by much greater face spalling. The explanation for both of these probably lies in the presence of nine minor faults in the workings about 70 m above the face.

The most interesting data are in Fig. 7.22, which shows the effect of setting pressures on *roof flaking and face spalling*, defined earlier and redefined in the figures. It can be seen that the frequency of roof cavities decreases significantly with increasing setting pressure. Similarly face spalling reduces by the same sort of magnitude. The reasons for this are probably interrelated and also related to the reduction in convergence and in lateral face movement. Some of these may be reiterated:

(a) The reduction in convergence will reduce bending stresses in the roof beam or plate where it 'hinges' at or behind the face line with a resultant reduction in flaking (see Fig. 7.1).

(b) The reduction in lateral expansion of the face will reduce the tendency to spalling along expansion fractures.

(c) The reduction in spalling by giving greater support to the roof and reducing the canopy tip to face distance also reduces the unsupported roof area.

These and other points may be illustrated by examining the strain contours obtained from borehole measurements above the face in Figs 7.23 to 7.25. These have been selected to compare for each face the strain distribution at two mean setting load densities, one at about $220 \, \mathrm{kN \, m^{-2}}$, equivalent to that exerted by the normal power support ring main pressure; the other at about $420 \, \mathrm{kN \, m^{-2}}$ equivalent to that exerted by the maximum boosted positive set pressure.

In Figs 7.23 to 7.25 strain has been computed from the relative deformation of anchor wire extensometers (Figs 7.23 and 7.24) and magnetic extensometers (Fig. 7.25) installed in a borehole initially drilled in front of the support canopy tip. The experimental procedure varied. On F25 and K36 faces (Figs 7.23 and 7.24) following each advance of the face, support pressures for 10 m each side of the borehole were raised through the full range of available pressures. The data in Figs 7.23 and 7.24 are selected from this range.

On F27 face two boreholes were drilled and all the supports on the face were maintained at a pressure sufficient to give the two setting load densities quoted in Fig. 7.25. Although the techniques differ, the necessary act of support lowering during longwall mining is sufficient to allow valid comparisons to be made.

Divergent views have been expressed as to the height of the immediate roof which is affected by roof support (see for instance Alder, 1968; Habernicht, 1972; Wagner and Steijn, 1979). The thickness of lower strata which cave in the goaf after the removal of support has been quoted in the literature as varying from 1.5 times to 13 times the seam thickness (Ilstein, 1960). Few, if any, observations have been reported which explain in detail the mechanics of deformation of the immediate roof strata supported by the face supports.

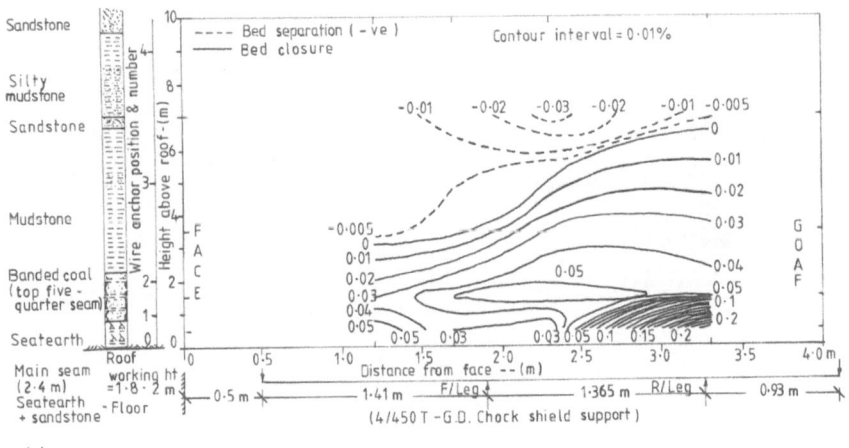

(a)

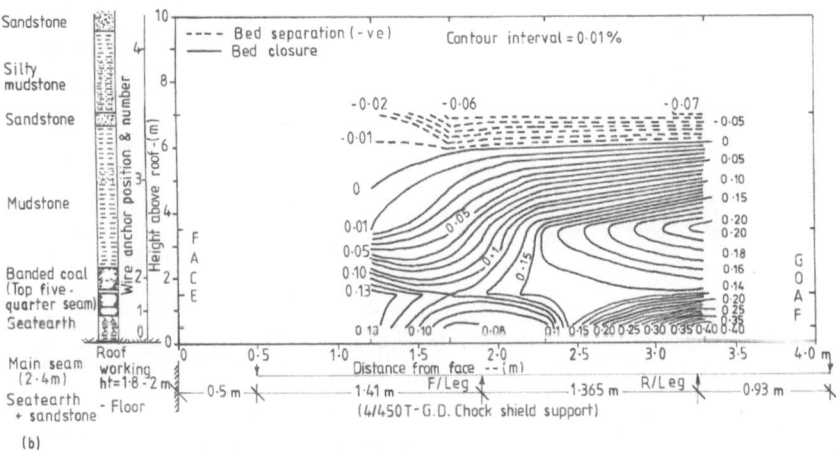

(b)

Fig. 7.23 Contours of vertical strain above supports at F25 face, Westoe Colliery: (a) when the mean setting load density was raised to $260 \, kN \, m^{-2}$ and (b) when the mean setting load density was raised to $420 \, kN \, m^{-2}$. Strains were computed from differential roof displacements between anchors connected to a wire extensometer system (after Gupta and Farmer, 1983).

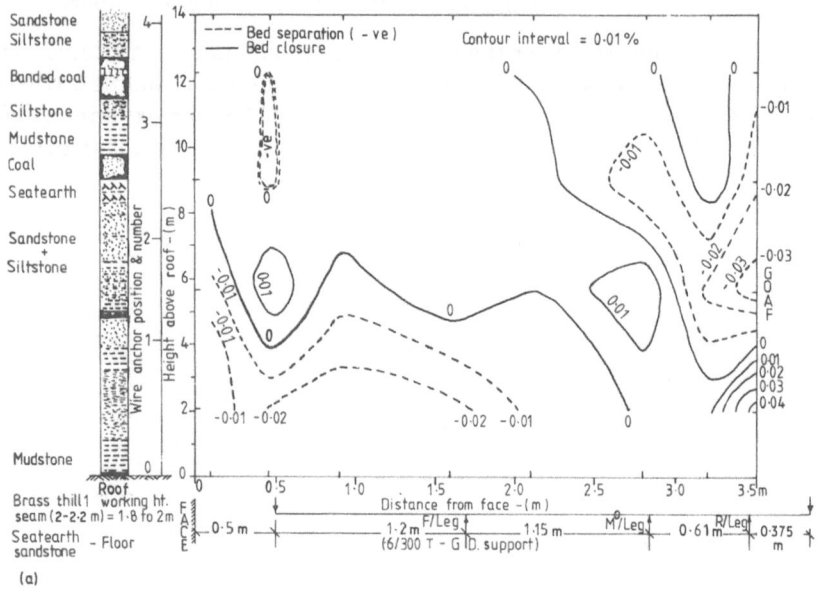

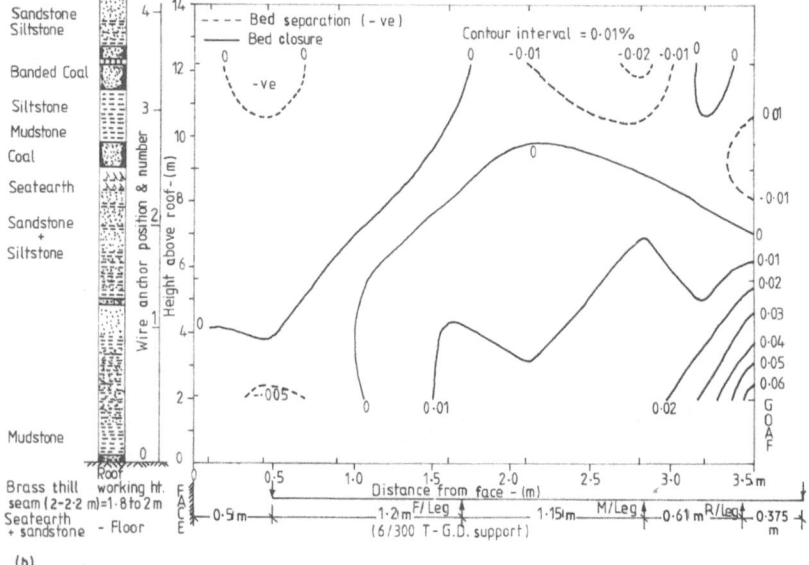

Fig. 7.24 Contours of vertical strain above supports at K36 face, Westoe Colliery: (a) when the mean setting load density was raised to 220 kN m^{-2} and (b) when the mean setting load density was raised to 420 kN m^{-2}. Strains were computed from differential roof displacements between anchors connected to a wire extensometer system (after Gupta and Farmer, 1983).

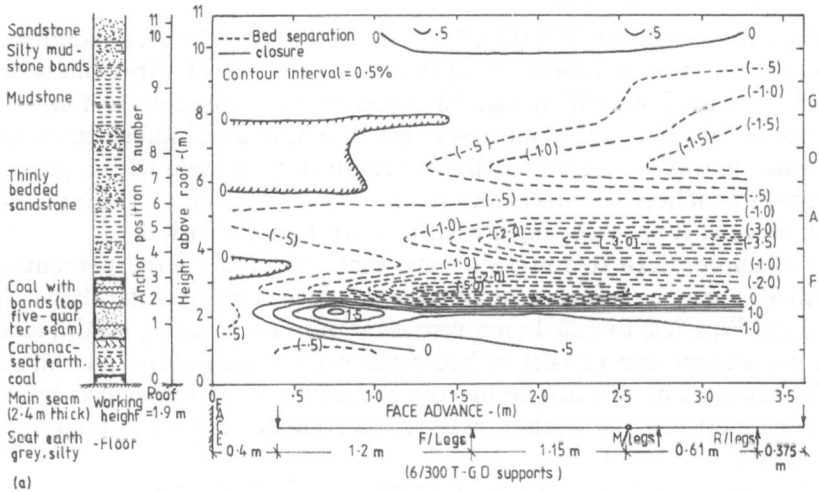

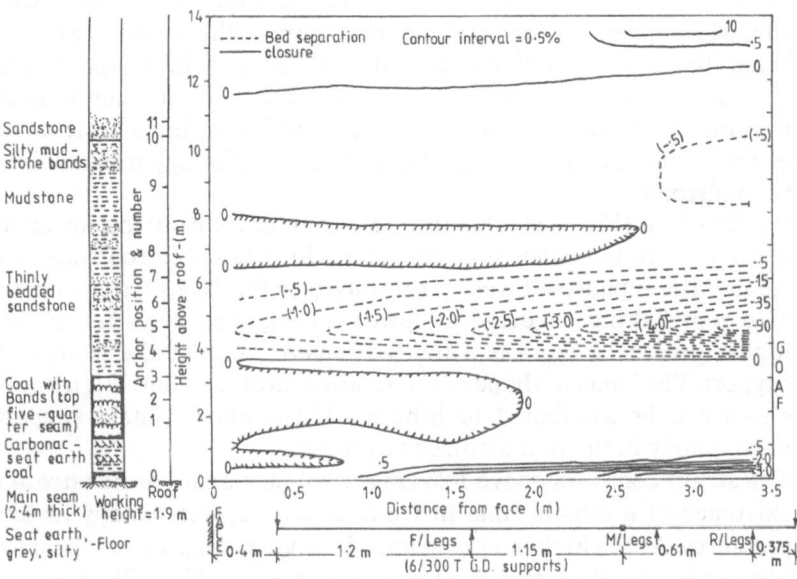

Fig. 7.25 Contours of vertical strain above supports at F27 face, Westoe Colliery: (a) when the mean setting load density was raised to $220\,\mathrm{kN\,m^{-2}}$ and (b) when the mean setting load density was raised to $450\,\mathrm{kN\,m^{-2}}$. Strains were computed from differential roof displacements between magnetic anchors monitored using a reed switch extensometer system (after Gupta and Farmer, 1983).

The experiments on F25 (Fig. 7.23) and K36 (Fig. 7.24) faces confirmed that with a wire extensometer system the presence of wires limits the number of anchors which can be installed in a borehole to four and introduces many possible sources of error. While useful information was obtained with the system, the higher resolution possible with a greater number of anchors was necessary in order to study in detail the influence of increasing support setting pressure on strata deformation.

It is appropriate to reiterate here that the plots of the percentage vertical strain contours against face advance, and heights above the seam, in Figs 7.23 to 7.25 do not represent pure engineering strain, since the movements can consist of bed separation, rock failure, and elastic deformation. The magnitude of strain appears to be influenced by the distance between the anchors with higher anchor distances tending to give lower strain values.

The more important observations which can be made are:

(a) On F25 face (Fig. 7.23) the data are confined to the section above the support. The main change with increased pressure is from a wedge shaped compression zone above the support at the lower pressure to a beam shaped compression zone, with significantly increased compression. It is interesting to note that the change from compression to tension about 7 m above the roof level, and indicating a possible zone of detachment, coincides with the change in lithology from mudstone to sandstone.

(b) On K36 face (Fig. 7.24) the data are confused but at the lower load density there is evidence of considerable tension or roof sag at the front of the supports. It was on this face (see Fig. 7.22(b)) that considerable roof flaking occurred at lower setting load densities. Even at higher pressures there is a limited tension zone at the front of the support. The 'domed' shape of the strain contours at both high and low stress can be attributed to hinging of the chock shield support – particularly in the roof disturbed by flaking.

(c) The strain contours above F25 (Fig. 7.25) face again demonstrate the existence of a tensile zone in front of the support canopy at lower pressures. At the higher setting load density, a compression 'beam' 4 m high and extending to the face line is created. This illustrates the basic mechanism of beam building in the roof above face supports at higher setting pressures.

In order to investigate the potential effect of abutment stresses on deformation ahead of the coal face, at K36 face boreholes were drilled horizontally into the face and circular ring magnets were stuck to the coal surface of the borehole (Fig. 7.26). The movement of the anchors relative

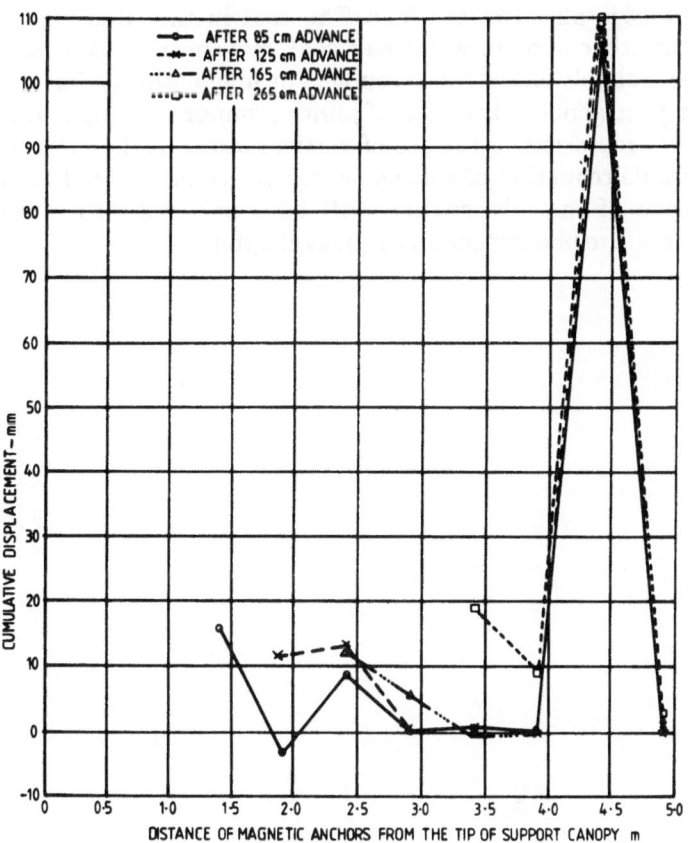

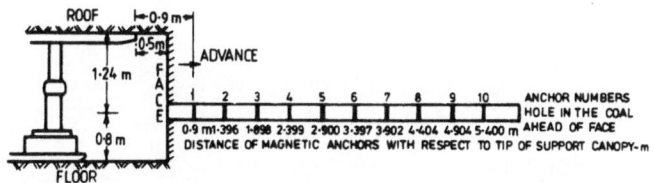

Fig. 7.26 Horizontal displacement of magnetic anchors located in a horizontal borehole drilled 6 m ahead of the centre of K36 coal face, Westoe Colliery (after Gupta, 1982).

to the canopy tip was determined by passing a reed switch attached to a rod and tape along a plastic tube located centrally in the boreholes.

The system indicated the existence of breaks ahead of the face. Figure 7.26 shows a particular example of a break 110 mm wide which originally occurred 4.5 m ahead of the face. The coal in this area was relatively strong. In other boreholes in weak coal, movements at this and greater depths completely closed the borehole. The implication is that this break – following the general direction of jointing/minor cleating parallel to the face line – represents a release of vertical stress at the abutment and a considerable reduction of vertical stress in the immediate face area. The importance of this observation will be considered with other stress-related fracture phenomena in the next chapter.

8

BURSTS AND
INSTABILITY

The ultimate manifestations of the process of stress distribution from excavations to abutments or pillars induced by mining are *rockbursts*. These are usually associated with hard rock mining but, provided the conditions for strain softening brittle fracture can be created as in Figs 1.1 to 1.3 and 1.5, similar phenomena can occur in coal mining.

Brittle fracture of the rock is accompanied by dilation, and the formation of fracture or shear planes on to which the strain energy stored by compression is released rapidly. Some of this strain energy, and possibly some potential energy lost from the rock mass, is released into the surrounding rock as *seismic energy* and can be detected and sometimes felt as precursive and major *seismic events*.

Rockbursts are usually defined (Salamon, 1983) as a seismic event which causes damage to mine workings. On this basis most of the collapses in coal mines would not qualify as rockbursts and the term *bumps* is probably preferable.

Seismic events are, however, generated by the same mechanisms in coal mines as in hard rock mines, and it is useful to summarize the conditions listed by Salamon (1983) for initiation of such events. These are:

1. A region of the rock mass must be in a state of *unstable equilibrium*. This will require the presence of an appropriately stressed discontinuity or a volume of rock in which stresses are changing in such a way as to induce fracture.
2. Induced stresses must affect the region and must become large enough to *trigger* the instability.
3. A substantial amount of *strain energy* must be stored in the rock in the affected region which will act as a source of kinetic energy when released by the instability.

In coal mines, induced stresses are largest at the abutments of faces or roadways (Fig. 6.1) or in pillars (Fig. 2.6). Seismic events are generated, with increasing severity, at depths in excess of 800 m – when the stresses in an abutment or pillar exceed the strength of the rock, so that it fails in a brittle manner. There are several specific types of burst or instability which occur in coal mines and which are recognizable as generic phenomena, accompanied by seismic activity. These include:

(a) *Rockbursts* of a classical type where an exposed abutment ahead of an advancing or retreating longwall face collapses rapidly with violent expulsion of coal or rock, often in quantities of 100–200 tonnes
(b) Roof collapse during *caving* above longwall mining operations, where cantilevering of strong beds can occasionally cause serious problems
(c) *Pillar* collapse in the same or over- or underlying seams due to interaction of superimposed mining induced stresses from current and previous workings
(d) *Outbursts* of gas and coal in anthracite mines
(e) Activation of existing *fault* planes by mining activities adjacent to and particularly underneath major cross-measures faults

With the exception of the latter, there are case history data available on these types of occurrence. It is, of course, always dangerous to attempt to generalize phenomena which are often motivated by specific criteria. For this reason it is often better to describe and explain in detail particular occurrences, rather than to develop general hypotheses.

8.1   Rockbursts in coal mines

Collapse of abutments, recognizable as a classic type of rockburst, where release of energy accompanied by ejection of coal can create a significant hazard to, as well as interrupting, a mining operation, have occurred in various countries. They have been described *inter alia* by Jeremic (1980) in Canada; Crouch and Fairhurst (1974) in United States; Moore and Hanes (1980) in Australia; Kidybinski (1980) in Poland; Bräuner (1981) in Germany and Josien, Breniaux, Daumalin, Doligez and Georgel (1982) in France. In Britain where such occurrences are less common, they have been described by the South Staffordshire and Warwickshire Institute of Mining Engineers (1933).

The best description of the mechanics of bursts in coal is given by Crouch and Fairhurst (1974), and expanded by Board and Fairhurst (1983), although it owes a lot to earlier work on conventional rockbursts summarized by Cook (1983) and Salamon (1983). It is worth considering in some detail – using the results of tests in triaxial compression illustrated in Figs 1.1 to 1.5. A rock specimen failing in triaxial compression follows a

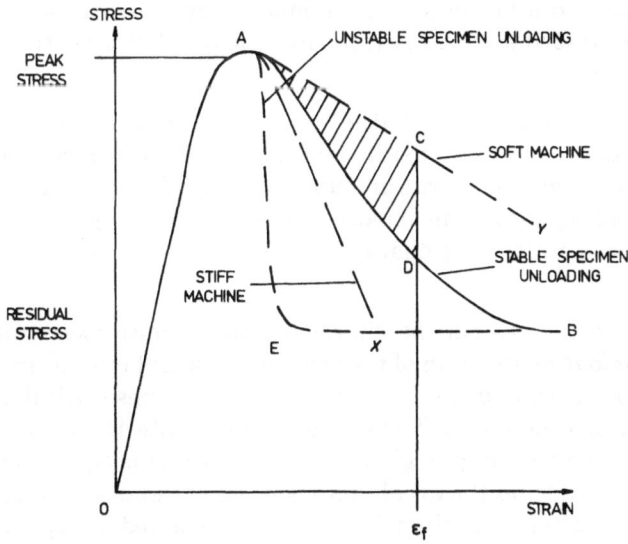

Fig. 8.1 Unloading characteristics of a rock specimen loaded in triaxial compression during brittle failure compared with the unloading characteristics of a stiffer and less stiff testing machine.

particular unloading deformation characteristic (Fig. 8.1) after fracture. During this regime (AB) the strength of the specimen reduces from peak to residual and it only remains *stable* provided the testing machine behaves in a stiff manner and has a stiffer (*stiffness* being defined as stress divided by strain) unloading characteristic (AX) than the specimen. This can be achieved by various means (summarized by Hudson, Brown and Fairhurst (1972)) in the laboratory, most commonly by a servo-control system. If the testing machine behaves in a *soft* manner and has a less stiff unloading characteristic (AY) than the specimen, then the *strain energy* at a given strain (say ε_f) and represented by the area ACD will be transformed into *kinetic energy* accelerating the disintegration of the specimen.

As can be seen from Figs 1.1 to 1.5 (where stiffness of the machine was not high and unstable deformation is apparent in most of the stronger rocks) the unloading stiffness of the rocks decreases with increasing confinement and (usually) decreasing strength. If the deformation becomes unstable then the unloading path follows AEB with rapid release of energy stored during the stable loading process as kinetic energy on to discontinuity surfaces. Since strain energy is stored in the rock volume (dimension L^3) and dissipated along a surface (dimension L^2) the energy released during fracture will increase in proportion to the dimension of the failed rock.

Translated to a larger scale, the laboratory observations can be summarized to describe the conditions for release of stored strain energy as kinetic energy:

(a) The rock being loaded (the coal seam in coal mining) must be subjected to a stress of sufficient magnitude over a sufficiently large volume to release a large amount of energy if it fractures.
(b) The loading conditions imposed by the surrounding rock must be such that their unloading characteristic is less stiff than the fracturing rock.

These two factors can be characterized in hard rock mining as the *energy dissipation function* of the stressed rock and the *energy release rate* of the rock mass, and the interaction between these will determine the likelihood of rockbursts. The energy release rate is the rate of energy released by extraction of a given area of rock and is equal to the product of the mean force on the areal increment before extraction and the mean convergence after extraction. If no rock is extracted or support is applied of sufficient magnitude to prevent convergence the energy release rate is zero. The same will apply if the energy is dissipated by fracture of the rock – as in successful caving. On the other hand if all the force is transferred to the rock mass convergence will increase and the energy release rate will increase. Dempster, Tyser and Wagner (1983) show a strong correlation between seismic energy released and energy release rate. An optimum low and uniform energy release rate should be sought.

The *energy dissipation function* of the stressed rock is related primarily to its ability to yield or fracture in small volumes, releasing energy in a controlled manner. It is to changes in this function that most rockburst control measures are aimed. The objective is to soften the rock so that it can absorb a greater amount of energy without producing kinematic effects. This can be achieved by blasting to induce fractures, water injection or drilling.

In coal mining rockbursts generally require a combination of a particularly strong roof and floor, a strong, usually anthracitic coal and a depth in excess of 800 m. There are also differences in rockburst phenomena. Because coal in an unconfined state (Fig. 1.5) has low strength, pillar edges and coal faces tend to yield easily and the coal tends to unload in a stable manner, fracturing parallel to the face (see Figs 7.4 and 7.5) and concentrating higher stresses in the centre of the pillars (see Figs 2.4 and 2.6) or abutments. Thus at some distance ahead of the face a peak stress/ peak strength anomaly may occur exacerbated by uneven yielding of the surrounding coal and by the sedimentary and tectonic structures which occur in the Coal Measures.

The extreme manifestation of this type of structure is described by

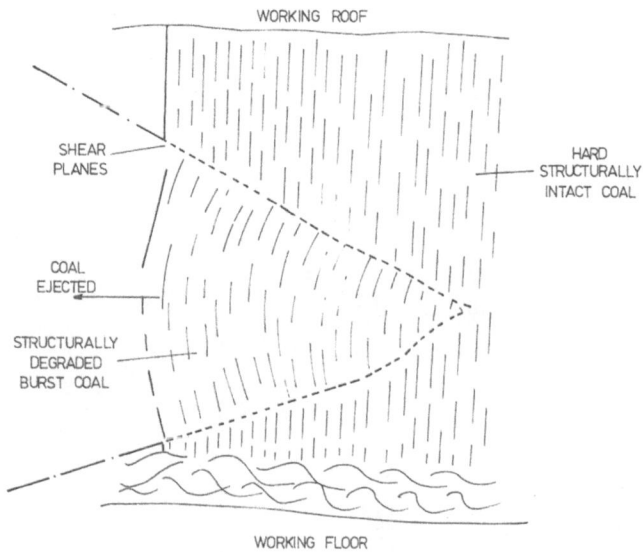

WORKING ROOF

SHEAR PLANES

HARD STRUCTURALLY INTACT COAL

COAL EJECTED

STRUCTURALLY DEGRADED BURST COAL

WORKING FLOOR

Fig. 8.2   Section through a typical rockburst cone (after Moore and Hanes, 1980).

Moore and Hanes (1980) in prime coking and semi-anthracite coals at Leichardt Colliery, Central Queensland. The prime factor here appears to be the strength of the 3 m seam section overlain by 15 to 39 m of sandstones (up to 70 %) shales and siltstones. Where rockbursts occur mining induced stresses are concentrated in the immediate abutment and there is little evidence of abutment yielding. Although the average seam depth is only 400 m large scale bursts have occurred. Failure in the coal face varies from a progressive crush or yielding where coal is heavily sheared and cleated, to explosive bursts of up to 300 tonnes in bright, hard structurally intact coal. In section (Fig. 8.2) the burst zones are typically conical with boundaries defined by induced fractures. The long axes of these fractures are typically elliptical and parallel to the face direction. The coal is not always completely ejected, but is always intensely sheared. The intact coal is always hard and contains little if any structural deformation or degradation.

Case histories of monitoring of *seismic activity* in coal mines are not common. In the Polish Upper Silesian coalfield, Trombik and Zuberek (1979) describe measurements in a seam at a depth of 700 m in an area highly susceptible to coal bumps. The longwall faces were retreating at 0.75 m per cut at a height of about 3.0 m in mainly bright strong coal, leaving about 1.5 m of roof beneath mixed sandstone and shale strata. The waste was backfilled with sand fill.

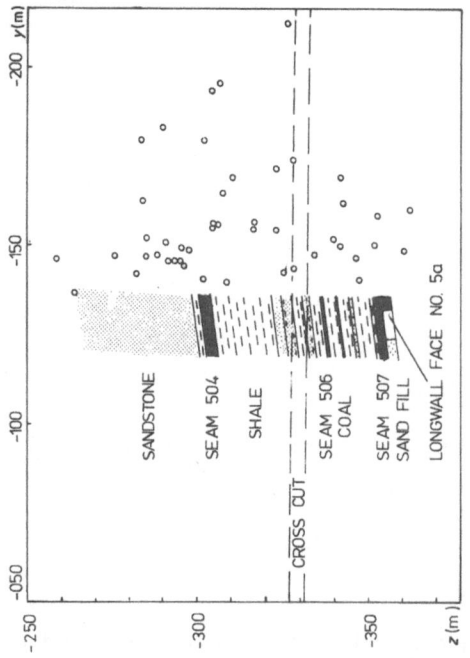

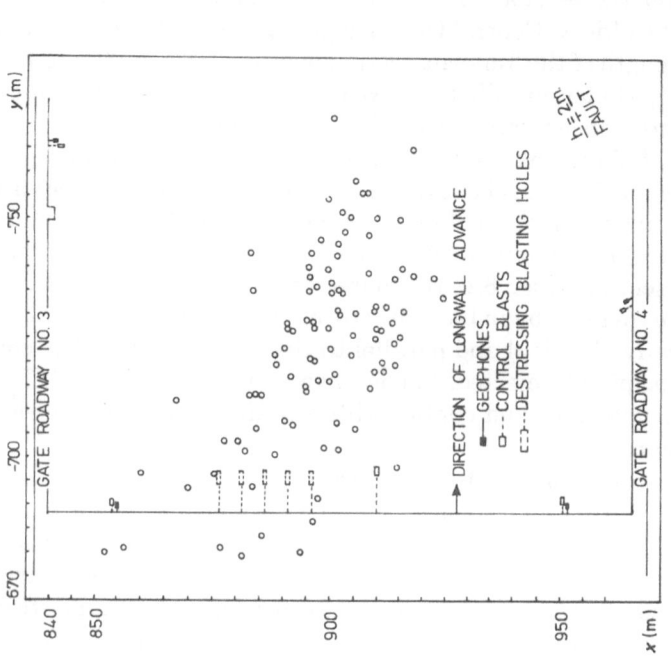

Fig. 8.3 Sources of seismic events (a) in the plane of the seam and (b) in a vertical section through the centre line of a retreating longwall face immediately following destress blasting at Szombierki coal mine, Upper Silesia (after Trombik and Zuberek, 1978).

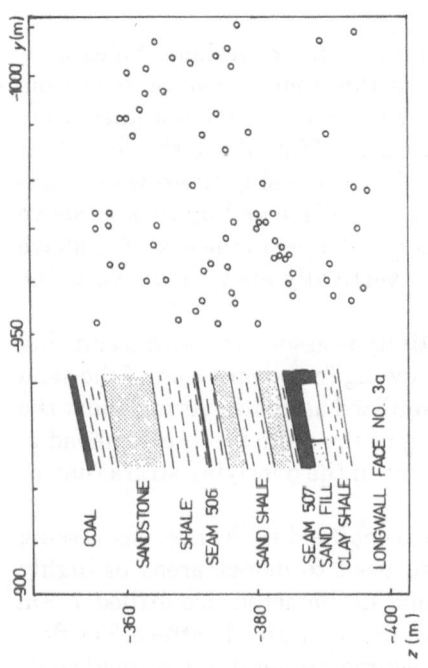

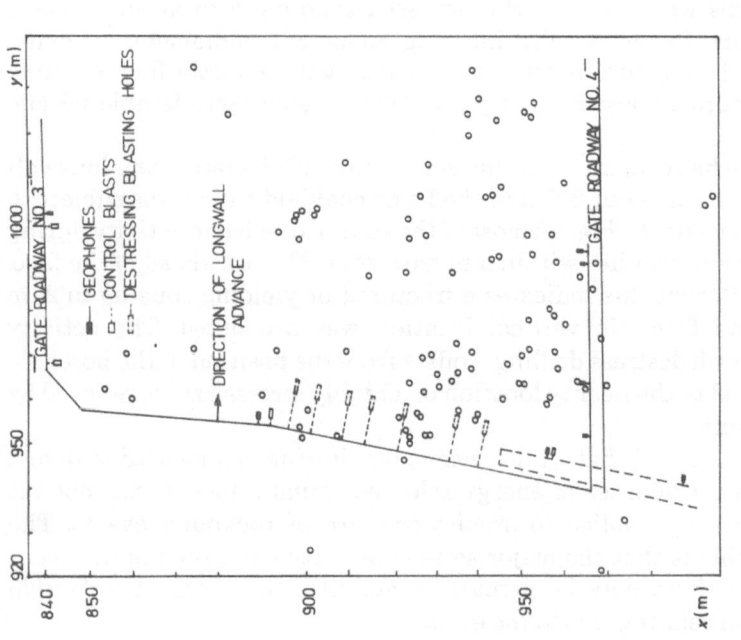

Fig. 8.4 Sources of seismic events, following destressing blasting: (a) in the plane of the seam and (b) in a vertical section through the centre line of a retreating longwall face (after Trombik and Zuberek, 1978).

In order to dissipate some of the energy in the coal, blast holes were drilled 10 m ahead of the face to destress this zone. It can be seen from source locations of seismic events in Figs 8.3 and 8.4 obtained from geophones at the seam level and in a crosscut 70 m above the seam that this was successful. Immediately after the destressing there were, however, numerous events 10 m–80 m ahead of the face and up to 90 m above (Fig. 8.3) the face, although these were concentrated in the first 40 m above the face. There were virtually no events vertically above and behind the face.

The implication that most seismic activity is associated with fracturing of the roof rocks might be expected. Allowing for compaction of the sand fill by 40–50 %, the roof rocks can be visualized as cantilevering about the top of the coal seam. Sudden destressing of the seam up to 10 m ahead of the face is more likely to induce movement in the overlying strata than at the seam level.

In the Ruhr coalfield and the Provence coalfield in France, destressing is achieved by drilling – which is also used to detect areas of highly stressed coal. Boreholes, 43 mm or 50 mm in diameter, are drilled 7–8 m ahead of the face and from roadways to a predetermined pattern. The flow rate of cuttings from the borehole is measured. Generally in normal coals drilling a 43 m *borehole test* yields a cutting flow of 2–3 l min^{-1} which can rise to 10 l min^{-1} in stressed zones to very high levels in outburst zones. The test aims to detect a highly stressed zone close to an undamaged abutment, and in the test drilling programme 95 mm diameter relieving holes are drilled up to 25 m long at 2–3 m intervals along the face and from advance heading. These relieving holes often induce considerable seismic activity.

Will (1979) has examined seismic activity ahead of a retreating longwall face at a depth of about 800 m in the Ruhr coalfield which was subject to rockburst activity. In Fig. 8.5 most of the source of seismic activity during a 24 h mining period lies within a narrow zone 20–30 m ahead of the face. Will suggests that this indicates a fractured or yielding zone up to 20 m ahead of the face. No vertical location was attempted. The activity associated with destress drilling – offset from the position of the borehole – is attributed to the nearby location of a highly stressed zone, relieved by the destressing.

It should be noted that while seismic monitoring can be used to detect and explain the sources of energy release around a face, it has not yet been successfully applied to predict or warn of rockburst events. The reason for this is that the major seismic activity monitored is the event itself. While there may be percursive activity, there are considerable difficulties in relating this to the event.

In predicting the onset of rockbursts in coal mines it is more important

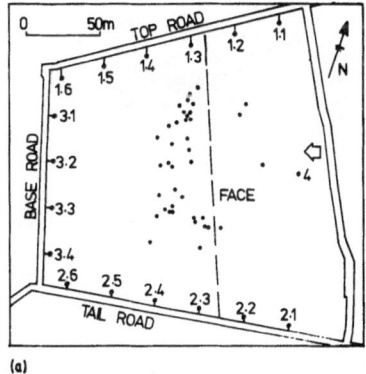
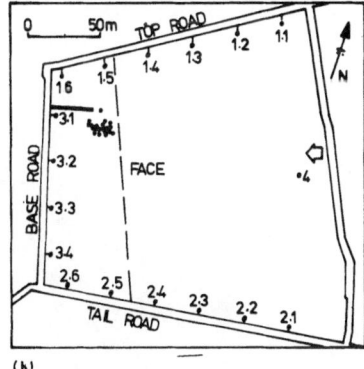

Fig. 8.5   (a) Localized sources of seismic events during a 24 hour mining period induced ahead of a retreating longwall face; (b) localized sources of seismic events during drilling of a destressing borehole ahead of the same retreating longwall face in the Ruhr coalfield, Germany (after Will, 1979).

to isolate some of the specific factors which are associated with them. These include – as mentioned previously – a high *energy release rate* and a low *energy dissipation function*. These in turn may be related to some, or all, of the following:

(a) A structurally intact, usually high rank coal seam
(b) Structurally continuous, high stiffness rocks (usually sandstone and limestone) forming the roof and floor.
(c) An even stress distribution through the roof and floor rocks to a stiff deformation resistant coal seam – usually inferring a flat undisturbed seam structure
(d) A high geostatic stress – minimum depths of 200 m for rockbursts have been quoted by Kidybinski (1980) for Polish mines, but normally depths in excess of 700 m would be expected for serious events.

8.2   Outbursts in coal

While there may be some gas associated with or released during rockburst activity in coal mines, it is not a major factor in initiating or propagating the rockburst. Outbursts of coal and gas – invariably methane gas – are quite distinct phenomena. They are characterized by a rapid release of pulverized coal and gas into a mine roadway or face.

Outbursts often occur at the ends of faces or in headings although they can also occur in mid-face. They are preceded by considerable noise

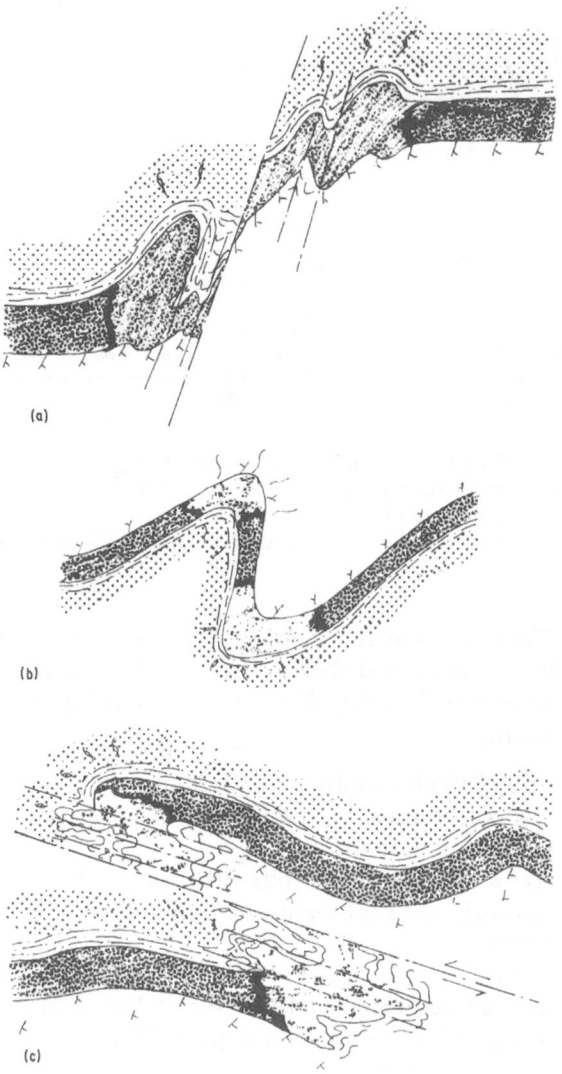

Fig. 8.6 Idealized sections (after Godden, 1981) of outburst structures in anthracite mines of the South Wales coalfield. Outbursts are associated with (a) normal faults, (b) folds and (c) reverse faults. Each type of structure has been deformed by a combination of high active tectonic stresses and high triaxial confinement. The roof strata are typically strong massive and competent sandstones or siltstones with tension gashes infilled with quartz. The floor strata are strong hard seatearths, usually sheared in the zone of disturbance. The anthracite changes from a normal strong continuous seam (shaded black) through a transition zone (interleaved) to a sheared pulverized outburst type anthracite (shaded grey) in the parts of the seam subjected to maximum disturbance.

followed by a burst of the coal face-wall which releases large quantities of gas and coal, often from quite a small opening of fracture. The amount of coal released – virtually as a fluidized bed propelled by desorbed gas, often at a rapid rate – can be several hundred tonnes.

There are several geological and tectonic features which are invariably associated with outbursts. These include:

(a) The strata have been tectonically disturbed often by a thrust fault or fold.
(b) The tectonic disturbance has been accompanied by shearing of the coal which changes its state to a finely sheared mass.
(c) Significant amounts of adsorbed gases are contained on the surfaces of the sheared coal particles.
(d) The roof and floor rocks and the coal abutment on all sides of the sheared zone are strong and intact.

Typical structures associated with outbursts in the South Wales coalfield have been illustrated by Godden (1981) and are reproduced in Fig. 8.6. In the South Wales coalfield – and in other coalfields throughout the world (see Hargraves, 1980; Shepherd *et al.*, 1981) – outbursts tend to occur in high rank coals of high vitrinite content. They also occur in some evaporite deposits (Gimm and Pforr, 1964), particularly potash. In Britain outbursts occur at the one remaining mine in the Gwendraeth Valley – Cynheidre Colliery – at infrequent intervals. It is symptomatic of a common approach to mining engineering problems that they have never been subjected to more research than at present.

A surface network of geophones has been installed at Cynheidre by Long, Kusznir, Blenkinsop and Smith (1984) to investigate seismic activity associated with outburst phenomena. The seismic activity monitored over a two day period in 1983 leading to an outburst in the vicinity of a major fault is illustrated in Fig. 8.7. It is interesting to compare the location of events ahead of the face and above the face with those in Figs 8.3 and 8.4. They show evidence of considerable activity ahead of and above the face – little activity behind the face.

In Fig. 8.8 events leading up to an outburst, expressed in terms of event frequency, are illustrated. They illustrate several useful points:

(a) Precursive seismic activity occurs prior to the occurrence of outbursts.
(b) This activity is of microseismic magnitude.
(c) The precursive activity is probably in *addition* to that associated with general longwall mining.
(d) It has not yet been possible to identify the source mechanism of the seismic activity.

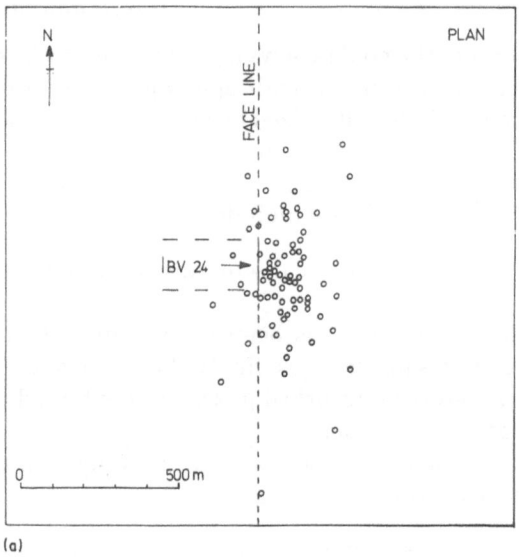

(a)

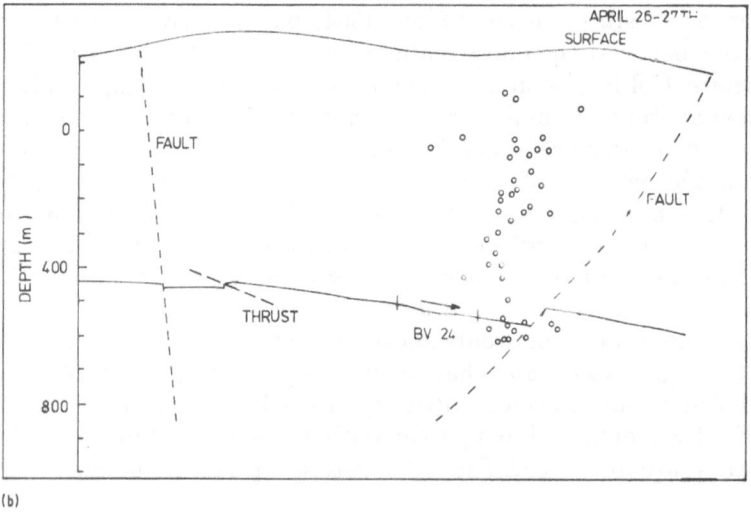

(b)

Fig. 8.7 Sources of seismic events during a two day period in the vicinity of a face at Cynheidre mine, South Wales: (a) in the plane of the seam, (b) in a vertical section through the centre line of an advancing longwall face (after Long *et al.*, 1984).

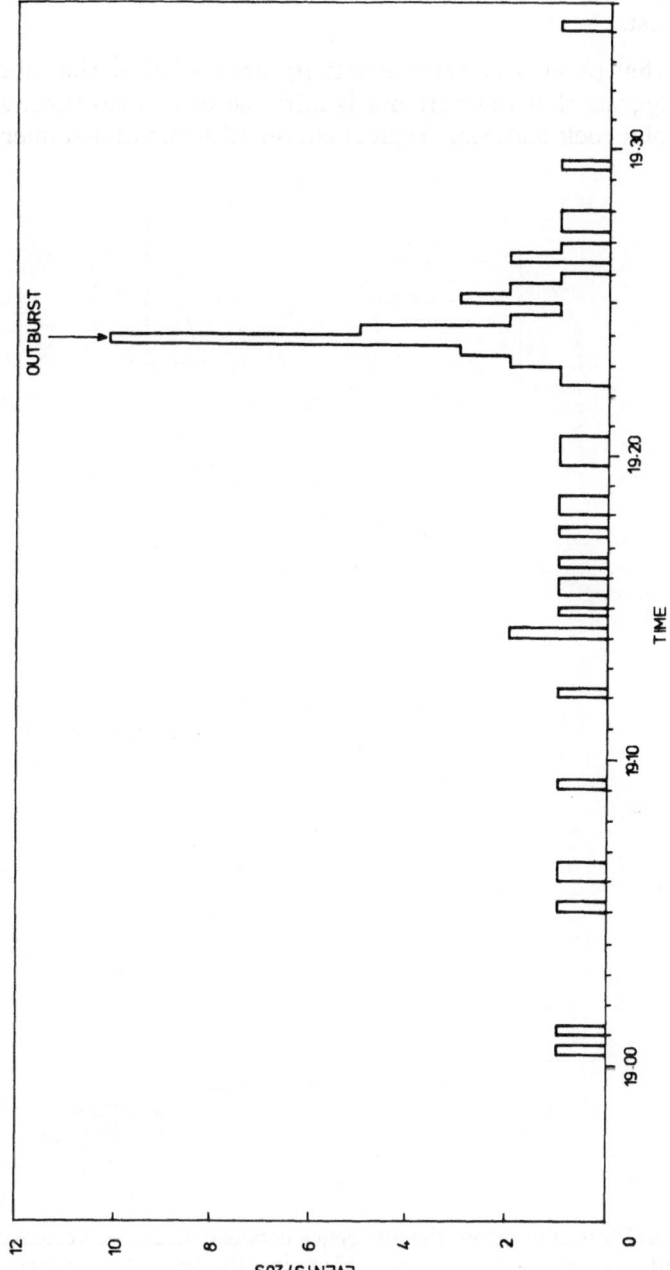

Fig. 8.8 Precursive and associated seismic activity leading to and during an outburst at Cynheidre mine, South Wales (after Long *et al.*, 1984).

271

(e) The seismic events occur with increasing frequency prior to the outburst.

The fact that precursive seismic activity occurs before the spontaneous events suggests that the outburst is initiated or evolves through brittle fracture of a rock material. Typical curves of accumulated microseismic

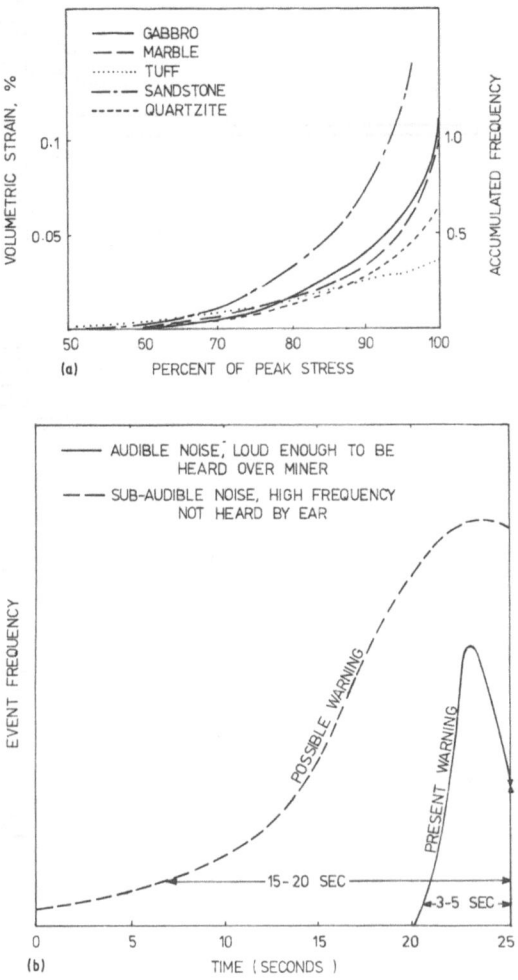

Fig. 8.9 (a) Plots of dilation (the difference between measured volumetric strain and computed 'elastic' volumetric strain) against the percentage of fracture stress and a dimensionless representation of accumulated frequency of microseismic events for five rocks (after Scholz, 1968); (b) a hypothetical relation between event frequency and time before an outburst (after Leighton, 1982).

activity during breakdown of a brittle material have been obtained by Scholz (1968) and others (see Fig. 8.9). It is likely that the activity envelope for anthracite will follow this pattern. It is significant that Leighton (1982) in a speculative approach to outburst initiation and prediction proposes a similar relation (Fig. 8.9) between the rock noise rate and time.

If the precursory seismic activity is associated with rock dilatancy and the growth of microcracks it is tempting to speculate on the role of the desorbing gas on the stability and growth dynamics of the microcracks. The gas if desorbed at pressure into the cracks may accelerate the crack growth process, so precipitating the explosive disintegration of the rock and the violent release of gas and coal fragments.

The indications are that outbursts are a stress related phenomenon in which a confining abutment of high rank coal collapses rapidly and in a brittle (see Fig. 1.5) manner. The collapse is assisted by, and allows the release of, gas which desorbs rapidly from the surfaces of pulverized coal – sheared by tectonic action, and constrained by strong and impermeable roof and floor rocks and the anthracite abutment. Outbursts may occur spontaneously or may be induced by precautionary firing – and possibly by destress drilling. Where they occur spontaneously they may be accompanied by precursive seismic activity. The difficulty is that this activity (Fig. 8.9) is so difficult to assess and the events are so rapid in their occurrence that it may not be possible to use this activity to predict the onset of outbursts.

8.3   Interseam interaction

Residual stresses in pillars (as illustrated in Fig. 2.6) are a major problem in coal mining. The effects of pillar interaction in the North Staffordshire coalfield have been described by Scurfield (1970) and Billington and Jacomb-Hood (1976). Usually tunnels or workings directly beneath or above pillars are seriously affected by the transmitted residual stress. A distance of 200 m is common for this interaction and up to 500 m has been postulated. Even allowing for the high magnitude of the residual stresses this may appear surprising. The phenomenon has been studied by Maury (1970) and Gaziev and Erlikham (1971) who have shown that in a layered and jointed rock mass the penetration of maximum stresses varies with the angle between the applied stress and the stratification, and the frictional resistance along the stratification. In Coal Measures strata containing bedding plane shear zones (Chapter 1) at sandstone – seatearth contacts, this may be low. The general picture proposed by Gaziev and Erlikham (1971) is illustrated in Fig. 8.10. A specific example proposed by Maury (1970) is illustrated in Fig. 8.11.

A case history of inter-seam interaction – in this case involving seismic activity from pillar collapse – in the North Staffordshire area at Hem Heath Colliery has been studied by Kusznir, Ashwin and Bradley (1980a) and Kusznir, Farmer, Ashwin and Bradley (1980b). These events were felt at the surface and caused considerable public alarm. Consequently a

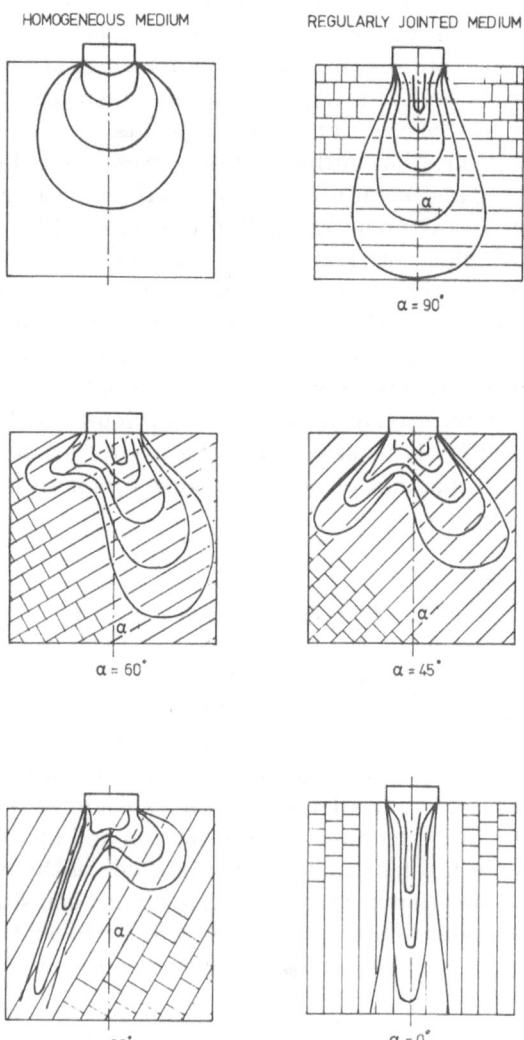

Fig. 8.10 Stress distribution beneath a rigid loading plate applied to homogeneous strata and jointed media inclined at various angles (after Gaziev and Erlikham, 1971).

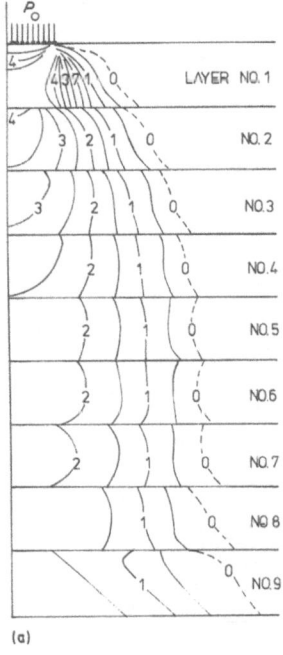
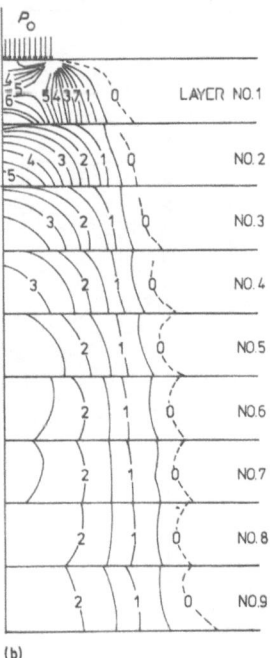

(a)

(b)

Fig. 8.11 Model test results (after Maury, 1970) showing the effect on stress transmission beneath a rigid plate above layered strata on a rigid base of the interface friction angle. The numbers refer to the ratio of P_0 transferred. In (a) $\phi = 36°$ in (b) $\phi \rightarrow 0°$.

surface seismometer array was installed to determine the exact location of the events.

The coal workings, in the area under investigation, are shown in Fig. 8.12. Panels 204, 205 and 206 of the Ten Foot seam were being mined when seismic activity occurred at the surface. Also shown in Fig. 8.12 is the fault distribution adjacent to the Ten Foot panels and the extent of previous workings in the Ten Foot seam. Figure 8.13 shows the extent of previous workings in the adjacent Moss, Bowling Alley, Rowhurst and Great Row seams. A sketch of a geological section (Fig. 8.14) drawn approximately N–S along the line of advance of panel 205, shows the approximate depth of the Ten Foot seam and the other previously worked seams. The dip was to south at approximately 8°. The geology comprised a typical cyclical sequence of coal seams with mudstone roofs, fireclay floors and intermediate sandstones, which were generally thin lenticular shaped bodies with limited lateral continuity. The area lay between two large faults – the Newcastle and Apedale faults. These faults showed no sign of recent

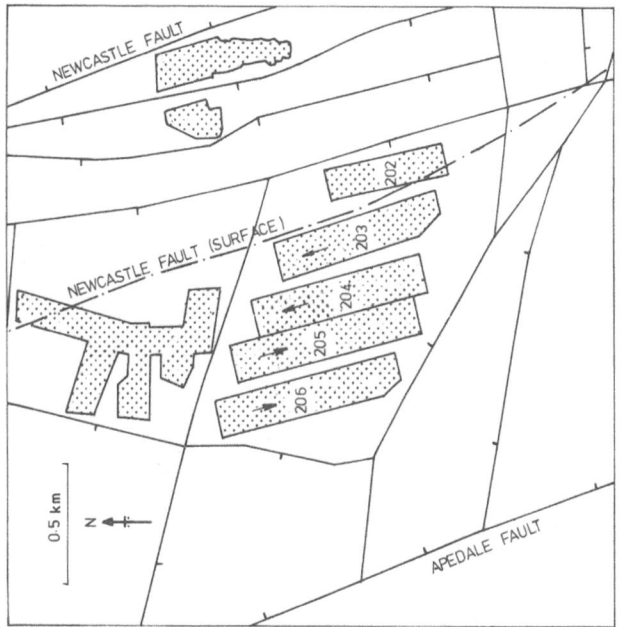

Fig. 8.12   Plan of longwall workings in the Ten Foot seam at Hem Heath Mine, Staffordshire, Britain, with fault locations at the seam level.

movement and the area is not known to be naturally seismically active. The cleat direction was normal to the line of major faulting and sub-parallel to the face line.

The average depth of the Ten Foot seam in the area was approximately 900 m. It was underlain by workings in the Bowling Alley seam at a depth of 15 m below the Ten Foot seam and overlain by workings in the Moss seam at a height of 170 m above the Ten Foot seam. There were also workings in the Rowhurst and Great Row seam, but these were considered too high above the Ten Foot seam to cause interaction problems. There was also minor faulting at the Ten Foot seam level, particularly between 203 and 204 faces and to the west of 206 face. Two main pillars had been left in the Bowling Alley seam (Fig. 8.13(a)), one 25 m wide intersecting 204, 205 and 206 faces, and one 40 m wide intersecting the line of 202 and 203 faces. Both these pillars were affected by minor faulting. In the Moss seam there was one major pillar (Fig. 8.13(c)) crossing the line of 204, 205 and 206 faces and approximately 200 m wide.

The coal in the Ten Foot seam was worked from five longwall faces (202 to 206) whose face width and height were approximately 200 m and 2.8 m respectively. Faces 205 and 206 were worked from north to south (down dip) as retreating faces while 204 face was worked from south to north (up dip) as an advancing face. The faces were advanced in 0.5 m strips by an Anderson Mavor shearer loader and supported by Gullick 6/240 self-advancing supports, and the roof was fully caved. The average advance rate for 206 face was 30 m per week. Pillars of coal were left between adjacent Ten Foot panels with the exception of faces 205 and 204, a 30 m pillar being left between 205 and 206. The mining conditions were generally good. However, some deterioration of 205 and 206 face access roadways occurred when passing under or above pillars left in the Moss and Bowling Alley seam workings.

The seismometer network, operated during the mining of 205 and 206 faces, enabled the computation of tremor hypocentres which are shown in relation to the active mine workings, previous workings and faults in Fig. 8.15. The seismometer network, the data processing and the calculation of tremor hypocentre locations and magnitudes have been described by

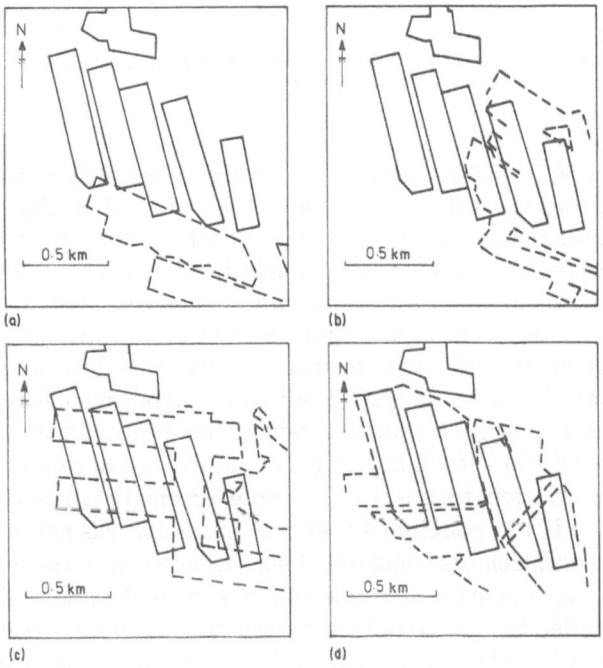

Fig. 8.13   Plan of longwall workings in the Ten Foot seam at Hem Heath Mine on to which have been superimposed plans of workings in (a) Great Row seam, (b) Rowhurst seam, (c) Moss seam and (d) Bowling Alley seam.

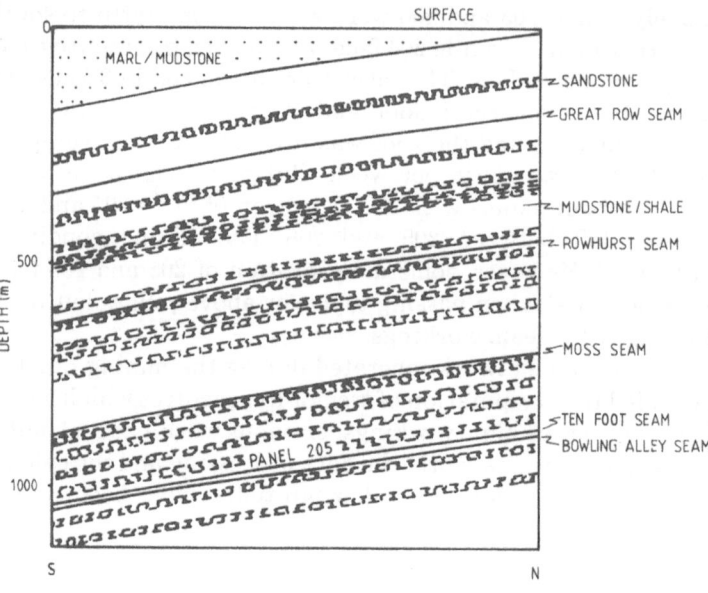

Fig. 8.14   Vertical section parallel to the direction of advance of 205 face (see Figs 8.12, 8.13).

Westbrook (1977) and Kusznir *et al.* (1980a). Plan (Fig. 8.15(a), (b)) and vertical sections normal to (Fig. 8.15(c), (d)) and parallel (Fig. 15(e), (f)) to the coal face are shown. Tremors which occurred during the mining of 205 and 206 faces are shown separately. In addition to the tremor hypocentres, the magnitude (M_L) of each event was calculated and each event is represented on the plans and sections by a circle whose radius is proportional to its magnitude. The accuracy of location has been estimated (Kusznir *et al.*, 1980a) to be ± 200 m. From the hypocentre location shown in Fig. 8.15 it can be seen that the tremors are situated within the mined area and are within a few hundred metres of the active panel in each case. No obvious relation to any of the large or small faults can be seen. Consequently it was concluded that the seismicity was not controlled by faults. The earth tremors occurred at depths between 400 m and 100 m, the majority being located above the active Ten Foot panels. The vertical section parallel to the coal face shows that many hypocentres were situated slightly to the east of the active panel so that they lay over the previously extracted panel.

Many of the larger events were felt by local inhabitants on the surface. These events corresponded usually to those with a magnitude greater than 2.5. Most seismicity occurred in the period from Monday to Friday

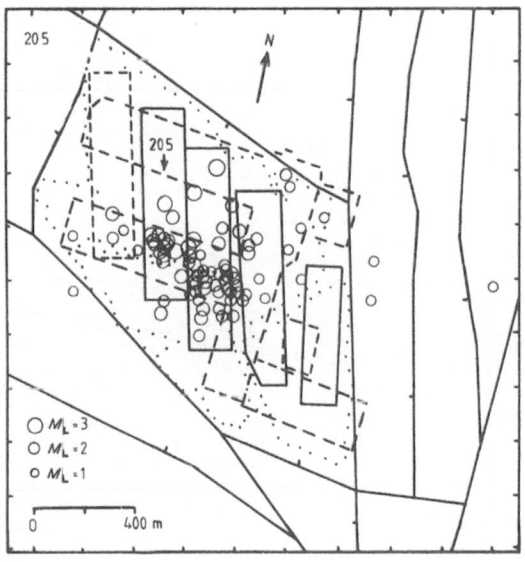

(a)

(b)

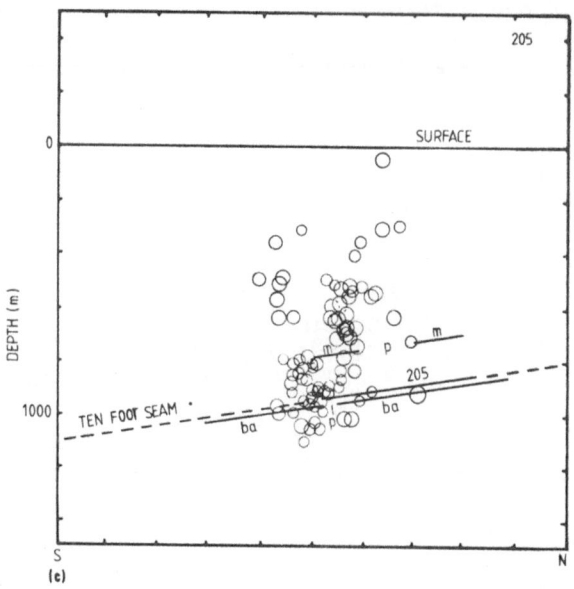

(c)

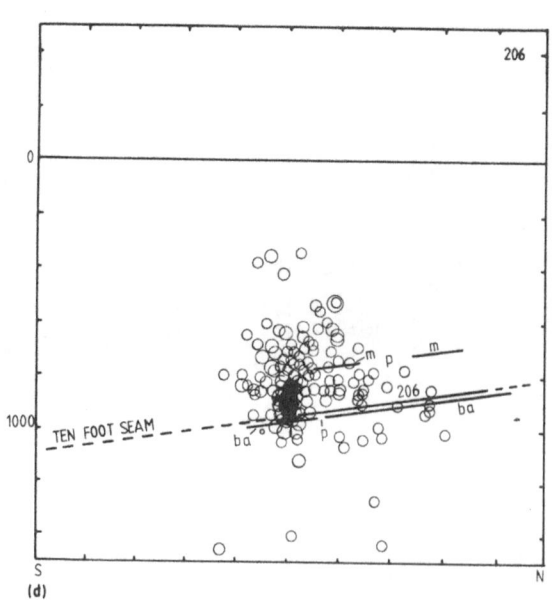

(d)

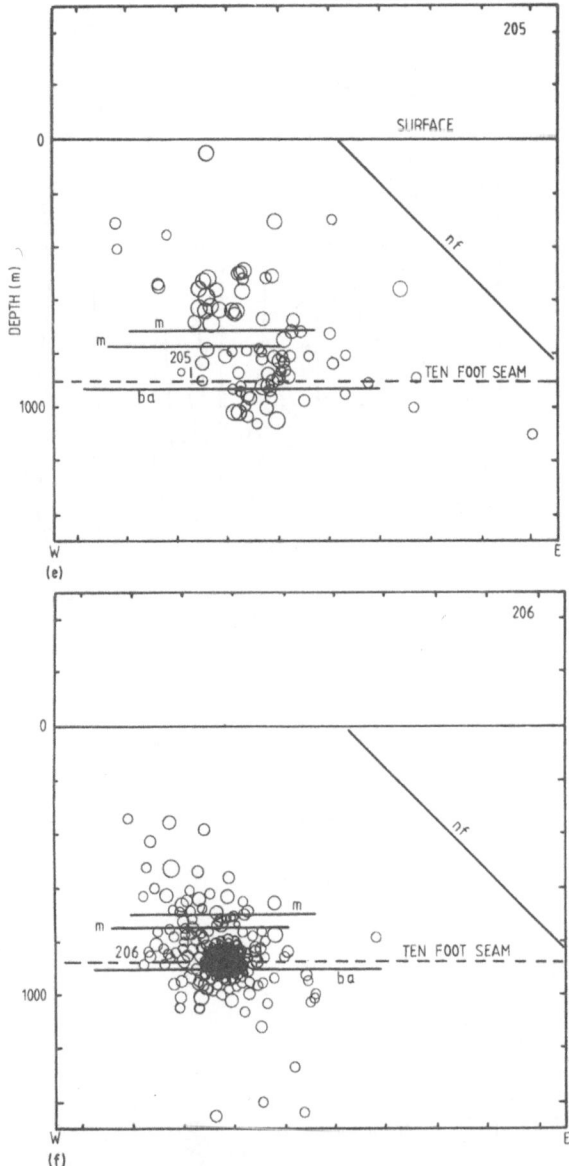

Fig. 8.15 Position of tremor hypocentres detected from a surface seismometer array during the working of 205 and 206 faces at Hem Heath Mine: (a) plan view during working of 205 face; (b) plan view during working of 206 face; (c) vertical section parallel to direction of face advance during working of 205 face; (d) vertical section parallel to direction of face advance during working of 206 face; (e) vertical section parallel to face axis – 205 face; (f) vertical section to face axis – 206 face.

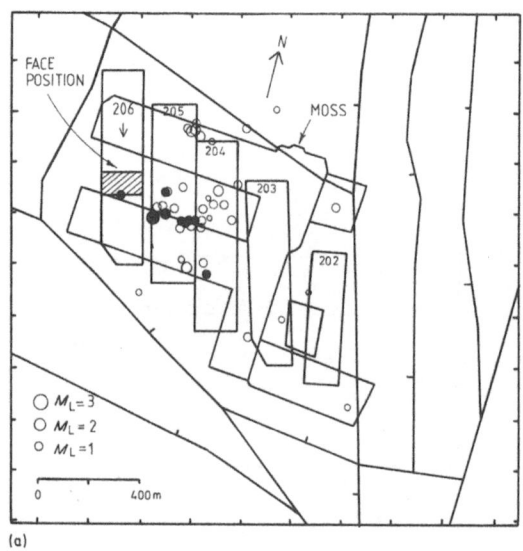

(a)

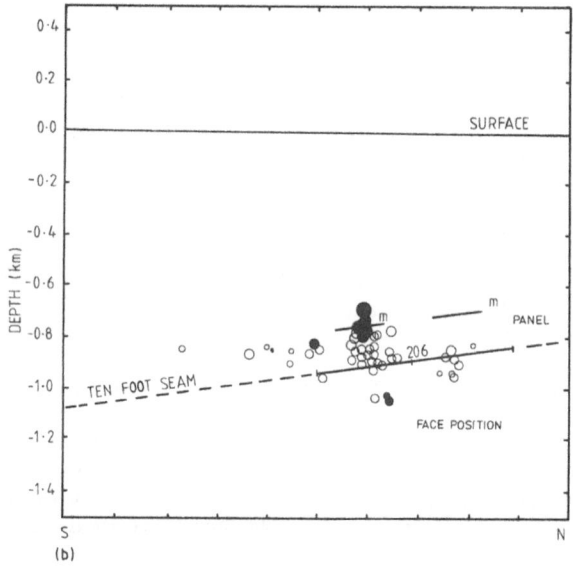

(b)

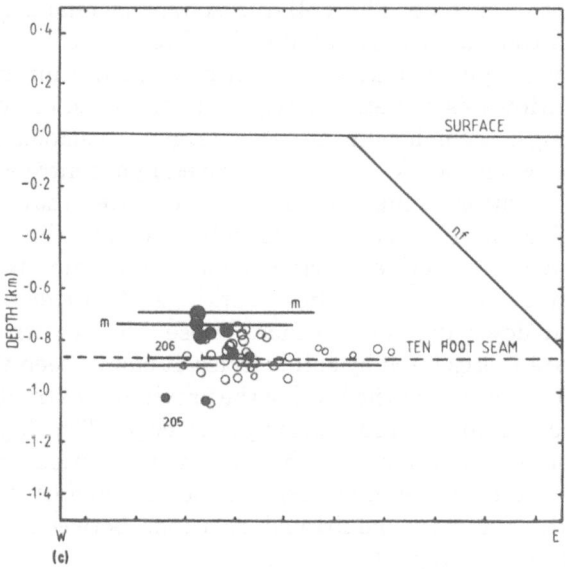

Fig. 8.16 Position of hypocentre locations detected from an improved seismometer layout, during working of 206 face. The position of the face during the month duration of these observations is shown by hatched ornament. Seismic events with a shear mechanism are represented by solid circles. (a) Plan view; (b) vertical section parallel to direction of face advance; (c) vertical section parallel to face axis.

with much less seismic activity at the weekends. This coincided with the mine working pattern.

Detailed examination of the tremor hypocentre locations and the position of the active coal face at the time of the tremors showed that:

(a) The tremors occurred adjacent to and advanced in unison with the coal face.

(b) The *larger* magnitude events, usually felt at the surface, only occurred when the face passed either below the pillars left by previous workings in the Moss seam, or above the pillar in the Bowling Alley seam.

(c) The *smaller* events were not strongly affected by the position of pillars in the Moss and Bowling Alley seams.

(d) Few seismic events occurred before the face had advanced 250 m.

Analysis of first motions of the tremors suggested that two different source mechanisms occurred – *shear* and *collapse* source mechanisms. The shear events were those of greater magnitude and were those felt by local residents at the surface. The collapse source mechanism events have smaller magnitude and were not felt by local residents.

For a period of approximately one month during the mining of 206 face a higher resolution seismometer network was installed directly above the mine workings, complementing the larger seismometer networks. Together the two networks greatly improved hypocentre estimates – the accuracy of location being estimated to be better than $+50$ m. These hypocentre locations, shown in Fig. 8.16, indicate that most tremors occurred above the active workings but all within an approximate vertical distance of 250 m from these workings. Most tremors occurred in advance (and down dip) of the face. Those events with a shear type mechanism are represented by solid circles. It can be seen that the larger of such events occurred at the level of the previously extracted Moss seam and were located in the Moss seam *pillar* edges. The higher accuracy locations confirm that most of the events, in particular the collapse events, occurred over the previously extracted panel (in this case panel 205). Thus the following additional relations between seismicity and mining can be added to (a)–(d) above:

(e) The larger magnitude events (also felt) possessed a shear source mechanism and were situated in the edges of pillars of previous workings – principally those in the overlying Moss seam.
(f) The smaller unfelt events, lying unusually within 150 m vertically above the level of the active panel, have a collapse or implosional source mechanism.
(g) Tremor hypocentres appear to be displaced about 200 m to the east of the active Ten Foot panel, i.e. over the previous panel.

From the tremor observations it can therefore be suggested that two types of tremor exist. Type A have magnitudes less than about 2.5, possess a collapse/implosional source mechanism and are situated within approximately 150 m in depth of the Ten Foot panels, usually above them. Type B events have larger magnitudes, were felt at the surface, have a shear source mechanism, and occur in the pillars of previous workings when the active Ten Foot panel face passes under or above the pillars.

The tremors result from energy released by fracture of rock. The largest recorded earth tremors (Fig. 8.16) had local magnitudes (M_L) at the surface of the order of 3. Seismic energy can be calculated from M_L using Richter's (1958) equation:

$$\text{Log } E = 9.9 + 1.9M_L - 0.02M_L^2 \tag{8.1}$$

which gives an energy output in the case of the largest event of 250 Mj. The energy output from the small events would be proportionately less, but nevertheless quite significant.

An estimate of the volume of rock which would be required to collapse to release 250 Mj of energy can be obtained by a comparison with energy released from collapse of an unconfined laboratory specimen, say for instance the Coal Measures sandstone of Fig. 1.1. Collapse of the unconfined specimen in the laboratory occurred at an axial strain of 0.6%. Strain energy, W, is given by the equation:

$$W = \tfrac{1}{2}(\sigma_1 \varepsilon_1 + \sigma_2 \varepsilon_2 + \sigma_3 \varepsilon_3) \qquad (8.2)$$

Consequently the strain energy stored in the unconfined rock immediately prior to failure was of the order of 0.3 Mj m^{-3} and most of this energy would have been released on failure. The volume of rock which would have to be subjected to shear fracture to cause this energy output would be 800 m^3, if all the stored strain energy were converted to seismic energy. Assuming a spherical seismic source this is equivalent to a shear plane of the area of about 86 m^2 which corresponds to a relatively minor part of the Moss seam pillar edge. It is, however, unreasonable to suppose that the seismic efficiency of the fracture process is 100% and that all the strain energy would be converted to seismic energy. A more realistic figure for seismic efficiency would be a figure of the order of 1% or even less (Walsh, 1977). If this is so then the local magnitude of 3.0 events corresponds to the fracturing of rock volume of 80 000 m^3 and shear plane area 1800 m^2.

While seismic events of magnitude 3.0 were perceptible at the surface they did not cause damage to buildings. However, any increase in magnitude above this value would greatly increase the hazard. Normally in weak Coal Measures rocks with a cyclic sequence of weak fireclay, shales, mudstones, thin sandstones, and coal seams, it is unusual, as discussed in Section 8.1, for the conditions for rapid energy release to occur. However, with increasing depths of working and overburden pressures and more rapid advances of longwall faces, all of which were present in this case and will increasingly occur in the future, more detailed attention to mine layouts, particularly when mining vertically adjacent seams, will be required if seismic events associated with coal mining are not to become a more frequent occurrence.

8.4   Roof collapse during caving

An interesting observation on the genesis of roof collapse during caving can be obtained from the previous case history. Type A events generally occur in the wastes of faces in the Ten Foot seam adjacent to those being worked. Some occur behind the working face and some appear to occur in the wastes of the old workings in the Moss seam. They are dilational or tensile in origin and must be attributed to collapse of roof layers above the worked areas. The events in a mechanical sense can be attributed to reduction of the compressive principal stress in the goaf area, resulting in tensile fracture of layers subjected to reduced confinement. The location of many collapse/dilational events over adjacent panels (Figs 8.15, 8.16) is unlikely to be explained by location error. An alternative explanation is that as the overall span of the workings increases, so bending and tensile stresses and strains in any layers of rock spanning the workings will increase. The stresses will be at a maximum in the centre of a hypothetical beam or plate. They will be reduced to a certain extent by pillars (as between 205 and 206 faces) at the seam level, but only to a certain extent. Sandstones will be particularly able to resist these stresses but failure along a fault or sandstone lens boundary, caused by increasing stresses from an adjacent working, will cause rupture with a release of energy. But the amount of energy released, except in exceptional cases, will be less than shear failures in pillars, because by the nature of the material (rock has a compressive strength equal to ten times its tensile strength), the strain energy capable of being stored in compression is much higher than that capable of being stored in tension. Type A events may therefore be described as resulting from the collapse of strata over the wastes – principally above panels 204, 205 and 206. They are the inevitable and normal noise and vibration associated with caving strata, and do not appear to constitute any form of hazard.

Hazards may however be associated with delayed or incomplete caving of longwall faces under strong sandstone roofs. These may under certain circumstances create rockburst conditions, and some of the conditions associated with *caving* are worth investigation. As has been seen, caving – the collapse of the roof behind advancing longwall supports – is essential to relieve abutment stresses. After a longwall face commences retreating or advancing from its starting point the immediate roof layer starts to sag and separate and a process of slabbing gradually moves upwards. However, stresses will not be relieved – and stored strain energy on the abutments reduced – until the caving reaches a height where the fractured rock can support a substantial overburden stress (see Fig. 6.1). The point at which this happens depends on the ease of caving of the overlying strata, the type of face support and the method of mining. If the strata do not cave easily the first major collapse or *weighting* can release large amounts of energy. In British mines with soft roofs the first weighting usually takes place after 20 to 30 m (see for instance Tubby and Farmer (1981)) from the barrier. In India (Singh and Sarkar, 1976) under roofs of strong shales and sandstones it can occur after 35 to 45 m advance. The first weighting and, to a lesser extent, subsequent periodic weightings are preceded by an increase in *face convergence* which often leads to severe damage to face supports and equipment. Singh and Sarkar (1976) describe an increase in convergence from 20 to 25 mm m^{-1} advance to 80–130 mm m^{-1} immediately prior to the weight. Zoltan and Bodonyi (1972) describe an increase in face convergence and support loads of 30–40% after an advance of 17–25 m. These are classic examples of increases in energy release rate immediately prior to a rockburst.

It is difficult to predict the position of the first weight since it does not involve the proximate or nether roof, but higher roof layers. It is more likely to be severe if the roof layers are thick and strong and contain a high proportion of sandstone or limestone. Empirical concepts of caveability discussed in Chapter 5 are not always helpful, nor roof strength computations based on arch beam theory (see for instance Denkhaus, 1964). Lampl (1983), after examining case histories in China and the United States where supports were damaged by heavy initial weighting, suggested, as have many South African investigators, that most benefits can be obtained by improvements to face supports. These would aim to increase their capacity to the maximum available; to increase setting pressures, using guaranteed sets, to levels as close to yield pressures as possible; to have large volume power set legs with an accumulator effect, and to introduce a *rapid yield* valve of the highest practical flow capacity. This would be set about 10% above the nominal yield load and be additional to normal yield valves. These suggestions developed empirically in response to rockburst conditions intuitively approach the problem using a methodology capable of being developed from the fundamental characteristics of rock deformation summarized in Fig. 8.1. It is an excellent example of practice preempting theory in response to an extreme engineering problem.

REFERENCES

Alder, L. (1968), Roof control in longwall mining. *Min. Congr. J.*, March, 58–67.

Altounyan, P.F.R. (1977), *Use of computer assisted numerical techniques for the design of underground structures*, University of Newcastle upon Tyne Internal Report.

Altounyan, P.F.R., Bell, M.J., Farmer, I.W. and Happer, C.J. (1982), Temperature stress and strain measurements during and after construction of concrete linings in frozen sandstone. *Proc. 3rd Int. Symp. Ground Freezing*, pp. 343–8.

Altounyan, P.F.R. and Farmer, I.W. (1981), Tunnel lining pressures during groundwater freezing and thawing. *Proc. 5th Rapid Excavation and Tunelling Conference, San Francisco*, Society of Mining Engineers of AIME, pp. 784–800.

Altounyan, P.F.R., Shelton, P.D. and Wang Hao (1983), Shaft lining pressures during sinking through deep aquifer rock. *Proc. 5th Congr. Int. Soc. Rock Mechanics, Melbourne*.

American Society for Testing and Materials (ASTM) (1967), *Standard method of test for triaxial compressive strength of undrained rock core specimens without pore pressure measurements*, D2664–67.

Arioglu, E. (1976), *Factors affecting the design of support systems for use in roadways associated with longwall faces in Coal Measures strata*, PhD Thesis, University of Newcastle upon Tyne.

Ashwin, D.P., Campbell, S.G., Kibble, J.D., Haskayne, J.D., Moore, J.F.A. and Shepherd, R. (1970), Some fundamental aspects of face powered support design. *Min. Engr*, **129**, 659–71.

Atkinson, J.H. and Bransby, P.L. (1978), *The Mechanics of Soils; an Introduction to Critical State Soil Mechanics*, McGraw-Hill, London.

Attewell, P.B. (1977), Ground movements caused by tunelling in soil. *Proc. Conf. Large Ground Movements and Structures, Cardiff*, Pentech, London, pp. 812–948.

Attewell, P.B. and Farmer, I.W. (1974), Ground deformation resulting from shield tunelling in London Clay. *Can. Geotech. J.*, **11**, 380–95.

Attewell, P.B. and Farmer, I.W. (1976), *Principles of Engineering Geology*, Chapman and Hall, London.

Attewell, P.B. and Farmer, I.W. (1977), Investigations of a local anomalous stress distribution in a sub-aqueous rock tunnel. *Proc. Conf. Rock Engng, Newcastle upon Tyne*, British Geotechnical Society, London, pp. 613–22.

Attewell, P.B., Farmer, I.W. and Glossop, N.G. (1978), Ground deformation caused by tunnelling in a silty alluvial clay. *Ground Engng*, 11 (8), 32–41.

Attewell, P.B., Farmer, I.W. and Wickson, J.L. (1976), Measurements of a ground-lining interaction pressure in an underwater tunnel in Coal Measures rock. In *Tunnelling '76*, Institution of Mining and Metallurgy, London, pp. 255–64.

Auld, F.A. (1979), Design of concrete shaft linings. *Proc. Instn Civ. Engrs*, 67, 817–32.

Auld, F.A. (1982), Ultimate strength of concrete shaft linings and its influence on design. In *Strata Mechanics* (ed. I.W. Farmer), Elsevier, Amsterdam, pp. 134–40.

Babcock, C., Morgan, T., and Haramy, K. (1981), Review of pillar design equations including the effects of constraint. *Proc. 1st Conf. Ground Control in Mining, W. Virginia University, Morgantown*, pp. 23–34.

Badger, C.W., Cummings, A.D. and Whitmore, R.L. (1956), The disintegration of shales in water. *J. Inst. Fuel*, 29, 417–23.

Barla, G.B. and Boshkov, S. (1978), *Investigation of differential strata movements and water table fluctuations during longwall workings at Somerset mine*, National Technical Information Service Report, FE-9041-1.

Barton, N., Lien, R. and Lunde, J. (1974), Engineering classification of rock masses for the design of tunnel supports. *Rock Mech.*, 6, 189–236.

Batchelor, A.S. (1972), *The correlation of roadway displacement with the stress redistribution and strata movements caused by longwall mining*, PhD Thesis, University of Nottingham.

Bates, J.J. (1978), An analysis of powered support behaviour. *Min. Engr*, 137, 681–92.

Bates, J.J., Butler, J.W., Smith, G. and Waring, B. (1975), *Design and analysis of powered support hydraulic supply and distribution system: suggestions for standardisation of system and components*, Report No. 56, January 1975, Mining Research and Development Establishment, National Coal Board.

Bell, M.J. (1982), The design of shaft linings in Coal Measures rocks. In *Strata Mechanics* (ed. I.W. Farmer), Elsevier, Amsterdam, pp. 160–6.

Berry, D.S. (1977), Progress in the analysis of ground movements due to mining. *Proc. Conf. Large Ground Movements and Structures, Cardiff*, Pentech, London, pp. 781–811.

Bieniawski, Z.T. (1968), The effect of specimen size on compressive strength of coal. *Int. J. Rock Mech. Min. Sci.*, 5, 325–35.

Bieniawski, Z.T. (1973), Engineering classification of jointed rock masses. *Trans. S. Afr. Inst. Civ. Engrs*, 15, 335–44.

Bieniawski, Z.T. (1976), Rock mass classification in rock engineering. *Proc. Symp. Exploration for Rock Eng., Johannesburg*, Balkema, Rotterdam, pp. 97–106.

Bieniawski, Z.T. (1981), Improved design of coal pillars for US mining conditions. *Proc. 1st Conf. Ground Control in Mining, W. Virginia University, Morgantown, Va.*, pp. 13–22.

Billington, C.J. and Jacomb-Hood, E.W. (1976), New approaches to steel supports for tunnels in mines. In *Tunnelling '76*, Institution of Mining and Metallurgy, London, pp. 349–58.

Blades, M.J. (1975), Developments in monolithic packing systems. Paper presented to Institution of Mining Engineers, South Midlands and South Staffs Branch, December 1975.

Bloor, A. (1982), Deformation of a circular concrete roadway lining in response to strata movements. In *Strata Mechanics*, Elsevier, Amsterdam, pp. 223–9.

Board, M.P. and Fairhurst, C. (1983), Rockburst control through destressing – a case example. In *Rockbursts; Prediction and Control*, Institution of Mining and Metallurgy, London, pp. 91–101.

Bonell, R.A. (1980), *Shield type supports – experience in the North Yorkshire Area*, National Coal Board Internal Report.

Brabbins, M.W. (1978), Longwall mining in USA and Australia. *Colliery Guardian International*, pp. 57–65.

Braun, W.M. (1980), Application of the NATM in deep tunnels and difficult rock formations. *Tunnels Tunnelling*, **12 (2)**, 17–9.

Brauner, G. (1973), *Subsidence due to underground mining*, Parts 1 and 2. US Bureau of Mines, Information Circulars 8571–2.

Bräuner, G. (1981). *Gebirgsdruck und Gebirgsschlage*, Verlag Gluckauf, Essen.

Bridgeman, P.W. (1947), Volume changes in the plastic states of simple compression. *J. Appl. Phys.*, **20**, 141–51.

British Standards Institution (1972), *The structural use of concrete*, CP 110, Part 1, BSI, London.

Brown, E.T. (ed.) (1981), *Rock Characterisation, Testing and Monitoring*, Pergamon, Oxford.

Brown, E.T. (1981), Putting the NATM into perspective. *Tunnels Tunnelling*, **13 (10)**, 13–6.

Brown, E.T. and Hoek, E. (1978), Trends in relationships between measured in-situ stresses and depth. *Int. J. Rock Mech. Min. Sci.*, **15**, 211–5.

Bryan, A., Bryan, J.G. and Fouche, J. (1964), Some problems of strata control and support in pillar workings. *Min. Engr*, **123**, 238–54.

Bunting, D. (1911), Chamber pillars in deep anthracite mines. *Trans. Am. Inst. Min. Engrs*, **42**, 235–45.

Carr, F. and Lewis, S. (1973), *The problem of support in high speed drivages*, National Coal Board Internal Report.

Carr, J.L., Shepherd, R., Walton, J.T. and Clarke, E.B. (1967), Advance heading practice and problems. *Min. Engr.*, **127**, 63–72.

Carver, J., Luxmoore, S., Howieson, I.A. and Jones, H.D. (1976), Safety and health in coal mine tunnel drivage. In *Tunnelling '76*, Institution of Mining and Metallurgy, London, pp. 85–95.

Carver, J., Cowan, J. and Binns, P.D. (1977), Strata control aspects of steep seam longwall mining in the United Kingdom. *6th Int. Conf. Strata Control, Banff, Canada*, Paper 30.

Chugh, Y.P. and Missavage, R.A. (1981), Effects of moisture on strata control in mines. *Engng Geol.*, **17**, 241–55.

Clarke, A.M. (1963), A contribution to the understanding of washouts, swalleys, splits and other seam variations and the amelioration of the effects of mining in South Durham. *Min. Engr*, **122**, 667–706.

Clarke, A.M. (1976), Why modern exploration has little to do with geology and much more to do with mining. *Colliery Guardian*, **224**, 325–36.

Coates, D.F. (1970), *Rock mechanics principles*, Mines Branch Monograph 874, Department of Energy, Mines and Resources, Ottawa.

Cook, N.G.W. (1983), Origin of rockbursts. In *Rockbursts; Prediction and Control*, Institution of Mining and Metallurgy, London, pp. 1–9.

Crouch, S.L. (1970), *The influence of failed rock on the mechanical behaviour of underground excavations*, PhD Thesis, University of Minnesota.

Crouch, S.L. and Fairhurst, C. (1974), *The mechanics of coal mine bumps and the interaction between coal pillars, mine roof and floor*, Report Research Contract H0101778 to US Bureau of Mines, Washington, DC.

Daemen, J.J.K. and Fairhurst, C. (1972), Rock failure and tunnel support loading. *Proc. Int. Symp. Underground Openings, Lucerne*, Swiss Society for Soil Mechanics and Foundation Engineering, Zurich, pp. 359–69.

Daemen, J.J.K. and Hood, M. (1981), Subsidence profile functions derived from mechanistic rock mass models. *Proc. Workshop on Surface Subsidence due to underground mining*. West Virginia University, Morgantown, Va.

Davies, R.A. (1977), *A catalogue of strength properties of some Coal Measures rocks*, National Coal Board Internal Report.

Deere, D.U. (1968), Geological considerations. In *Rock Mechanics in Engineering Practice* (eds K.G. Stagg and O.C. Zienkiewicz), Wiley, London, pp. 1–20.

Deere, D.U., Peck, R.B., Monsees, J.E. and Schmidt, B. (1969), Design of tunnel support systems. *Final Report on Contract No 3-0152*, US Department Transportation, Washington, DC.

Deere, D.U. (1979), Applied rock mechanics, the importance of weak geological features. *Proc. 4th Int. Congr. Int. Soc. Rock Mechanics, Montreux*, **3**, pp. 22–5.

Deere, D.U., Hendon, A.J., Patton, F.D. and Cording, E.J. (1966), Design of surface and near surface construction in rock. *Proc. 8th US Symp. Rock Mechanics, Minneapolis*, pp. 237–303.

Deere, D.U. and Miller, R.P. (1966), *Engineering classification and index properties for intact rock*, US Air Force Weapons Lab. Report AFWL-TR-65-16, Kirkland, New Mexico.

De la Cruz, R.V. and Goodman, R.E. (1970), Theoretical basis of the borehole deepening method of absolute stress measurement. *Proc. 11th US Symp. Rock Mechanics, Berkeley*, pp. 353–79.

Dempster, E.L., Tyser, J.A. and Wagner, H. (1983), Regional aspects of mining induced seismicity: theoretical and management considerations. In *Rockbursts; Prediction and Control*, Institution of Mining and Metallurgy, London, pp. 37–52.

Denkhaus, H.G. (1964), Critical review of strata movement theories and their application to practical problems. *J. S. Afr. Inst. Min. Metall.*, **64**, 310–332.

Dietz, H.K.O. (1982), Aspects of advanced grouting during shaft sinking in South Africa. *Int. J. Mine Water*, **2**, 19–28.

Dudley, W.R. (1977), *A field investigation into the influence of anhydrite packing on longwall gate road stability*, PhD Thesis, University of Newcastle upon Tyne.

Evans, I. and Pomeroy, C.D. (1966), *Strength Fracture and Workability of Coal*, Pergamon, Oxford.

Evans, W.H. (1941), The strength of undermined strata. *Trans. Inst. Min. Engrs*, **50**, 475–532.

Evans, W.H., Hogan, M.A. and Vallis, E.H. (1941), An investigation of the load on packs at moderate depths. *Trans. Inst. Min. Engrs*, **50**, 339–77.

Everling, G. (1977), Discussion (Bonell, R.A. A review of strata control experience and current trends). *6th Int. Conf. Strata Control, Banff, Canada.*

Farmer, I.W. (1982), Deformation of access roadways and roadside packs in mines. In *Strata Mechanics* (ed. I.W. Farmer), Elsevier, Amsterdam, pp. 207–12.

Farmer, I.W. (1983), *Engineering Behaviour of Rocks*, Chapman and Hall, London.

Farmer, I.W. and Altounyan, P.F.R. (1980), The mechanics of ground deformation above a caving longwall face. *Proc. 2nd Conf. Ground Movements and Structures, Cardiff*, Pentech, London, pp. 75–91.

Farmer, I.W. and Attewell, P.B. (1973), The effect of particle strength on the compression of crushed aggregate. *Rock Mech.*, **5**, 237–48.

Farmer, I.W. and Attewell, P.B. (1975), A note on the similarities between ground movement around soft ground tunnels and longwall mining operations. *Min. Engr*, **134**, 397–403.

Farmer, I.W. and Glossop, N.H. (1983), Design constraints for full face tunnelling machines in coal mines, *Int. J. Min. Engng*, **1**, 57–70.

Farmer, I.W. and Pooley, F.D. (1967), A hypothesis to explain the occurrence of outbursts in coal, based on a study of West Wales outburst coal. *Int. J. Rock Mech. Min. Sci.*, **4**, 189–93.

Farmer, I.W., Price, A.M. and Youdan, D.G. (1980), Design of tunnels in Coal Measures rocks. In *Eurotunnel '80, Basle*, Institution of Mining and Metallurgy, London, pp. 13–7.

Farmer, I.W. and Robertson, J.T. (1975), The effect of pack construction on roadway stability behind working faces. *Min. Engr*, **134**, 599–606.

Farmer, I.W. and Shelton, P.D. (1980), Factors that affect underground rock-bolt reinforcement systems design. *Trans. Inst. Min. Metall., Sect. A*, **89**, 68–83.

Faulkner, R. and Phillips, D.W. (1935), Cleavage induced by mining, *Trans. Inst. Min. Engrs*, **134**, 599–606.

Fenner, R. (1938), Untersuchungen zur Erkenntnis des Gebirgsdruckes, Gluckauf, **74**, pp. 691–6; 705–15. *Study of ground pressures*, Tech. Translation 515, NRC Division of Building Research, Ottawa.

Forrest, W. (1978), The Selby Project, *Min. Engr*, **138**, 237–46.

Forrest, W. and Black, J.C. (1979), Hydrogeological analysis, ground treatment and special construction techniques at Selby: Gascoigne Wood surface drift mine. In *Tunnelling '79*, Institution of Mining and Metallurgy, London, pp. 256–63.

Franklin, J.A., Broch, E. and Walton, G. (1971), Logging and mechanical character of rock. *Trans. Inst. Min. Metall.*, **80**, A1–9.

Gamble, J.C. (1971), *Durability–plasticity classification of shales and other rocks*, PhD Thesis, University of Illinois.

Garritty, P. (1981), *Effects of mining on surface and sub-surface water bodies*, PhD Thesis, University of Newcastle upon Tyne.

Garritty, P. (1982), Water percolation into fully caved longwall faces. In *Strata Mechanics* (ed. I.W. Farmer), Elsevier, Amsterdam, pp. 25–9.

Garritty, P. (1983), Water flow into undersea mine workings, *Int. J. Min. Engng*, **1**, 237–51.

Gaziev, E.G. and Erlikham, S.A. (1971), Stresses and strains in anisotropic rock foundation. *Int. Symp. Rock Mechanics, Nancy*, Paper 2-1.

Gilbert, M.J. (1982), Pressures and displacements around a lined shaft in weak rock. In *Strata Mechanics* (ed. I.W. Farmer), Elsevier, Amsterdam, pp. 174–7.

Gilbert, M.J. and Farmer, I.W. (1981), A time dependent model for lining pressure based on strength concepts. *Int. Symp. Weak Rock, Tokyo*, **1**, pp. 137–42.

Gimm, W.A.R. and Pforr, H. (1964), Breaking behaviour of rock salt under rock bursts and gas outbursts. *Proc. 4th Int. Conf. Strata Control and Rock Mechanics, Columbia University, NY*, pp. 434–49.

Glossop, N.H. (1982), Some observations on the use of stochastic methods for the prediction of subsidence over longwall panels. In *Strata Mechanics* (ed. I.W. Farmer), Elsevier, Amsterdam, pp. 59–62.

Godden, S.J. (1981), *A special geological report on the problems of outbursts of coal and gas*, National Coal Board Unpublished report.

Godden, S.J. (1982), *Bursts and outbursts in coal mines*, Unpublished MSc Thesis, University of Newcastle upon Tyne.

Goodman, R.E., G-h. Shi and Boyle, W. (1982), Calculation of support for hard jointed rock using the key block principle. *Proc. 23rd US Symp. Rock Mechanics, Berkeley*, pp. 883–98.

Gorrie, C. and Scott, G. (1970), Some aspects of caving on powered support faces, *Min. Engr*, **129**, 677–87.

Gotze, W. and Kammer, W. (1976), Die Auswirkungen von Strecken furung und Ausbau auf die Querschnitts verminderung von Abbaustrecken. *Gluckauf*, **112**, 946–53.

Graham, J.J. (1978), A review of some recent powered support developments, *Min. Engr*, **137**, 665–78.

Greenwald, H.P., Howarth, H.C. and Hartman, I. (1939), *Experiments on the strength of small pillars of coal in the Pittsburgh bed*, US Bureau of Mines, Technical Paper **605**.

Gupta, R.N. (1982), *The influence of setting pressure and strata behaviour on longwall faces*, PhD Thesis, University of Newcastle upon Tyne.

Gupta, R.N. and Farmer, I.W. (1981), A magnetic ring extensometer system for strata deformation measurement in coal mines. *Min. Engr*, **141**, 303–5.

Gupta, R.N. and Farmer, I.W. (1982), Relations between strata deformation and support performance on longwall faces. In *Strata Mechanics* (ed. I.W. Farmer), Elsevier, Amsterdam, pp. 74–81.

Gupta, R.N. and Farmer, I.W. (1983), Strata deformation and support performance at a longwall coal face. *Proc. 5th Congr. Int. Soc. Rock Mechanics, Melbourne*.

Habernicht, H. (1972), Systematic development of powered supports for faces in weak rock. *Proc. 5th Int. Conf. Strata Control, London*, Paper 6.

Hargraves, A.J. (1980), A review of instantaneous outburst data. *Proc. Symp. Occurrence, Prediction and Control of Outbursts, Victoria*, pp. 1–18.

Harris, G.W. (1974), A sandbox model used to examine the stress distribution around a simulated longwall coal face, *Int. J. Rock. Mech. Min. Sci.*, **11**, 325–35.

Harwood, S. (1980), *Design of underground openings in layered strata*, MSc Thesis, University of Newcastle upon Tyne.

Hassani, F.P. and Scoble, M.J. (1981), Properties of weak rocks with special reference to the shear strength of their discontinuities as encountered in British surface coal mining. *Proc. Symp. Weak Rock, Tokyo*, Vol. 1, pp. 355–64.

Hayward, D. (1982), Robbins races Titan in Selby tunnels. *New Civ. Engr*, 5 August, pp. 18–20.

Hazen, G. and Artler, L. (1976), Practical coal pillar design problem. *Min. Congr. J.*, June, 86–92.

Hedley, D.G.F. (1969), Design criteria for multi-wire borehole extensometer systems. *Proc. 1st Can. Symp. Mine Surveying and Rock Deformation Measurement, New Brunswick, Canada*.

Hedley, D.G.F. and Grant, F. (1972), Stope and pillar design for the Elliot Lake Uranium mines. *CIM Bull.*, July, 37–44.

Hess, H. (1972), Roof control by powered supports in the West German coal mining industry. *5th Int. Conf. Strata Control, London*, Paper 1.

Hinde, C.G. (1978), *An investigation of hydraulic powered support performances in various mining conditions*, PhD Thesis, University College, Cardiff.

Hobbs, D.W. (1968), Scale model studies of strata movement around mine roadways – III. *Int. J. Rock Mech. Min. Sci.*, **5**, 245–51.

Hobbs, D.W. (1969), Strata movement around mine roadways – results of scale model studies. *Min. Engr*, **128**, 461–71.

Hobbs, D.W. (1970), The behaviour of broken rock under triaxial compression, *Int. J. Rock Mech. Min. Sci.*, **7**, 125–48.

Hobbs, N.B. (1975), The prediction of settlement of structures on rock. *Conf. Settlement of Structures, Cambridge*, Pentech, London, pp. 579–610.

Hobst, L. and Zajic, J. (1983), *Anchoring in Rock and Soil*, Elsevier, Amsterdam.

Hodkin, D.L. (1978), *Interaction between longwall and pillared workings at Lynemouth Colliery*, PhD Thesis, University of Newcastle upon Tyne.

Hodkin, D. (1982), Interaction between pillar workings at Ellington Colliery. In *Strata Mechanics* (ed. I.W. Farmer), Elsevier, Amsterdam, pp. 275–7.

Hoek, E. and Bray, J.W. (1976), *Rock Slope Engineering*, Institution of Mining and Metallurgy, London.

Hoek, E. and Brown, E.T. (1981), *Underground Excavations in Rock*, Institution of Mining and Metallurgy, London.

Holland, C.T. (1964), Strength of coal in mine pillars, *Proc. 6th US Symp. Rock Mechanics, University of Missouri, Rolla*, pp. 450–66.

Holmes, P. (1979), *The engineering significance of sheared Coal Measures strata with specific reference to the stability of underground structures*, Internal Report, University of Newcastle upon Tyne.

Holmes, P. (1983), *Geotechnical factors affecting longwall mining of the Barnsley Seam*, PhD Thesis, University of Newcastle upon Tyne.

Howell, R.C., Wright, F.D. and Dearinger, J.A. (1977), Ground movement and pressure changes associated with shortwall mining. *Proc. 17th US Symp. Rock Mechanics, Utah*, pp. 40–5.

Hudson, J.A., Brown, E.T. and Fairhurst, C. (1972), Soft, stiff and servocontrolled testing machines; a review with reference to rock failure. *Engng Geol. (Amsterdam)*, **6**, 155–89.

Ilstein, A. (1960), Influence of the resistance of the support system on the manifestation of rock pressure in longwall faces. *Proc. 3rd Int. Conf. Strata Control, Paris*, pp. 127–32.

Jackson, B.H. (1968), The development and use of powered support in Western Germany. *Min. Engr*, **128**, 567–74.

Jackson, D.J.H. (1979), Testing of shield supports. *Min. Engr*, **138**, 763–71.

Jacobi, O. (1966), The increase of roof flaking in longwall faces as a result of workings under pillar edges and of abutment pressure of adjacent workings. *Int. J. Rock Mech. Min. Sci.*, **3**, 221–30.

Jacobi, O. (1976), *Praxis der Gebirgsbeherrschung*, Verlag Gluckauf, Esen.

Jacobi, O., Everling, G. and Irresberger, H. (1964), Research with a view to the development of powered face supports, *Proc. 4th Int. Conf. Strata Control and Rock Mechanics, New York*, pp. 166–85.

Jaeger, J.C. and Cook, N.G.W. (1979), Fundamentals of Rock Mechanics, 3rd edn, Chapman and Hall, London.

Jamison, D.B. and Cook, N.G.W. (1979), in *Fundamentals of Rock Mechanics*, 3rd edn (eds J.C. Jaeger and N.G.W. Cook), Chapman and Hall, London.

Jenkins, J.D. and Szeki, A. (1964), *The properties of some rock materials and their behaviour in pillars*, University of Newcastle upon Tyne Unpublished Report.

Jennings, D.H. (1981), *Relations between seepage flow and radial pressures on concrete shaft linings in saturated strata*, Unpublished MSc Thesis, University of Newcastle upon Tyne.

Jeremic, M.L. (1980), Influence of shear deformation structures in coal – selecting method of mining. *Rock Mech.*, **13**, 23–38.

Johnson, G. (1963), *Study of rock pressures round mining excavations*, PhD Thesis, University of Durham.

Johnson, G. (1973), Rock mechanics – a nomograph for the assessment of roadway conditions. *Colliery Guardian*, **221**, 16–20.

Jones, R.T. (1980), Internal Report, Mining Research and Development Establishment, National Coal Board.

Josien, J., Breniaux, C., Daumalin, M., Doligez, M., Georgel, P. (1982), The dynamic effects of strata pressure: rockbursts. Paper to *7th Int. Conf. Strata Control, Liège*.

Josien, J.P. (1972), The functioning of supports and their effect on roof behaviour on the faces. *Proc. 5th Int. Conf. Strata Control, London*, Paper 7.

Kenny, P. (1969), The caving of the waste on longwall faces. *Int. J. Rock Mech. Min. Sci.*, **6**, 541–55.

Kenny, P. and Wilson, A.H. (1963), Strata movements on faces equipped with conventional and powered supports. *Min. Engr*, **122**, pp. 524–38.

Kenny, T.C. (1967), *Shearing resistance of natural quick clays*, PhD Thesis, University of London.

Kent, P.E. (1966), The structure of the concealed Carboniferous rocks of North East England. *Proc. Yorks. Geol. Soc.*, **35**, 323–52.

Kidybinski, A. (1980), Significance of in-situ strength measurements for prediction of outburst hazard in coal mines in Lower Silesia. *Proc. Symp. Occurrence Prediction and Control of Outbursts in Coal Mines, Victoria*, pp. 193–203.

King, H.J., Whittaker, B.N. and Batchelor, A.S. (1972), The effects of interaction in mine layouts, *Proc. 5th Int. Conf. Strata Control, London*, Paper 17.

Kirmani, F.A.K. (1972), *Experimental and theoretical studies of strata conditions around two mine roadways with special reference to fracture phenomena*, PhD Thesis, University of Newcastle upon Tyne.

Klein, J. (1982), Present state of freezer shaft design in mining. In *Strata Mechanics* (ed. I.W. Farmer), Elsevier, Amsterdam, pp. 147–53.

Knill, J.L. (1983), Quoted in *New Civ. Engr*, No. 557, 15 September, p. 10.

Kusznir, N.J., Ashwin, D.P. and Bradley, L.G. (1980a), Mining induced seismicity in the North Staffordshire coalfield, England. *Int. J. Rock Mech. Min. Sci.*, **17**, 45–53.

Kusznir, N.J., Farmer, I.W., Ashwin, D.P., Bradley, L.G. and Al Saigh (1980b), Observations and mechanics of seismicity associated with coal mining in North Staffordshire. *Proc. 21st US Rock Mechanics Symp., University of Missouri, Rolla*, pp. 632–40.

Ladanyi, B. (1974), Use of the long-term strength concept in the determination of ground pressure on tunnel linings. *Proc. 3rd Congr. Int. Soc. Rock Mechanics, Denver*, **2**, Part B, pp. 1150–6.

Lain, M.J. (1974), An investigation into some aspects of the mineralogy, behaviour and physical properties of underclays, University of Newcastle upon Tyne Internal Report.

Lampl, F. (1983), Effect and limitations of abnormal loads on roof supports. In *Rockbursts; Prediction and Control*, Institution of Mining and Metallurgy, London, pp. 123–32.

Lang, T. (1971), Rock reinforcement. *Bull. Assoc. Engng Geol.*, **9**, 213–9.

Leigh, R.D. (1963), *Strata pressures and rock mass movements induced by longwall mining*, PhD Thesis, University of Durham.

Leighton, F. (1982), The search for a method to provide warning of coal and gas outbursts. *Proc. 2nd Conf. Ground Control in Mines, Morgantown, West Virginia*, pp. 113–7.

Linden, Walter von der (1977), Summary of performance of shield type supports in the US. *Trans. Soc. Min. Engrs AIME*, **262**, pp. 156–64.

Litwiniszyn, J. (1964), On certain linear and non-linear stratas theoretical models. *Proc. 4th Int. Conf. Strata Control and Rock Mechanics, New York*, pp. 384–96.

Long, R.E., Kusznir, N.J., Blenkinsop, T.G. and Smith, M.J. (1984), Contribution to discussion. *Trans. Inst. Min. Metall.*, **93**.

McLintock, F.A. and Walsh, J. B. (1962), Friction on Griffith cracks in rocks under pressure. *Proc. 4th Nat. Congr. Appl. Mech., Berkeley, 1962*, pp. 1015–21.

Macrea, J.C. and Lawson, W. (1954), The incidence of cleat fracture in some Yorkshire coal seams. *Proc. Leeds Geol. Assoc.*, **6**, 224–7.

Mainil, P. (1965), Contribution to the study of ground measurements under the influence of mining operations. *Int. J. Rock Mech. Min. Sci.*, **2**, 225–43.

Mallory, R. (1980), Internal Reports, Mining Research and Development Establishment, National Coal Board.

Marr, J.E. (1959), A new approach to estimation of mining subsidence. *Trans. Inst. Min. Engrs*, **118**, 697–707.

Marr, J.E. (1975), The application of the zone area system to the prediction of mining subsidence. *Min. Engr*, **135**, 53–60.

Martos, F. (1958), Concerning an approximate equation of the subsidence trough and its time factor. *Proc. Int. Congr. Strata Control, Leipzig*, pp. 191–205.

Maury, V. (1970), Distribution of stresses in discontinuous layered systems. *Water Power*, **22**, 195–202.

Moore, R.D. and Hanes, J. (1980), Bursts at Leichardt Colliery, Central Queensland and the apparent benefits of mining by shotfiring. *Proc. Symp. Occurrence, Prediction and Control of Outbursts in Coal Mines, Victoria*, pp. 71–84.

Morgan, D. (1982), Longwall coal face roof supports for the 1980s. *Min. Engr*, **141**, 25–9.

Morris, A.H. (1978), Cross-measure drifting – equipment and techniques. *Min. Engr*, **138**, 65–72.

Morrison, R.G.K., Corlett, A.V. and Rice, H.R. (1956), *Report of special committee on mining practices at Elliot Lake*, Ontario Department of Mines, Bulletin 155.

Moseley, J.T.B. (1980), Internal Reports, Mining Research and Development Establishment, National Coal Board.

Mottahed, P. and Szeki, A. (1982), The collapse of room and pillar workings in a shaley gypsum mine due to dynamic loading. In *Strata Mechanics* (ed. I.W. Farmer), Elsevier, Amsterdam, pp. 260–3.

Muir Wood, A.M. (1979), Ground behaviour and support for mining and tunnelling. *Trans. Inst. Min. Metall.* (Sect. A: Min. industry), **88**, A23–34.

National Coal Board (1968), *Working under the sea*, Mining Department Instruction, PI 1968/69 (revised 1971), National Coal Board.

National Coal Board (1972), *Design of mine layouts with reference to geological and geometrical factors*, Working Party Report, Mining Department, National Coal Board.

National Coal Board (1975), *Subsidence Engineers' Handbook*, 2nd Edn, Mining Department, National Coal Board, London.

Obert, L. and Duvall, W.I. (1967), *Rock Mechanics and Design of Structures in Rock*, Wiley, New York.

Obert, L., Windes, S.L. and Duvall, W.A. (1946), *Standardised tests for determining the physical properties of mine rocks*, US Bureau of Mines, Report Invest. 3891.

Orawecz, K. (1977), Analogue modelling of stresses and displacements in board and pillar workings in coal mines. *Int. J. Rock Mech. Min. Sci.*, **14**, 7–23.

Peng, S.S. and Harthill, M. (eds.) (1981), *Workshop on surface subsidence due to underground mining*, West Virginia University, Morgantown, Va.

Peng, S.S. and So, W.H. (1983), The causes of cyclic excessive convergence at the longwall tail entry. *Int. J. Min. Engng*, **1**, 27–42.

Phillips, D.W. (1944), Rockbursts or 'bumps' in coal mines. *Trans. Inst. Min. Engrs*, **104**, 55–84.

Phillips, D.W. and Jones, T.J. (1941), Strata movements ahead of and behind longwall faces. *Trans. Inst. Min. Engrs*, **101**, 346–62.

Potts, E.L.J. (1964), Current investigations in rock mechanics and strata control. *Proc. 4th Int. Conf. Strata Control, New York*, pp. 29–44.

Price, A.M. (1979), *The effect of confining pressure on the post-yield deformation characteristics of rocks*, PhD Thesis, University of Newcastle upon Tyne.

Price, A.M. and Farmer, I.W. (1979), Application of yield models to rock. *Int. J. Rock Mech. Min. Sci.*, **16**, 157–9.

Price, A.M. and Farmer, I.W. (1980), A general failure criterion for rocks. *Proc. 21st US Symp. Rock Mechanics, University of Missouri, Rolla*, pp. 256–64.

Price, A.M. and Farmer, I.W. (1981), The Hvorslev surface in rock deformation. *Int. J. Rock Mech. Min. Sci.*, **18**, 229–34.

Price, N.J. (1966), *Fault and Joint Development in Brittle and Semi-brittle Rock*, Pergamon, Oxford.

Price, R.J. (1980), Application of higher setting loads to powered supports in the South Nottinghamshire Area. Paper presented to the Institution of Mining Engineers, Notts and N. Derbys. Branch, February.

Price, R.J. and Pickering, M.H.B. (1981), Application of higher setting loads to powered supports in South Notts Area. *Min. Engr*, **140**, 841–8.

Rabciewicz, L. (1969), Stability of tunnels under rock load. *Water Power*, **21**, 266–73.

Richter, C.F. (1958), *Elementary Seismology*, Freeman, San Francisco.

Salahey, M.R., Money, M.S. and Dearman, W.R. (1977), The occurrence and engineering properties of intraformational shears in Carboniferous rocks. *Proc. Conf. Rock Engineering, Newcastle upon Tyne*, pp. 311–28.

Salamon, M.G.D. (1964), Elastic analysis of displacements and stresses induced by the mining of seam or reef deposits, Parts I–IV, *J. S. Afr. Inst. Min. Metall.*, **64**, 129–49, 197–218, 468–500; and **65**, 319–28.

Salamon, M.G.D. (1983), Rockbursts hazard and the fight for its alleviation in Southern African goldmine. In *Rockbursts; Prediction and Control*, Institution of Mining and Metallurgy, London, pp. 11–36.

Salamon, M.G.D. and Monro, A.H. (1967), A study of the strength of coal pillars. *J. S. Afr. Inst. Min. Metall.*, **68**, 55–67.

Saxena, N.C. and Singh, B. (1982), Subsidence behaviour Coal Measures over bord and pillar workings. In *Strata Mechanics* (ed. I.W. Farmer), Vol. 32, Developments in Geotechnical Engineering, Amsterdam, Elsevier, pp. 283–5.

Schmidt, B. (1969), *Settlements and ground movements associated with tunnelling in soil*, PhD Thesis, University of Illinois, Urbana.

Schofield, A.N. and Wroth, C.P. (1968), *Critical State Soil Mechanics*, McGraw-Hill, London.

Scholz, C.H. (1968), Microfracturing and the inelastic deformation of rock in compression. *J. Geophys. Res.*, **73**, 1417–31.

Scurfield, R.W. (1970), Staffordshire mining layout for the mid-1970s. *Min. Engr*, **130**, 73–84.

Shadbolt, C.H. (1977), Mining subsidence – historical review and state of the art. *Proc. Conf. Large Ground Movements and Structures*, Pentech, London, pp. 705–48.

Shelton, P.D. and Farmer, I.W. (1984), Deformation of a deep mine inset during construction in Coal Measures rock. (To be published.)

Shepherd, J., Rixon, L.K. and Griffiths, L. (1981), Outbursts and geological structures in coal mines: a review. *Int. J. Rock Mech. Min. Sci. & Geomech. Abstr.*, **18**, 267–83.

Shepherd, R. (1964), Study of strata control on mechanised coal faces. *Proc. 4th Int. Conf. Strata Control and Rock Mechanics, New York*, pp. 230–47.

Shepherd, R. (1977), Powered roof supports; today and tomorrow. *Mine and Quarry*, **6** (3), 17–26.

Shepherd, R. and Ashwin, D.P. (1968), Measurement and interpretation of strata behaviour on mechanised faces. *Colliery Guardian*, **216**, 795–804.

Shepherd, R. and Wilson, A.H. (1960), Measurement of strain in concrete shafts and roadway linings. *Trans. Inst. Min. Engrs*, **119**, 561–74.

Singh, B. and Sarkar, S.K. (1976), The investigation into strata control failures at caved longwall faces in India and a new approach for support planning to avoid such occurrences. *Proc. 6th Int. Conf. Strata Control, Banff, Canada.*

Singh, K.H. (1968), *Experiments in strata control with reference to rapidly advancing and retreating longwall systems*, PhD Thesis, University of Newcastle upon Tyne.

Singh, T.N. (1981), *Laboratory model simulation of mine strata deformation*, MSc Thesis, University of Newcastle upon Tyne.

Singh, T.N. and Singh, B. (1967), Fundamentals of mine modelling. *J. Mines, Metals and Fuels*, **25**, 254–9.

Skempton, A.W. (1961), Effective stress in soils, concrete and rocks. In *Pore Pressure and Suction in Soils*, British Geotechnical Society, Butterworth, London, pp. 4–16.

Skempton, A.W. (1964), Long-term stability of clay slopes. *Géotechnique,* **14**, 77–101.

Skempton, A.W. (1966), Some observations on tectonic shear zones. *Proc. 1st Int. Congr., Int. Soc. Rock Mechanics, Lisbon*, Vol. 1, pp. 329–35.

Smart, B.G.D., Isaac, A.K. and Hinde, C.G. (1980), Investigations into the relationship between powered support performance and the working environment with particular reference to strata convergence. *Proc. 21st US Symp. Rock Mechanics, University of Missouri, Rolla*, pp. 332–44.

Smart, B.G.D., Rowlands, N. and Isaac, A.K. (1982), Progress towards establishing relationships between the mineralogy and physical properties of Coal Measures rocks. *Int. J. Rock Mech. Min. Sci. & Geomech. Abstr.,* **19**, 81–9.

Snowdon, R. (1979), *Performance of a full face tunnelling machine in Coal Measures rocks at Dowdon Colliery*, MSc Thesis, University of Newcastle upon Tyne.

Snowdon, R.A., Glossop, N.H. and Farmer, I.W. (1983), Performance of a full face tunnelling machine in Coal Measures strata. In *Eurotunnel '83, Basle*, Institution of Mining and Metallurgy, London, pp. 129–36.

Sorensen, W.K. and Pariseau, W.G. (1978), Statistical analysis of laboratory compressive strength and Young's modulus data for the design of production pillars in coal mines. *Proc. 19th US Rock Mechanics Symp., Lake Tahoe, Nev.*, pp. 40–7.

South Staffordshire and Warwickshire Institute of Mining Engineers (1933), The occurrence of bumps in the thick coal seam of South Staffordshire. *Trans. Inst. Min. Engrs*, **85**, 116–47.

Speck, R.C. (1981), Influence of certain geological and geotechnical factors on coal mine floor stability – a case study. *Proc. 1st Conf. Ground Control in Mining, W. Virginia University, Morgantown, Va.*, pp. 44–9.

Spruth, F. (1951), Distribution of pressure up, on and near the coal face. *Conf. Int. sur le pression de terrains, Liège*, pp. 121–39.

Spruth, F. (1966), Nochmals: Zur Frage der streckendamme. *Gluckauf*, **102**, 241–5.

Stassen, P. and van Duyse, H. (1977), Development of supports in stone roads in the Campine Coalfields. *Proc. 6th Int. Conf. Strata Control, Banff, Canada*, Paper 15.

Stateham, R.M. and Ratcliffe, D.E. (1976), *Humidity – a cycle effect in coal mine roof stability*, RI 8291, US Bureau of Mines, Washington.

Stimpson, B. and Walton, G. (1970), Clay mylonites in English Coal Measures – their significance in opencast stability. *Proc. 1st Int. Congr. International Association Engineering Geologists, Paris, 1970*, **2**, pp. 1388–93.

Streat, F.A. (1954), Strength and stability of pillars in coal mines. *J. Chem. Met. Min. Soc. S. Africa*, **68**, pp. 55–67.

Stunya, V. and Meyer, C. (1960), Caving of rigid roof rock. *Colliery Engng*, **37**, 208–17.

Styler, A.N. (1980), *Interburden strata deformation and interaction effects associated with longwall mining*, PhD Thesis, University of Newcastle upon Tyne.

Sweet, A.L. and Bogdanoff, J.L. (1965), Stochastic model for predicting subsidence. *J. Engng Mech. Div., Am. Soc. Civ. Engrs*, **91** (EM2), pp. 21–45.

Sziwlski, A.B. and Whittaker, B.N. (1975), Control of strata movement around face ends. *Min. Engr*, **134**, 515–25.

Terzaghi, K. (1943), *Theoretical Soil Mechanics*, Wiley, New York.

Terzaghi, K. (1946), Rock defects and loads on tunnel supports. In *Rock Tunnelling with Steel Supports* (eds R.V. Proctor and T.L. White), The Commercial Shearing and Stamping Co., Youngstown, Ohio, pp. 17–99.

Terzaghi, K. and Peck, R.B. (1967), *Soil Mechanics in Engineering Practice*, 2nd Edn, Wiley, New York.

Thomas, L.J. (1964), An interim assessment of strain measurements in concrete lined shafts and insets at Wolstanton Colliery. *Int. J. Rock Mech. Min. Sci.*, **1**, 547–61.

Trombik, M. and Zuberek, W. (1980), Location of microseismic sources at the Szambierki coal mine. *Proc. 2nd Conf. Acoustic Emission and Microseismic Activity in Geologic Structure and Materials, Penn State University*, pp. 171–90.

Tsur-Lavie, Y. and Denekamp, S.A. (1982), Size and shape effect on pillar design. *Proc. Symp. Strata Mechanics, Newcastle upon Tyne*, Elsevier, Amsterdam, pp. 245–8.

Tubby, J. and Farmer, I.W. (1981), Stability of undersea workings at Lynemouth and Ellington Collieries. *Min. Engr*, **141**, 87–96.

Tuffs, G.W. (1982), Current trends in shaft sinking in Britain. *Min. Technol. (London)*, **64**, 323–7.

Van Eekhout, E.M. and Peng, S.S. (1975), The effect of humidity on the compliance of Coal Mine shales. *Int. J. Rock Mech. Min. Sci.*, **12**, 335–40.

Van Heerden, W.L. (1974), *In-situ determination of complete stress–strain characteristics of 1.4 m square specimens with width to height ratios up to 3.4*, CSIR Research Report ME 1265.

Varo, L. and Passaris, E.K.S. (1977), The role of water in the creep properties of halite, *Rock Engineering*, British Geotechnical Society, London, pp. 85–100.

Voight, B. and Pariseau, W. (1970), State of the predictive art in subsidence engineering. *J. Soil Mech. Fdn Engng Div., Am. Soc. Civ. Engrs*, **89**, No. SM4, 91–126.

Wagner, A.A. (1957), The use of the Unified Soil Classification System by the Bureau of Reclamation. *Proc. 4th Int. Conf. Soil Mech. Fdn Engng, London*, **1**, pp. 125–34.

Wagner, H. (1974), Determination of the complete load-deformation characteristics of coal pillars. *Proc. 4th Congr., Int. Soc. Rock Mechanics, Denver*, Vol. 2B, pp. 1076–81.

Wagner, H. and Steijn, J.J. (1979), Effect of local strata conditions on the support requirements of longwall faces. *Proc. 4th Int. Cong. Int. Soc. Rock Mechanics, Montreux*, **1**, pp. 557–64.

Walsh, J.B. (1977), Energy changes due to mining. *Int. J. Rock Mech. Min. Sci.*, **14**, 25–34.

Wang, F.D., Skelley, W.A. and Wolgamott, J. (1977), *In-situ* coal pillar strength study. *Proc. 18th US Symp. Rock Mechanics, Golden, Col.*, pp. 2B5–1.

Wang, Hao (1982), *Design of deep mine structures in weak rocks*, MSc Thesis, University of Newcastle upon Tyne.

Ward, W.H. (1978), Ground supports for tunnels in weak rocks. *Géotechnique*, **28**, 133–71.

Ward, W.H., Coates, D.J. and Tedd, P. (1976), Performance of tunnel supports in the Four Fathom Mudstone. In *Tunnelling '76*, Institution of Mining and Metallurgy, London, pp. 329–40.

Waring, B. (1977), Application of increased setting pressures. *Conf. Powered Support Hydraulics System Design, Performance and Condition Testing, MRDE, Gresley Old Hall*, Section 4.

Westbrook, G.K. (1977), Investigation of earth tremors at Stoke on Trent. In *Instrumentation for Ground Vibration and Earthquakes*, Institution of Civil Engineers, London, pp. 76–82.

Wheeler, A. (1974), Support on the face. *Proc. Symp. Mining Methods, Harrogate*, Institution of Mining Engineers, London, pp. 65–78.

Whickham, G.E., Tiedeman, H.R. and Skinner, E.H. (1972), Support determination based on geological predictions. *Proc. North American Rapid Excavation and Tunnelling Conf., Chicago*, Vol. 1, pp. 43–64.

Whittaker, B.N. (1974), An appraisal of strata control practice. *Min. Engr*, **134**, 9–23.

Whittaker, B.N. (1975), Contribution to discussion. *Min. Engr*, **134**, 488–90.

Whittaker, B.N. (1976), Recording, treatment and interpretation of roadway deformation surveys. *Min. Engr*, **135**, 607–17.

Whittaker, B.N. (1979), Evaluation of the design requirements and performance of gate roadways. *Min. Engr*, **138**, 535–48.

Wiggill, R.B. (1963), The effects of different support methods on strata behaviour around stoping excavations. *J. S. Afr. Inst. Min. Metall.*, **63**, 391–426.

Wilkinson, J.H. and Evans, J.W. (1963), Some experiences with self advancing supports at Lea Hall Colliery. *Min. Engr*, **122**, 541–56.

Will, M. (1979), Seismic activity and mining operations. *Proc. 2nd Conf. Acoustic Emission and Microseismic Activity in Geologic Structures and Materials, Penn State University*, pp. 191–208.

Wilson, A.H. (1960), *The measurement and interpretation of strain in the concrete linings of insets at Abernant*, MRE Report No. 2155, National Coal Board, London.

Wilson, A.H. (1964), Conclusions from recent strata control measurements made by the Mining Research and Development Establishment. *Min. Engr*, **123**, 367–80.

Wilson, A.H. (1975), Support load requirements on longwall faces. *Min. Engr*, **134**, 479–91.

Wilson, A.H. (1977), The stability of tunnels in soft rocks at depth. *Proc. Conf. Rock Engng, Newcastle upon Tyne*, pp. 511–27.

Wilson, A.H. (1978), Various aspects of longwall roof support. *Colliery Guardian International*, pp. 50–6.

Wilson, A.H. (1980), *The stability of underground workings in the soft rocks of the Coal Measures*, PhD Thesis, University of Nottingham.

Wilson, A.H. (1981), Stress and stability in coal ribsides and pillars. *Proc. 1st Conf. Ground Control in Mining, W. Virginia University, Morgantown, Va.*, pp. 1–12.

Wilson, A.H. (1983), The stability of underground workings in the soft rocks of the Coal Measures. *Int. J. Min. Engng*, **1**, 91–181.

Wilson, A.H. and Ashwin, D.P. (1972), Research into the determination of pillar size. *Min. Engr*, **131**, 409–30.

Woodley, J.N.L. and Osborne, B.A. (1980), MRDE experience in pump packing. *Min. Engr*, **140**, 437–43.

Worsey, P.N. (1978), Internal Report, University of Newcastle upon Tyne.

Wright, F.D. (1973), Roof control through beam action and arching, *SME Mining Engineers Handbook*, Amer. Inst. Min. Met. Pet. Engrs, New York, **1**, pp. 13.80–13.96.

Zern, E.N. (1926), *Coal miners' pocket book*, McGraw-Hill, New York.

Zoltan, T. and Bodanyi, J. (1972), Roof control in seams with weak surrounding rocks, *Proc. 5th Int. Strata Control Conf., London*, Paper 16.

AUTHOR INDEX

SUBJECT INDEX